NONLINEAR DYNAMICS

WILEY SERIES IN NONLINEAR SCIENCE

Series Editors: ALI H. NAYFEH, Virginia Tech
                  ARUN V. HOLDEN, University of Leeds

Abdullaev   Theory of Solitons in Inhomogeneous Media
Bolotin   Stability Problems in Fracture Mechanics
Kahn and Zarmi   Nonlinear Dynamics: Exploration through
                          Normal Forms
Moon (ed.)   Dynamics and Chaos in Manufacturing Processes
Nayfeh   Method of Normal Forms
Nayfeh and Balachandran   Applied Nonlinear Dynamics
Nayfeh and Pai   Linear and Nonlinear Structural Mechanics
Ott, Sauer, and Yorke   Coping with Chaos
Pfeiffer and Glocker   Multibody Dynamics with Unilateral Contacts
Qu   Robust Control of Nonlinear Uncertain Systems
Vakakis et al.   Normal Modes and Localization in Nonlinear Systems
Yamamoto and Ishida   Linear and Nonlinear Rotor Dynamics: A Modern
                          Treatment with Applications

NONLINEAR DYNAMICS
Exploration through Normal Forms

PETER B. KAHN
Department of Physics
State University of New York at Stony Brook
Stony Brook, New York

YAIR ZARMI
Center for Energy and Environmental Physics
The Jacob Blaustein Institute for Desert Research
Ben-Gurion University of the Negev
Sede Boqer Campus, Israel

A Wiley-Interscience Publication
JOHN WILEY & SONS, INC.
New York • Chichester • Weinheim • Brisbane • Singapore • Toronto

This text is printed on acid-free paper.

Copyright © 1998 by John Wiley & Sons, Inc.

All rights reserved. Published simultaneously in Canada.

Reproduction or translation of any part of this work beyond that permitted by Section 107 or 108 of the 1976 United States Copyright Act without the permission of the copyright owner is unlawful. Requests for permission or further information should be addressed to the Permissions Department, John Wiley & Sons, Inc., 605 Third Avenue, New York, NY 10158-0012.

Library of Congress Cataloging in Publication Data:

Kahn, Peter B.
    Nonlinear dynamics : exploration through normal forms / Peter B. Kahn and Yair Zarmi.
       p. cm.
    Includes bibliographical references and index.
    ISBN 0-471-17682-6 (cloth : alk. paper)
    1. Dynamics. 2. Nonlinear theories. 3. Normal forms (Mathematics) I. Zarmi, Yair, 1942– . II. Title.
QA845.K34   1997
531'.11—dc21   97-9275

Printed in the United States of America

10 9 8 7 6 5 4 3 2 1

CONTENTS

Preface   xi

1 The Text: Its Scope, Style, and Content   1

   1.1 Introduction, 1
       1.1.1 Near-identity transformations, 6
       1.1.2 Basic ideas behind the method of normal forms, 7
       1.1.3 Departures from linear theory: Background, 8
   1.2 Goals: What the Book Is All About, 13
       1.2.1 Introductory comments regarding the normal form expansion and freedom, 13
       1.2.2 Reference models, 17
   References, 20

2 Basic Concepts   23

   2.1 Definitions, 23
       2.1.1 Fixed point, 23
       2.1.2 Hyperbolic fixed point, 24
       2.1.3 Center fixed point, 25
       2.1.4 Stability of fixed points, 28
       2.1.5 Stability of solutions, 29
       2.1.6 Limit cycle, 33
       2.1.7 The O symbol, 35
       2.1.8 Asymptotic series, 36
       2.1.9 Near-identity transformation, 38
   2.2 Fundamental Theorems, 38
       2.2.1 Existence and uniqueness of solutions: Lipschitz condition, 38
       2.2.2 Hartman–Grobman theorem, 39
       2.2.3 Poincaré–Lyapunov theorem, 42

2.2.4 Difference between nonlinear and linear problems, 44
2.2.5 One-dimensional examples of Poincaré–Lyapunov theorem, 45
2.2.6 Gronwall's lemma, 46
2.3 Stable, Unstable, and Center Manifolds, 48
2.3.1 Stable and unstable manifolds, 48
2.3.2 Center manifold, 51
References, 54

3 Naive Perturbation Theory (NPT)   55

3.1 Definition and Basic Aspects of NPT, 56
3.1.1 Eigenvalues with negative real part, 59
3.1.2 Pure imaginary eigenvalues, 62
3.1.3 An alternative approach, 64
3.2 Discussion, 69
References, 71

4 Formalism of Perturbation Expansion   73

4.1 Freedom in Perturbation Expansion of Solutions of Dynamical Systems, 73
4.1.1 Contrasting the naive perturbation expansion and Poincaré's normal form expansion, 79
4.2 Resonant Combinations of Eigenvalues, 84
4.2.1 Poincaré domain, 85
4.2.2 Siegel domain, 87
4.2.3 Small-denominator problem, 91
4.3 The Method of Normal Forms, 92
4.4 The Last Freedom: Free Resonant Terms in x_n, 98
4.5 Intuitive Meaning of Lie Bracket, 99
4.6 Convergence of Normal Form Transformation, 102
4.6.1 General case, 102
4.6.2 Two-dimensional case, 104
4.7 Eigenvalue Update: Case c in Bruno's Analysis, 108
4.7.1 Energy-conserving oscillations: Behavior near a center, 108
4.7.2 Behavior near a saddle fixed point, 109
4.8 Time Validity of the Truncated Expansion, 111
4.8.1 Naive perturbation theory, 112
4.8.2 Normal form error estimates for planar oscillatory systems, 117
References, 127

CONTENTS vii

5 Problems with Eigenvalues That Have Negative Real Part 129

5.1 Basic Result: Review, 130
5.2 One Coordinate with a Negative Real Eigenvalue:
    No Possible Resonances, 134
    5.2.1 Procedure, 134
5.3 Analysis of a Specific Example, 136
    5.3.1 Case 1: Free functions set = 0, 136
    5.3.2 Comparison with an expansion of the exact solution, 138
    5.3.3 Implementation of the initial conditions, 138
    5.3.4 Case 2: Free functions, 139
5.4 Finite-Time Blowup, 143
    5.4.1 Illustrative examples, 145
5.5 Concluding Remarks, 157
References, 158

6 Normal Form Expansion for Conservative Planar Systems 159

6.1 The Perturbed Simple Harmonic Oscillator, 159
6.2 The Duffing Oscillator, 160
    6.2.1 Normal form expansion, 161
    6.2.2 Further discussion, 163
    6.2.3 Choice of the free functions, 166
    6.2.4 Implementation of initial conditions: Equivalence of
          different choices of free functions, 171
    6.2.5 Numerical comparisons, 171
    6.2.6 Quantitative analysis, 174
6.3 Precision and Duration of the Validity of the Expansion, 178
    6.3.1 Dependence on accuracy with which the updated
          frequency is known, 178
    6.3.2 Precision and duration of validity of MNF
          expansion, 179
6.4 Oscillator with Quadratic Perturbation, 182
    6.4.1 Surprise with "zero–zero" choice, 183
6.5 The Pendulum, 185
6.6 The "Altered" Duffing Oscillator, 187
6.7 Additional Examples, 191
    6.7.1 Mixing of scales, 192
    6.7.2 Example where the minimal normal form is not useful, 193
6.8 Boundary-Value Problems: Bifurcations and
    Normal Forms, 196
References, 207

7 Dissipative Planar Systems 209

7.1 Oscillators with Cubic Damping, 211
  7.1.1 Linear oscillator with cubic damping, 211
  7.1.2 Pendulum with cubic damping, 215
  7.1.3 Duffing oscillator with cubic damping, 218
7.2 Limit Cycle Systems, 221
  7.2.1 Definition of a limit cycle, 221
7.3 First-Order Normal Form Analysis for Limit Cycle Systems, 223
7.4 Qualitative Aspects of the Expansion, 225
  7.4.1 Structure of the expansion, 225
  7.4.2 Observations, 227
  7.4.3 Limit circle radius renormalization, 228
  7.4.4 The Van der Pol oscillator revisited, 230
  7.4.5 The Rayleigh oscillator revisited, 233
7.5 Multiple Limit Cycles: Identification of the Expansion Parameter, 234
  7.5.1 Qualitative considerations, 235
  7.5.2 Perturbation analysis, 239
References, 246

8 Nonautonomous Oscillatory Systems 247

8.1 Introduction, 247
8.2 An Example from the Theory of Accelerator Design, 249
  8.2.1 The case $p = 1$, 251
  8.2.2 The case $p = 2$, 255
8.3 The Mathieu Equation, 265
  8.3.1 The case $n = 1$, 266
  8.3.2 The case $n = 2$, 268
  8.3.3 The general case, 270
  8.3.4 Do we have to convert the equations into autonomous ones? 270
8.4 Fortuitous "Explosive Instabilities": Examples of the Nonlinear Mathieu Equation, 271
  8.4.1 One-parameter problem, 271
  8.4.2 Two-parameter problem, 276
8.5 Harmonic Oscillators with Periodically Modulated Friction, 282
  8.5.1 Linear friction, 282
  8.5.2 The modified Rayleigh oscillator, 284
8.6 Normal Form Perturbation Analysis of NMR Equations, 288
  8.6.1 The model, 289
  8.6.2 The normal form expansion, 291

CONTENTS

8.6.3 Results, 295
8.6.4 Do we need to substitute $z \equiv \exp(-i\omega_R t)$? 297
References, 298

9 Problems with a Zero Eigenvalue 301

9.1 Analysis Using the Normal Form Expansion, 302
    9.1.1 Illustrative examples, 304
9.2 Center Manifold Method: Connection with Method of Normal Forms, 309
    9.2.1 Method of normal forms, 310
    9.2.2 Method of the center manifold, 313
9.3 Additional Examples, 315
9.4 Concluding Remarks, 319
References, 320

10 Higher-Dimensional Hamiltonian Systems 323

10.1 Introduction, 323
10.2 Betatron Oscillations, 324
    10.2.1 Qualitative arguments, 325
    10.2.2 The unperturbed system, 326
    10.2.3 First-order perturbation analysis, 327
    10.2.4 Second-order analysis, 330
10.3 Action–Angle Variables, 332
    10.3.1 Single harmonic oscillator, 334
    10.3.2 Nearly integrable systems, 335
10.4 Coupled Harmonic Oscillators, 337
    10.4.1 Structure of the normal form equations, 340
    10.4.2 "Constant" of motion, 342
    10.4.3 Canonical transformation choice, 343
    10.4.4 Constraint on free functions, 345
    10.4.5 Minimal normal forms? 346
10.5 Example of Two Coupled Oscillators, 348
    10.5.1 Perturbation analysis, 349
    10.5.2 Numerical results, 352
10.6 Comments Regarding Symplectic Maps, 359
References, 363

11 Higher-Dimensional Dissipative Systems 365

11.1 Limit Cycles in Multidimensional Dissipative Systems, 365
    11.1.1 Center manifold with a limit cycle for leading coordinates, 365

11.1.2 Bifurcation in a dissipative system—a scaling example, 371
11.1.3 Limit cycle in a three-dimensional system—no center manifold structure, 379

11.2 Onset of Oscillations in Hopf Bifurcations, 383
11.2.1 Derivation of the normal form, 385
11.2.2 Usual choice (no free functions), 386
11.2.3 Special choice of free functions—limit circle radius "renormalization", 388

11.3 Phase Locking in Coupled Oscillators, 390
References, 394

**Appendix Conservative System with One Degree of Freedom: 395
Calculation of the Period**

Index **399**

PREFACE

As soon as one starts thinking about nonlinear dynamics, the richness of the subject emerges. It encompasses almost all "not linear" phenomena and investigators have an enormous range of subject material from which to choose the focus of their investigation. It has been a field of particularly intense study in recent years, due, in great part, to the availability of modern computers and computer software. Many books have been written and there are now journals dedicated to the subject.

With these thoughts in mind, we have come to write this book to bring together a variety of ideas that we have developed in our investigations of a small portion of nonlinear dynamics. In the text, we have concentrated on constructing an exposition of the method of normal forms and its application to ordinary differential equations through perturbation analysis. In particular, we use the inherent freedom in the expansion to obtain expressions that are compact and have computational advantages, as is illustrated in a variety of applications.

The test begins, in Chapter 1, with a prologue that introduces basic ideas associated with perturbation methods as applied to problems in nonlinear dynamics, particularly the idea of a near-identity transformation and the role it plays in the normal form expansion.

Chapter 2 is concerned with establishing error estimates and a discussion of the time-validity of the perturbation expansions that we develop. We introduce some fundamental theorems in dynamical system analysis (i.e., the Poincaré–Lyapunov and the Hartman–Grobman theorems) as well as Gronwall's lemma, which is an essential tool in the estimation of and in the finding of bounds for errors in perturbative approximation schemes.

In Chapter 3 we introduce and discuss *naive perturbation theory* (NPT). The basic idea in this method is that one uses the solutions of the unperturbed problem as generators of a series expansion for the full problem. This is successful in some cases, but for a wide class of conservative problems it leads to spurious *secular* terms. These terms in the perturbation expansion, associated with a system that executes periodic motion, are an artifact of the structure of a finite number of terms in the expansion and *not* of the full solution. [In our expansions *secular* or *aperiodic* terms are those that appear as products of powers of the time multiplied by a trigonometric function. For example, such terms as $t \sin(t)$ or $t^2 \sin(t)$. We will encounter them as genuine and spurious terms in expansions. Illustrative examples are given.]

In Chapter 4, we develop the formalism of the perturbation expansion associated with the method of normal forms. We introduce the fundamental theorems of Poincaré and Poincaré–Dulac and address particular attention to the concept of resonance and where the eigenvalues of the unperturbed problem lie in the complex plane. This is followed by a systematic development of the method of normal forms and a detailed discussion of the concept of *freedom of choice*. We introduce the Duffing oscillator to illustrate various aspects of the normal form expansion and to show explicitly how one computes various terms.

In Chapter 5, we discuss problems in which the eigenvalues of the unperturbed system have negative real part. We learn that, for such problems, naive perturbation theory yields a perturbation expansion that is equivalent to the normal form expansion. This result follows, since it turns out that only a finite number of so-called *resonant terms* are possible and, as a result, the generators of the expansion can be taken to be the unperturbed solutions. Then, it is a matter of taste (or convenience) which method is employed. We also discuss aspects of finite-time blowup that lead to finite-time validity of the perturbation expansion. In such problems, the perturbation expansion is more correctly a *formal transformation* since, even if one writes it as a closed-form expression, the expansion is valid only for times less than the blowup time.

In Chapters 6 and 7, we primarily study systems with one degree of freedom that model conservative and dissipative systems, respectively. These systems describe electrical and mechanical oscillations and vibrations where the force has a dominant linear term and a small nonlinear one. The unperturbed system has a pair of complex conjugate pure imaginary eigenvalues and the perturbation permits periodic motion, damped oscillations, or limit cycles. One finds that generally one has *eigenvalue update* and as a consequence, for this class of problems, naive perturbation theory leads to spurious secular terms. The normal form analysis provides a faithful perturbation description of the motion. The discussion is quite extensive and brings into focus a great variety of aspects of the normal form expansion. For example, in Chapter 6, we use the freedom in the normal form expansion to introduce the concept of *minimal normal forms,* which is associated with a regrouping of terms in the expansion so that the latter has a compact form. By contrast, in Chapter 7, we use the freedom to "renormalize" the expansion to obtain a very simple way to view the onset of limit cycle oscillations.

In Chapter 8 we consider a rich variety of nonautonomous problems. These systems arise naturally when one considers forced linear and nonlinear oscillatory motion such as arises, for instance, in a discussion of the lin-

PREFACE

ear and nonlinear Mathieu equations, the study of pendula, orbits in celestial mechanics, electric circuits, nuclear magnetic resonance and for resonant oscillations of charged particles due to multipole errors in guiding magnetic fields in particle accelerators. Here one encounters and identifies real and spurious *explosive instabilities*. Also, one finds that true secular behavior is possible for nonautonomous systems. We analyze nonautonomous weakly nonlinear oscillatory systems with periodic forcing of the general type

$$\frac{dx}{dt} = -y, \qquad \frac{dy}{dt} = x + \epsilon G(x,y,\cos \omega t, \sin \omega t;\epsilon)$$

We view $\cos \omega t$ and $\sin \omega t$ (of which at least one appears in the equations explicitly) as the coordinates of an additional, spurious, harmonic oscillation, treated on equal footing with the position x and velocity y of the original problem. This increases the dimension of our system from 2 to 4, as is seen from the resulting equations:

$$\frac{dx}{dt} = -y, \qquad \frac{dy}{dt} = x + \epsilon F(x,y,w,w^*;\epsilon)$$

$$\frac{dw}{dt} = -i\,\omega w, \qquad \frac{dw^*}{dt} = i\,\omega w^*$$

$$w \equiv \exp(-i\,\omega t)$$

$$F(x,y,w,w^*;\epsilon) \equiv G(x,y,\cos \omega t, \sin \omega t;\epsilon)$$

where x, y, w, w^* are scalar variables, F is a scalar forcing function, and ϵ is a small parameter. Note that the harmonic time dependence, present in the forcing function, is represented by the variables w, w^*. This apparent complication works to our advantage in the normal form analysis as will become evident in the discussion.

Chapter 9 is devoted to a discussion of techniques for the treatment of problems in which the unperturbed part has one or more zero eigenvalues and the remaining ones have negative real part. One finds that the method of normal forms, without modification, provides a faithful characterization of such systems. (Specifically, it is easy to accommodate both the transient behavior and finite initial amplitudes within the standard normal form analysis.) For problems of this class, some investigators have found the *method of the center manifold* appealing. It has a rigorous mathematical foundation and is straightforward to implement. However, one must always keep in mind that, in this method, one is restricted to both sufficiently small initial amplitudes and a neighborhood of the origin. It cannot follow the transient behavior. (This center manifold technique is widely used in bifurcation analysis where the reduction of dimension does not cause any loss of essential information regarding the bifurcation. It is also used in studying high-dimensional systems of equations as a preconditioning step to a perturbation analysis.) Finally, we have been able to show, for the class of problems under consideration, the connection between the normal form equations and those obtained by the method of the center manifold.

In Chapter 10, we discuss several harmonic oscillators with nonlinear coupling. After establishing some general results, we continue the development of the analysis through the discussion of a Hamiltonian system consisting of two oscillators with weak nonlinear coupling. There is a tremendous amount of freedom in the expansion and one has to choose how to organize the calculation most efficiently.

In Chapter 11, we study higher-dimensional dissipative systems. We demonstrate, through a sequence of examples, the evolution of limit cycles and center manifolds. We then analyze the onset of oscillations in systems that undergo a Hopf bifurcation and the phenomenon of phase locking.

Some technical details, concerning the calculation of the period of oscillation of a one dimensional system in a perturbed harmonic potential, are given in the Appendix.

Portions of our work have appeared as articles in journals, conference proceedings, and edited volumes. We wish to both acknowledge and thank the publishers for their kind permission to use in the book material from the following sources:

"Minimal Normal Forms in Harmonic Oscillations with Small Nonlinear Perturbations," *Physica D* **54**, 65–74 (1991) (Elsevier Science-NL Sara Burgerhartstraat 25, 1055 KV Amsterdam, The Netherlands).

"Freedom in Small Parameter Expansion for Nonlinear Perturbations" and "Radius Renormalization in Limit Cycles," *Proc. Roy. Soc.* (London) **A440**, 189–199; **A443**, 83–94 (1993).

"Computational Effectiveness of Various Normal Form Expansions" and "Normal Form Perturbation Analysis of NMR Equations," pp. 319–326 and 411–414, in *ICCP-2 Conference* (Sept. 13–17, 1993) *Proceedings*, De-Yuan, Da-Hsuan Feng, Michael R. Strayer, and Tian-Yuan Zhang, eds. (1995), International Press, Cambridge, MA.

"Computational Aspects of Normal Form Expansions," pp. 633–661 in *AIP Conference Proceedings* No. 326, Yiton T. Yan, John P. Naples, and Michael Syphers, eds. (1995), AIP Press, Woodbury, NY.

"Normal Form Analysis of Non-Autonomous Oscillatory Systems," in *Dynamical Systems and Their Applications,* Vol. 4, pp. 359–374, R. Agarwal, ed. (1995), World Scientific Press, Singapore.

We thank Juan Lin and Diana Murray for their thorough reviews of the manuscript. Many of their suggested corrections were included in its final version. Thanks are due to Lee Segel, who read portions of the book and made constructive comments. Our love to our wives, Vicki McLane and Shulamit Amir-Zarmi, who were so encouraging and patient.

1

THE TEXT: ITS SCOPE, STYLE, AND CONTENT

1.1 INTRODUCTION

In our study of nature we observe phenomena that are strikingly different. Thus, attempting to understand properties of some limited set of observations, we try to group "similar" things together and hopefully develop an adequate systemization. In this endeavor, science has been quite successful over the years, as judged by our understanding of the periodic structure of the elements, development of a classification of biological systems according a hierarchical structure, chemical reactions, models of the solar system, the structure of elementary particles, etc.

At the next level, in our development of models of many physical systems, we have been successful at incorporating the salient features of the system by both designing and analyzing a linear description that serves as the basis for a sound first approximation. Examples of areas of application are classical and quantum mechanics, electric circuits, Newtonian theory of gravitation, Maxwell's equations, etc.

In the study of these problems, the analysis has relied heavily on the linearity of the basic equations and uses the principle of superposition as a fundamental component. Furthermore, classification schemes were developed of the equations themselves since diverse physical phenomena are described by the same mathematical functions and techniques. (For example, in one area of mathematics of interest to scientists, this has led to serious investigations of the properties of the so-called *special functions*.) This is a successful procedure, in part, because the structure of the solutions to linear equations is present in the equations themselves, independent of the initial conditions. The success in the use of linear equations was so great that our traditional body of scientific knowledge and the associated curriculum that one studies both rely almost exclusively on a linear analysis.

The desire to obtain an approximate, but useful, analytic "picture" of natural processes, has often led scientists to ignore those aspects of the mathematics that are associated with the nonlinearities in the model system. The basic premise has always been that a thorough understanding of the approximate linear system would give insight into the associated nonlinear system.

Alas, this hope is rarely fulfilled. Nonlinear phenomena that are not adequately described by the linear approximation are encountered in all areas of the quantitative sciences. Almost always, the nonlinear system has fundamental aspects that cannot be faithfully approximated by a linear perturbation scheme. Perhaps most important is the loss of the *principle of superposition* that plays a critical role in almost all aspects of linear systems. By training, one gets used to seeking a solution to a problem by constructing a sequence of functions that, when added together appropriately, yield a satisfactory approximation. The loss of this important principle leads one to realize that the spectrum of behavior of nonlinear systems is appreciably richer than that of linear systems, and, furthermore there generally does not exist the possibility of a systematic classification scheme based on the structure of the equations. This latter point follows in part because the singularity structure of the solution of a nonlinear differential equation is affected by the initial conditions.

Today it is becoming progressively clearer that the linear world, for which mathematics has been developed over a period of 300 years, is but a tiny corner of a much richer world that is being slowly unraveled. Over the years, many attempts have been made to improve our understanding of systems governed by nonlinear equations. Already in the 18th century, Clairot, Lagrange, and Laplace developed perturbation methods for the treatment of nonlinear problems in celestial mechanics. Jacobi, Poincaré, and Lindstedt studied similar problems toward the end of the 19th century. Rayleigh studied "self-sustained" vibrations within the framework of his theory of sound. His equation for the self-sustained oscillations in organ pipes turns out to be equivalent to the equation developed by Van der Pol in the 1920s for the current in a triode. Both systems include nonlinear dissipative terms that drive the system toward a self-sustaining oscillatory mode denoted today as a *limit cycle*.

Nonlinear models arose in a variety of contexts. At the beginning of this century, Ross proposed a modified exponential growth model, now known as a *logistic equation*, to describe the spread of malaria. The same model was used by Verhulst and others to account for populations that were "growth-limited." This type of model has played a key role in many studies of biological systems and in the analysis of chemical reactions governed by the law of mass action [1]. Volterra introduced a set of coupled nonlinear equations that were used to describe oscillations in fish populations. It is a beautiful model that has played an important role in the development of mathematical

INTRODUCTION

biology. It is interesting that in the construction of these equations, Volterra was greatly influenced by concurrent developments in statistical mechanics. In particular, Volterra's discussion of the so-called encounters is reminiscent of methods used by Boltzmann to study the approach to equilibrium of an ideal gas, the celebrated *H theorem* [2]. A similar model was introduced by Lotka to study a class of chemical kinetics and rate equations. The equations are now known as the *Volterra–Lotka equations*, and are used extensively as an idealization of the interaction of two isolated species, a predator and a prey and interactions of "forces in conflict" (i.e., the Lanchester equations). The equations are often generalized to include dissipative terms similar to those in Rayleigh's and Van der Pol's equations, to describe sustained oscillating chemical reactions known as *Belousov–Zhabotinskii equations* [3].

In the period from the 1930s to the 1960s Krylov, Bogoliubov, and Mitropolskii [4,5] developed methods for analyzing oscillatory systems that contained small nonlinear perturbations. Later, interest arose in trying to develop analytic techniques that could be used to study nonlinear problems in which the nonlinearity is not a small perturbation. Important milestones were, among others, the Lorenz equations [6] which arise from a truncated Navier–Stokes equation for fluid motion together with heat and mass transfer, meteorology, chemical kinetics, etc.; May's work on discrete models and his role in drawing attention to the importance of transferring one's focus from the study of linear to the study of nonlinear phenomena [7]; Feigenbaum's analysis of discrete nonlinear maps and their relevance to the behavior of continuous nonlinear systems [8]; and finally the advent of the concepts of chaos and fractals.

In this work we primarily consider the modeling of *weakly* nonlinear systems by means of ordinary nonlinear differential equations. Our working hypothesis is that the associated linear system has been solved and that our efforts go toward constructing a reliable and systematic perturbation scheme, as a power series in a small parameter. The resulting approximation has a prescribed error, valid for a finite duration (that can be estimated) in time. When the perturbations are included, the basic equation takes the form

$$\frac{d\mathbf{x}}{dt} = \mathbf{A}\mathbf{x} + \varepsilon \mathbf{F}(\varepsilon, \mathbf{x}) \tag{1.1}$$

where $\mathbf{x}$ is an n-dimensional vector and $\mathbf{A}$ is a diagonal $n \times n$ constant matrix. (The diagonalization greatly simplifies the algebra.) The interaction or perturbation is written as $\mathbf{F}(\varepsilon, \mathbf{x})$, where ε is a small parameter, $|\varepsilon| \ll 1$, and $\mathbf{F}$ is an n-dimensional vector field that may contain linear and nonlinear terms. The nonlinear terms are polynomials or functions with Taylor series expansions in **x** of degree ≥ 2. One locates the fixed points, i.e., those values of **x**(*t*) such that $d\mathbf{x}/dt = 0$; transforms them so that they are at the origin; and studies the behavior of the solutions as a power series in the small parameter for a finite interval of time. Finding an approximation to the solution **x**(*t*) may begin with expanding **x** in a power series in ε:

$$\mathbf{x}(t) = \mathbf{x}_0 + \mathbf{x}_1 + \varepsilon^2 \mathbf{x}_2 + \cdots \qquad (1.2)$$

The *zero-order term* $\mathbf{x}_0(t)$, holds the key to the development of the perturbation expansion. To see this, visualize an *n*-dimensional sphere with radius $O(\varepsilon)$ (defined in some appropriate norm in the vector space) which contains the exact solution as it evolves in time. Then, any choice for the vector function $\mathbf{x}_0(t)$ is a valid one as long as it remains within $O(\varepsilon)$ of **x**(*t*). For example, one can construct $\mathbf{x}_0(t)$ entirely from the unperturbed part of the equation. This is called *naive perturbation theory*. It turns out that this procedure often has serious shortcomings and it is found that a scheme, such as the method of normal forms, that incorporates in the quantity $\mathbf{x}_0(t)$ significant aspects of the interaction terms leads to a more effective perturbation expansion.

The "path" connecting $\mathbf{x}_0((t)$ and **x**(*t*) (see Fig 1.1) depicts the progressive approach to **x**(*t*) due to the successive addition of higher order corrections $\varepsilon^n \mathbf{x}_n$ to $\mathbf{x}_0(t)$. If the series expansion of **x**(*t*) as given by Equation (1.2) converges, the "path" will converge to **x**(*t*) as $t \to \infty$. If the series is an asymptotic one, then the "path" will initially approach **x**(*t*), but as the successively higher-order terms are included, it will diverge away from it. The choice of $\mathbf{x}_0(t)$ and the "path" are interrelated, since all $\mathbf{x}_n$ must satisfy equations resulting from the dynamical equation, and the constraints that are derived from the imposed initial conditions. In the following chapters this program will be realized through the normal form expansion.

The Equation (1.1) describes an *autonomous system* if the right hand side does not explicitly contain the time. Then the rate of evolution of the system is entirely determined by its present state. In this regard, it is important to note that in autonomous systems with one degree of freedom (i.e., two variables, *x* and *y*), the motion takes place in the phase plane. The solution is unique and, as a result, trajectories cannot cross, except at fixed points. The analysis is straightforward and will form the basis of most of our discussion. The situation changes when Equation (1.1) includes interaction terms that have explicit time dependence. It then describes a *nonautonomous system*. The interaction term is changed from $\varepsilon \mathbf{F}(\varepsilon,\mathbf{x})$ to $\varepsilon \mathbf{F}(\varepsilon,\mathbf{x},t)$. For example our equation might be

INTRODUCTION

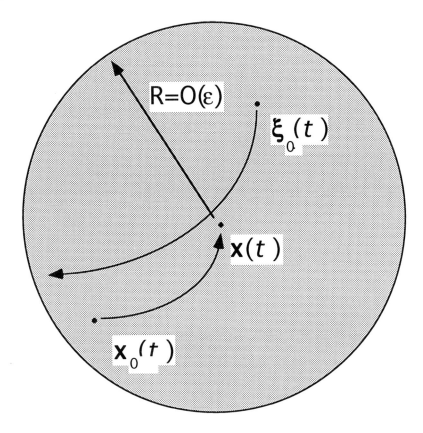

Figure 1.1 Pictorial representation of approximation schemes: $\mathbf{x}_0(t)$ and $\xi_0(t)$ are two valid zero-order approximations, $\mathbf{x}_0(t)$ yields a convergent perturbation series, and $\xi_0(t)$ yields an asymptotic series.

$$\frac{dx}{dt} = y \qquad \frac{dy}{dt} = -x + \varepsilon\{\alpha + \beta\exp(-t)\}x^2 y \qquad (1.3a)$$

One may study the equation as it is given or introduce a third dependent variable, z, by writing Equation (1.3a) as

$$\frac{dx}{dt} = y \qquad \frac{dy}{dt} = -x + \varepsilon\{\alpha + \beta\exp(-z)\}x^2 y \qquad \frac{dz}{dt} = 1 \qquad (1.3b)$$

We have converted Equation (1.3a), a nonautonomous system with two dependent variables, x and y, into to Equation (1.3b), an autonomous system in three variables (x, y, z). This leads us to conclude that, there is no way to distinguish between the variable we call z and the "time" t. It then follows that as soon as one considers systems with more than two variables, the distinction between autonomous and nonautonomous is no longer clear. [Generally, a nonautonomous system of degree n can be transformed into an autonomous one of degree $(n+1)$ by assigning the variable "time" to be the $(n+1)$st dimension.] One needs to keep in mind that the solution to Equation (1.3b) is unique. Hence, its trajectory in three-dimensional space cannot cross except at fixed points. However, the projection of the true motion onto a plane may yield crossing trajectories.

This situation arises in the study of conservative systems that are perturbed by time-dependent forces. The latter have the capacity to interact with the natural (or unperturbed) frequency of the system leading to conditions of resonance that may result in excitations and instability. Often one would like to obtain relationships among parameters characterizing such a problem that both describe and separate regions of stable and unstable motions. One learns that the richness of nonautonomous systems requires us to engage in an aspect of nonlinear analysis that is both inadequately explored and full of pitfalls. Thus, it is judicious to begin by having a firm grasp of autonomous systems with two variables before exploring nonautonomous ones.

Finally, it is crucial to realize that, in this work, we concentrate exclusively on problems that are amenable to a perturbation expansion that is a power series in a small parameter. By this restriction, we cannot treat problems associated with chaotic motion.

1.1.1 Near-identity transformations

Investigators have devised a variety of perturbation schemes to attack problems modeled by Equation (1.1). This is not surprising, as the procedure one introduces needs to be tailored to a certain extent to the questions one asks and the form one want these answers to take. For example, the *method of averaging* (MOA) is applied with success in the studies of perturbed oscillatory systems. The physical idea behind the method is that, for these systems it is possible to separate, by a *smoothing* or *averaging* technique the slowly varying component from those components that vary rapidly. Another perturbation scheme is the *method of multiple timescales* (MMTS). It is applied with success in the study of many types of equations, including perturbed oscillations, boundary-layer problems, and a class of partial-differential equations. The basic idea is to introduce artificially separated (or independent) timescales that characterize different aspects of solution. In this text, we have chosen to follow the approach known as the *method of normal forms* (NF) that is based on rather different underlying ideas than either MOA or MMTS. (Although

INTRODUCTION

derived from a different set of principles than MOA or MMTS, it yields in appropriate instances the same set of hierarchical equations.)

1.1.2 Basic ideas behind the method of normal forms

There are four aspects of the normal form expansion that we mention *briefly* at this point, to provide some perspective and motivation for the development of the method. We would like to convey to the reader the flavor of the method through a quick run through some aspects of the method. Critical details and concepts will not be fully developed or explained in this section. The fleshing out of all the essential details and applications will form central core of the book.

A. We begin by referring to Equation (1.1) and introduce a near-identity transformation from the vector **x** to the variable **u**:

$$\mathbf{x} = \mathbf{u} + \sum_n \varepsilon^n \mathbf{T}_n(\mathbf{u}) \qquad (1.4a)$$

The vector **u** is the zero-order approximation and the **T** functions are the generators of the transformation. In this work, they will be polynomials or functions that have a Taylor series expansion in **u**. On occasion, one has to write the transformation in component form:

$$x_i = u_i + \varepsilon T_{i,1}(u_1, u_2, \ldots, u_n) + \varepsilon^2 T_{i,2}(u_1, u_2, \ldots, u_n) + \cdots \qquad (1.4b)$$

Each linear component, u_i, is derived from the associated vector component, x_i. All the components are subsequently related to one another in the perturbation expansion. Furthermore, if one had not diagonalized the matrix **A**, the i component of **x** would have the possibility to be related linearly to all the components of **u**.

A critical aspect of the method is that, in general, the variable **u** *does not* obey the unperturbed equation but rather is "updated" or modified by the perturbation. Thus, its equation of motion must depend on the perturbation parameter ε and be of the general form

$$\dot{\mathbf{u}} = \mathbf{U}_0(\mathbf{u}) + \sum_{n \geq 1} \varepsilon^n \mathbf{U}_n(\mathbf{u}) \qquad (1.5)$$

The quantity $\mathbf{U}_0$ is the unperturbed operator, and the vector functions $\mathbf{U}_n$ are the "updating" of the equation of motion of the zero order or the fundamental component of the full solution.

B. The **T** functions are the generators of the expansion and serve as modifications that are added to the zero-order term **u**, so as to obtain a better approximation to the structure of the full solution. They depend on the **u** variables. In our analysis, we will see that the **T** functions may be chosen so as to yield for the vector **u** equations of a simple form by eliminating as many extraneous terms in the equations.

C. A primary goal of the near-identity transformation is to make the **u** equation as "linear" as possible by choosing the **T** functions so as to eliminate as many of the nonlinear components as possible without losing the effect of the perturbation on the zero-order approximation. The capacity of the **T** functions to perform this role is linked to the structure of the spectrum on the linear operator, **A**, of Equation (1.1). One may choose the **T** functions so that the **u** equation is solvable or incorporates a major part of the effect of the perturbation already in the zero-order approximation. We will find that an obvious choice for the *assumed* structure of the **T** functions is to mimic the form of the perturbation, leading to a set of *linear* equations for the desired coefficients. (This aspect will be clarified through examples.)

D. A central point of our approach is the exploitation of the *freedom* in choosing the structure of the **T** functions, to modify aspects of the expansion, or to satisfy a desired constraint. For example, in Hamiltonian systems, one might wish to have a canonical near-identity transformation. Such an expansion is realized by using the freedom associated with the choice of the **T** functions. Once that freedom has been exploited, the result is a unique set of equations.

Comment. We develop the normal form expansions by means of a Taylor series. It is possible to develop a series using exponential transformations, the *Lie transformations*. This is an equivalent alternative for obtaining the normal form expansion [9–12]. The results obtained by either method are easily related to one another.

1.1.3 Departures from linear theory: Background

1. Timescales. Often one is interested in harmonic oscillations that are slightly perturbed by a nonlinearity. The resulting equations may display stable (almost harmonic) periodic oscillations or unstable (damped or growing amplitude) ones. In either case, the timescale over which significant changes occur in the system (period of oscillations in the oscillatory case, or timescale for amplitude damping) is affected by the nonlinearity: it depends on the amplitude of the oscillations. This is unexpected since our experience and intuition derived from linear systems and harmonic oscillations leads us to assume that the period of oscillations is independent of the amplitude. But, in truth, this result should not surprise us at all, as we have

INTRODUCTION

already encountered relations between amplitude and period in periodic motion in earlier studies. In particular, Kepler's law for planetary motion, states that "the square of the period of a planet orbiting the sun, is proportional to the cube of the major axis of the ellipse describing the orbit." The general case for conservative planar systems is well demonstrated by analyzing homogeneous potentials [13]. Consider the motion of a point particle of mass m in a one-dimensional conservative potential, $V(x)$. The equation of motion is

$$m\frac{d^2x}{dt^2} + \frac{dV}{dx} = 0 \qquad (1.6)$$

The total energy, E, is given by

$$E = \tfrac{1}{2}m\left(\frac{dx}{dt}\right)^2 + V(x) \qquad (1.7)$$

Solving for dx/dt, one has for the period of the closed-orbit motion

$$\frac{dx}{dt} = \pm\sqrt{\frac{2(E-V(x))}{m}} \quad ; \quad T = \oint \sqrt{\frac{m}{2(E-V(x))}}\, dx \qquad (1.8)$$

For a homogeneous symmetric potential of the form $V(x) = c \cdot x^\alpha$, one has $E = c \cdot x_0^\alpha$, where x_0 is the amplitude of the motion. The period is given by

$$\begin{aligned} T &= 4\int_0^{x_0}\sqrt{\frac{m}{2c x_0^\alpha\left(1-(x/x_0)^\alpha\right)}}\, dx \\ &= 4\sqrt{\frac{m}{2c}}\, x_0^{1-\alpha/2}\int_0^1 \sqrt{\frac{1}{(1-\xi^\alpha)}}\, d\xi \end{aligned} \qquad (1.9)$$

Consequently

$$T \propto x_0^{1-(1/2)\alpha} \qquad (1.10)$$

Two observations follow from this result:

1. If the potential has the form $V(x)=x^2$ (without loss of generality, we omit the multiplicative constant for the quadratic potential), then $T=2\pi$, independent of the amplitude. If, on the other hand, $V(x)$ contains a power of x that is different from 2, then the period depends on the amplitude. For example, if $V(x) \propto x^4$, then, from Equation (1.8) we find $T \propto 1/x_0$.

2. If a perturbative term is added to a harmonic potential, e.g., $V(x) = x^2 + c \cdot x^4$, then a mixing of scales takes place. For small-amplitude oscillations the potential is dominated by the x^2 term, the motion is very close to harmonic, and the period becomes almost independent of the amplitude. For large x, the x^4 term dominates, making the period amplitude dependent.

One should remember that when the harmonic component of the system is used as the basis for one's intuition and analysis, the "clock" one is using to observe the motion is "mismatched" to the true "clock": The period of the true oscillations is different from the period of the approximate harmonic motion. So, if one thinks of viewing the system with a stroboscope tuned to the unperturbed frequency, then after a sufficiently long time, one will observe the system to drift away from the periodic harmonic motion and appear to be executing aperiodic or "secular" motion relative to the periodic motion. Recall that, in our expansions, *secular* or *aperiodic* terms are those that appear as products of powers of the time multiplied by a trigonometric function. In the example just considered, the timescale over which this behavior appears to occur depends on the size of the quartic term in the potential. However, it is important to realize that, no matter how small the nonlinearity, if one adjusts one's calculational scheme so as to use an amplitude independent period as the basis for observation, one will encounter aperiodic behavior that is *not* characteristic of the true motion but rather an artifact of the approximation.

Thus, to recapitulate, keeping in mind that the true motion is periodic, if one attempts to develop an approximation scheme based on an expansion in harmonic components using the unperturbed period $T = 2\pi$, one misses the true periodicity. (A common example of this phenomenon that most of us have experienced while watching movies is the impression that the wheels of a moving car suddenly seem to be turning around slowly forward or backward, depending on the speed of the car. The cause of this illusion is the mismatch between the frequency at which the pictures are replaced and the frequency of the rotation of the wheels. If the two are matched, then the wheels seem to be static. If the two are just *slightly* different, then the wheels seem to be moving slowly at a speed that is proportional to the difference between the two frequencies.)

INTRODUCTION

The attempt to develop an approximation scheme that faithfully describes the motion of nonlinear conservative systems over long times was the subject of extensive investigation by scientists studying celestial motion, led by Poincaré and others, who developed methods associated with their names.

If the system is confined to a one-dimensional potential, executing bounded periodic motion, it has a closed orbit in phase space. The period for any conservative one-dimensional potential can be obtained by quadratures. (A nice discussion regarding the relation between the period and the potential for symmetric and asymmetric potentials can be found in the text on Mechanics by Landau and Lifshitz [13]). One can evaluate the required integrals numerically or as a power series in a small parameter to any desired accuracy (see the Appendix). With this said, many perturbation schemes develop the approximation for the displacement, $x(t)$, concurrent with the calculation of the frequency or period. In many cases this is not a recommended procedure. The calculations rely on different aspects of the equation, and it may be advantageous to develop separate expansions. For example, the calculation of the period involves a power-series expansion for the evaluation of a definite integral, while the successive terms in the displacement are often found by solving a hierarchy of algebraic equations.

2. Frequency multiplication–higher harmonics.

Nonlinear problems have a tendency to multiply frequencies. A linear harmonic oscillator

$$\ddot{x} + \omega_0^2 x = 0 \qquad (1.11a)$$

has one frequency, ω_0, and one period, $T = 2\pi/\omega_0$. Consider a nonlinear problem whose solution is periodic with period T. The solution will not have one pure frequency, $\omega = 2\pi/T$. Instead, the nonlinear terms will give rise to terms in ω as well as integer multiples of ω. Which multiples appear depends on the nature of the nonlinearity. For instance, consider a cubic nonlinearity:

$$\ddot{x} + \omega_0^2 x = \alpha x^3 \qquad (1.11b)$$

This will modify the frequency of the basic harmonic from ω_0 to some ω. In addition, it is easy to see that feeding a $\cos(\omega t)$ term into the equation generates on the r.h.s. two terms: a $\cos(\omega t)$ term and a $\cos(3\omega t)$ term. Repetition of this process reveals that in this problem, the solution will have higher harmonics with frequencies $k \cdot (3\omega)$ with all integers $k \geq 1$.

If the nonlinearity is quadratic

$$\ddot{x} + \omega_0^2 x = \alpha x^2 \qquad (1.11c)$$

then all integer multiples of the updated frequency will appear.

At least for small nonlinearities, we shall see that frequency multiplication is just a cosmetic effect, and that the core of the dynamics of the problem lies in the effect of the nonlinearity on the basic frequency of the solution and on the amplitude.

3. Loss of principle of superposition. The second order linear differential equation

$$\ddot{y} + y = 0$$

has two linearly independent solutions: $\sin(t)$ and $\cos(t)$. The general solution of the equation is a linear combination of these two solutions. This is an example for a general and basic characteristic of linear differential equations. Let us formally write a differential equation as

$$Lx = 0 \qquad (1.12)$$

where L is an operator and x is the solution. For a linear operator, with any two functions x_1 and x_2 and any constants α and β, one has:

$$L(\alpha x_1 + \beta x_2) = \alpha L x_1 + \beta L x_2 \qquad (1.13)$$

For example, if

$$L \equiv \frac{d^2}{dt^2} + \cos t,$$

then

$$L(\alpha x_1 + \beta x_2) = \alpha \left(\frac{d^2}{dt^2} x_1 + \cos t \, x_1 \right) + \beta \left(\frac{d^2}{dt^2} x_2 + \cos t \, x_2 \right)$$

$$= \alpha L x_1 + \beta L x_2$$

As a result, if x_1 and x_2 are both solutions of Equation (1.12), then any linear combination $\alpha x_1 + \beta x_2$ is also a solution that equation.

This is not the case in nonlinear problems. Equation (1.13) does not

INTRODUCTION

hold for nonlinear operators. For instance, in the nonlinear equation

$$\frac{d^2x}{dt^2} + x + \varepsilon x^3 = 0 \qquad (1.14)$$

we have

$$L(\alpha x_1 + \beta x_2) = \frac{d^2(\alpha x_1 + \beta x_2)}{dt^2} + (\alpha x_1 + \beta x_2) + \varepsilon(\alpha x_1 + \beta x_2)^3$$

$$= \alpha L x_1 + \beta L x_2 + \varepsilon\{(\alpha x_1 + \beta x_2)^3 - (\alpha^3 x_1^3 + \beta^3 x_2^3)\}$$

Thus, when x_1 and x_2 are solutions of Equation (1.14), their linear combination is not! Hence, *the principle of superposition does not hold for nonlinear problems*. An immediate conclusion is that the notion of *linearly independent solutions* loses its meaning.

4. Dependence on integration constants and/or initial conditions. The solutions of linear equations are linear functions of the initial conditions or integration constants. This is not the case in nonlinear problems. As an example, consider the equation

$$x\ddot{x} + \dot{x}^2 = 0$$

which is solved by

$$x = \pm\sqrt{At + B}$$

The dependence on the two integration constants A and B is nonlinear and specific to the equation. In particular, if x is a solution of a nonlinear problem, then, in general, $\alpha \cdot x$ is not a solution for $\alpha \neq 1$.

1.2 GOALS: WHAT THE BOOK IS ALL ABOUT

We do not want to overload the reader, but we feel that it is essential to give some overview even though important ideas and themes will not be fully explained until they appear in the appropriate sections.

1.2.1 Introductory comments regarding the normal form expansion and freedom

We have two primary goals.

The first goal is to explain what the method of normal forms is, and to introduce the concept of *nonuniqueness* or *freedom* in the normal form expansion and elaborate upon it. The method of normal forms has its origin in the work of Poincaré at the beginning of the century and was further developed in the early years primarily by Birkhoff and subsequently by Arnold, Siegel, and others. The fundamental idea is that with proper attention given to the structure of the spectrum of the linear terms in a nonlinear differential equation one introduces a "formal" near-identity transformation, in the neighborhood of a fixed point, to reduce the nonlinear flow to its most basic elements, yielding equations that are more amenable to analysis.

A central aspect of the method of normal forms is the separation of the problems into two classes: Those with eigenvalues that lie in a Poincaré domain and those with eigenvalues in a Siegel domain. We learn that problems of the former class are readily amenable to analysis. That is to say, everything proceeds in a straightforward and systematic way. By contrast, problems associated with the Siegel domain can yield striking complexity and, on occasion, chaotic behavior. Their analysis is much richer and often delicate.

In this work, we concentrate almost exclusively on problems that permit us to obtain the "formal" transformation as a differentiable near-identity transformation in the form of a power series in a small parameter. One begins by looking at the spectrum of the matrix **A** in Equation (1.1) and separating systems according to the location of the eigenvalues of **A**.

Case i. If all the eigenvalues have negative (or positive) real part, they are said to lie in a *Poincaré domain*. Problems of this nature are discussed in Chapter 5. Using this type of classification enables us to group and subsequently treat together a set of systems for which the perturbation expansion takes a similar form. Everything proceeds in a straightforward manner.

Case ii. Problems for which the unperturbed system has only two eigenvalues: $+i$ and $-i$ are separated into two groups. *Group 1:* A conservative system with small initial displacements. For sufficiently weak perturbations, one obtains a closed orbit in the phase plane. In the analysis, we seek both the correct frequency and a faithful description of the trajectory. Such systems are discussed in Chapter 6. *Group 2:* If the full linear term has eigenvalues with a small [i.e., $O(\varepsilon)$] nonzero real part, the system is dissipative and treated as described in Chapter 7.

Comment. We point out in passing that systems of this class are said to have their unperturbed eigenvalues in a Siegel domain. (The term *Siegel domain* encompasses a rich variety of systems including, for

GOALS: WHAT THE BOOK IS ALL ABOUT

example, those that have eigenvalues that straddle the imaginary axis. Problems in this class can exhibit striking complexity and, on occasion, chaotic behavior that outside the scope of this book. A full discussion of the classification of the eigenvalues is given in Chapter 5.) The introduction of this terminology *is not* helpful at this point, but enables us, at a later stage, to trace difficulties in particular expansions to the fact the eigenvalues are in a Siegel domain.

Case *iii*. Matrices that have one zero eigenvalue and the rest have negative real part are the subject of Chapter 9.

Other cases are discussed, but what we wish to emphasize is that the initial steps in the analysis will traced back to the classification of the eigenvalues.

The *second goal* is to show how near-identity transformations and associated techniques can be used to study a broad spectrum of nonlinear systems, emphasizing how the freedom in the expansion can be exploited to realize compact expansions that have a simple structure as well as significant computational advantages over other schemes.

In all our discussions, we explore the concept of freedom in perturbation expansions through a systematic development of the method of normal forms (NF). We illustrate the effectiveness of the normal form expansion by using it to obtain perturbative solutions for a broad class of nonlinear differential equations. Before the development of symbolic algebra computer programs, calculations were generally carried out only through first- or second-order, and methods such as *averaging* and *multiple scales* were used extensively. Higher-order calculations involve significant algebraic complexity and really cannot be done by hand. However, with the advent of readily available computer algebra software such as **MAPLE**, **MACSYMA**, or **MATHEMATICA** [14], the calculations are straightforward to organize, and the fundamental simplicity and beauty of the normal form expansion lead to its appeal. Finally, it is possible to show the equivalence between normal form expansions and the method of averaging for a limited class of problems.

A critical component in our thinking derives from the theorem of Hartman and Grobman that describes how the qualitative features of the linearized system reflect the dynamics of the full nonlinear equations [15–17]. This theorem states that near a hyperbolic fixed point there exists a change of coordinates, that need not be differentiable, that *locally* transforms the phase space orbits (or flows) into linear ones. (A hyperbolic fixed point is one such that in its neighborhood, all the associated eigenvalues, of the linear terms, have nonzero real part.) So we are guaranteed that in some neighborhood of the hyperbolic equilibrium point under consideration the nonlinear dynamics are qualitatively similar to the linear dynamics.

Another set of ideas comes into play when there are zero eigenvalues or pairs of complex conjugate pure imaginary eigenvalues and the remaining ones all have negative real part. It is possible, for this class of problems, to use the method of normal forms, or to introduce an approximation scheme associated with the method of the center manifold [16–18]. The latter method relies on the possibility, if one is in a neighborhood sufficiently close to the origin, to reduce the dimension of the system after a time such that the transient regime has passed. However, it is essential to keep in mind that this method is not applicable if the initial amplitudes are "too large" since the system might then be outside the basin of attraction associated with the decay onto the center manifold. (The eigenvalues of the linear system coincide with the stable or neutral modes of the linear system. The eigenvectors span corresponding invariant subspaces.) The Hartman–Grobman theorem and its generalization by Shoshitaishvili for the case of a nonhyperbolic fixed point [19] tell us that the nonlinear manifolds are tangent to the linear manifolds in the neighborhood of the equilibrium point. This means, for example, that if our system has some eigenvalues with zero real part and some with negative real part, the flow in phase space will be stable in the immediate neighborhood of the origin, except perhaps on the center manifold. In such cases, if one is not interested in the transient behavior, one may wish to restrict one's attention to the flow on the lower-dimensional center manifold.

That the normal form transformation is not unique has been extensively discussed in the literature and implemented in an effective way for a large class of problems [20–23]. (Also note that one encounters and discusses the same nonuniqueness in other perturbative schemes such as Averaging and MMTS.) With this nonuniqueness central in our mind, we say that there is *freedom* or *freedom of choice* associated with particular implementations of the normal form expansion that we have exploited to obtain *minimal normal forms* (MNF) [24]. There is much activity in this general area. Also, attention has been given to establishing conditions under which the Lie equation that is central to the NF expansion has a unique solution.

Summing up. In this work we use the method of normal forms and the freedom associated with nonunique aspects of perturbation theory to study various systems that can be separated into five classes.

1. Systems in which all the eigenvalues of the unperturbed linear problem have negative real part. The eigenvalues are said to lie in a *Poincaré domain*, and the normal form and the perturbation expansion have a particularly simple character.
2. Systems in which the unperturbed problem is a single simple

harmonic oscillator. It then can be perturbed by a change in the potential, so that the system remains conservative. The phase curves are changed, but remain closed. Alternatively, the harmonic oscillator can be perturbed by dissipative terms leading to damped motion or limit cycles. (We devote considerable attention to problems of this class and they are treated in detail in Chapters 6 and 7.)

3. Problems in which there is one zero eigenvalue and the rest have negative real part. Also, one can discuss, using the same techniques, systems in which one has a pair of pure imaginary eigenvalues and the remaining ones have negative real part. These problems are amenable to a normal form analysis, and this technique may, for the investigator's purposes, prove to be much superior to the *method of the center manifold*. The latter method is only effective after the transients have decayed and is limited to initial amplitudes that are "very close" to the fixed points. By way of contrast, the method of normal forms with only slightly more algebraic complexity is applicable to problems that have finite initial amplitudes, and it can follow the transient behavior until the system enters a region of finite-time blowup or has decayed onto the center manifold.

4. Nonautonomous systems have the capacity to exhibit very complex behavior. We restrict ourselves to a discussion of these systems in a region where one still has regular motion that is describable by a perturbation treatment. This excludes, of course, chaotic motion. However, it does cover a broad range of behavior.

5. We briefly treat coupled nonlinear oscillators. The problems are inherently much more complex, and a thorough analysis is beyond the scope of this text. However, it is straightforward to obtain the first- and second-order approximations and, for particular systems, to obtain some general results.

1.2.2 Reference models

In the development of our discussion in subsequent chapters, we will refer to two simple models that illustrate different aspects of the method of normal forms. By referring again and again to the same systems, it is possible to show how that various aspects of the normal form expansion present themselves. The models are as follows.

Model 1: the Duffing oscillator. This is a conservative system, generated from a simple harmonic oscillator, that is governed by a linear force, by the addition of a small cubic nonlinearity. The equation of motion in its original form is

$$\frac{d^2x}{dt^2} + x + \varepsilon x^3 = 0; \quad 0 < \varepsilon \ll 1, \quad x(0) = A, \quad \dot{x}(0) = 0 \quad (1.15)$$

The trajectory in the phase plane formed by x and dx/dt is a closed curve. It is a circle on which "wiggles," derived from interference terms associated with the nonlinearity, are superposed. We begin by first computing the period or angular frequency. This calculation is easily performed using a computer algebra software package such as MATHEMATICA or MAPLE. (We provide details of the calculation in the Appendix.) With the frequency known, we introduce the normal form expansion and compute, again using a computer algebra program, the U functions that yield the dynamical equation for the zero-order component of the motion and the T uunction generators. [See Equations (1.4a,b) and (1.5).] One then obtains equations for each order in ε. For example, through $O(\varepsilon)$ one obtains

$$U_0 = -i\,u$$

$$U_1 = -\left(\tfrac{3}{8}\right) i\, u^2 u^*$$

$$T_1 = \left(\tfrac{1}{16}\right) u^3 - \left(\tfrac{3}{16}\right) u u^{*2} - \left(\tfrac{1}{32}\right) u^{*3} + \alpha u^2 u^*$$

The last term in T_1, proportional to $u^2 u^*$, is a free term that appears because there is freedom in the expansion; it is undetermined by the constraints generated by the method of normal forms. It affects the first-order approximation to the solution. The choice of the free terms and the associated implementation of the initial conditions are central to our development and will be discussed in each chapter, beginning with Chapter 4. Essential details for the Duffing oscillator are given in Chapter 6, where one learns how to use the T and U functions to obtain an approximation for $x(t)$ through $O(\varepsilon)$. With α real, one finds

$$x(t) = \left[\rho + \varepsilon\left(-\tfrac{3}{16} + \alpha\right)\rho^3\right]\cos\theta + \varepsilon\left(\tfrac{1}{32}\right)\rho\cos 3\theta\,; \quad \theta = \omega t \quad (1.16)$$

where ρ is related to the initial amplitude, A, in a simple manner and the phase, θ, is determined by the full frequency (which is affected by the perturbation). The $O(\varepsilon^2)$ term will include trigonometric functions of θ, 3θ and 5θ.

Comment. Although it is too early to show the role that the free term plays in providing an improved approximation to the solution, we just make a small stab at it. The approximate solution given by Equation

GOALS: WHAT THE BOOK IS ALL ABOUT

(1.16) satisfies the original differential equation with an error $O(\varepsilon^2)$ independent of the α-term.

(Substitute the approximate solution into the original equation and satisfy yourself that this is indeed the case.) Thus, one concludes that it may be convenient to choose the coefficient α so as to suppress the fundamental component ($\cos\theta$) of $x(t)$ in $O(\varepsilon^2)$. In this way, one has the capacity to improve the quality of our approximation by in effect "reaching" into the next order of the expansion and including part of it in the present order. Furthermore, it is precisely the *freedom* associated with the equation of motion of the fundamental component that allows us to obtain compact expansions.

Model 2: the Van der Pol oscillator. We now introduce a model system that brings out a different aspect of the freedom in the perturbation expansion. This equation was introduced by Van der Pol to model an electric circuit that contained a nonlinear component called a *triode*. The equation is equivalent to that developed by Lord Rayleigh to model sound vibrations. The basic equation is

$$\frac{d^2x}{dt^2} + x = \varepsilon(1-x^2)\frac{dx}{dt} \quad 0 < \varepsilon \ll 1 \quad (1.17a)$$

If one looks at the equation and begins with a small amplitude, the right-hand side is positive and as the motion develops, energy is "pumped" into the system, thereby increasing the amplitude. When the displacement, x, exceeds 1, the sign of the dissipative term becomes negative, leading to deceleration and a slowing down and, eventually, a halt to the amplitude growth. The amplitude then decreases and falls below the value 1, when once again the dissipative term brings energy into the system repeating this behavior.

This process eventually yields a *limit cycle*. Its zero-order component, called a *limit circle*, has a radius approximately equal to 2. (The *limit cycle* describes the full curve, and the *limit circle* describes the curve associated with the zero-order approximation.) The limit cycle is one of the phenomena characteristic of dissipative nonlinear systems. For $t \to \infty$, it goes to a limiting oscillatory (unharmonic) motion. For such systems, the limit cycle is a stable closed trajectory in the phase plane that is approached asymptotically from interior and exterior spiral orbits. A much more precise and detailed discussion including definitions is given in Chapter 7.

Now, how does one obtain an approximate limit cycle radius for the Van der Pol equation? In the traditional perturbational approach, one calculates a lowest-order approximation to the radius of the *limit circle*, and higher-order terms yield corrections to this value. The calculations are facilitated by the use of computer algebra. In order to get a feeling for what is happening, it may be helpful to give a few details. One writes the zero-order approximation, u, as

$$u = \rho \exp(-i\,\varphi)$$

The asymptotic form of u is the *limit circle*, as ρ goes to a constant.

When the freedom in the expansion is *not* exploited, one finds that the radius of the limit circle of the Van der Pol oscillator is given by $\rho = 2 + O(\varepsilon^2)$. That is, potentially the radius is updated in higher orders of the expansion. However, as we show in Chapter 7, for problems of this class, the freedom in the normal form expansion can be used to "renormalize" the radius so that it is not modified by the higher-order corrections. The equation of motion is found, using computer algebra as a computational tool, with the implementation of procedures derived in Chapter 7. One obtains

$$\frac{d\rho}{dt} = \tfrac{1}{2}\varepsilon\rho\left(1 - \tfrac{1}{4}\rho^2\right)\left\{\begin{array}{l}\left(1 + \varepsilon^2\left(\tfrac{13}{2048}\rho^6 + \tfrac{11}{512}\rho^4\right)\right) \\ + \varepsilon^4\left(\begin{array}{l}-\tfrac{231}{4096}\rho^4 + \tfrac{441}{8192}\rho^6 \\ -\tfrac{33{,}913}{1{,}572{,}864}\rho^8 + \tfrac{15{,}467}{2{,}359{,}296}\rho^{10} \\ -\tfrac{46{,}356}{3{,}355{,}432}\rho^{12} + \tfrac{15{,}717}{134{,}217{,}728}\rho^{14}\end{array}\right) \\ + O(\varepsilon^6)\end{array}\right\} \quad (1.18)$$

The freedom in the expansion has been used to incorporate all the higher-order corrections as a multiplicative factor that modifies the lowest-order term in the equation of motion of ρ. The asymptotic value of the radius remains at 2, its value at the onset of the perturbation (i.e., in the limit $\varepsilon \to 0$).

REFERENCES

1. Lotka, A. J., *Elements of Mathematical Biology*, Dover, New York (1956).
2. D'Ancona, U., *Struggle for Existence*, Brill, Leiden (1954).
3. Winfree, A.T., *The Geometry of Biological Time*, Springer-Verlag, New York (1980).
4. Krylov, N. M., and N. M. Bogoliubov, *Introduction to Nonlinear Mechanics*, Princeton University Press, Princeton, NJ (1947).
5. Bogoliubov, N. M., and Ya. Mitropolsky, *Asymptotic Methods in the Theory of Nonlinear Oscillations*, Gordon & Breach, New York (1961).

6. Lorenz, E. N., "Deterministic Non-Periodic Flow," *J. Atmos. Sci.* **20**, 130 (1963).
7. May, R., "Simple Mathematical Models with Very Complicated Dynamics," *Nature* **261**, 459 (1976).
8. Feigenbaum, M. J., "Quantitative Universality for a Class of Nonlinear Transformations," *J. Stat. Phys.* **19**, 25 (1978).
9. Hori, G. I., "Theory of General Perturbations for Noncanonical Systems," *Publ. Astron. Soc. Japan* **23**, 567 (1971).
10. Dragt, A. J., and J. M. Finn, "Lie Series and Invariant Functions for Analytic Symplectic Maps," *J. Math. Phys.* **17**, 225 (1976).
11. Meyer, K. R., and G. R. Hall, *Introduction to Hamiltonian Dynamical Systems and the N-Body Problem*, Springer-Verlag, New York (1992).
12. Nayfeh, A. H., *Perturbation Methods*, Wiley, New York (1973).
13. Landau, L. D., and E. M. Lifshitz, *Mechanics*, Pergamon Press, Bristol (1960).
14. Rand, R. H., and D. Armbruster, *Perturbation Methods, Bifurcation Theory and Computer Algebra*, Springer-Verlag, New York (1987).
15. Arnold, V. I., *Geometrical Methods in the Theory of Ordinary Differential Equations*, 2nd ed., Springer-Verlag, New York (1988).
16. Guckenheimer, J., and P. J. Holmes, *Nonlinear Oscillations, Dynamical Systems, and Bifurcations of Vector Fields*, Springer-Verlag, New York (1988).
17. Wiggins, S., *Introduction to Applied Nonlinear Dynamical Systems and Chaos*, Springer-Verlag, New York (1990).
18. Carr, J., *Applications of Centre Manifold Theory*, Springer-Verlag, New York (1981).
19. Crawford, J. D., "Introduction to Bifurcation Theory," *Rev. Mod. Phys.* **63**, 991 (1991).
20. Bruno, A. D., *Local Methods in Nonlinear Differential Equations*, Springer-Verlag, New York (1989).
21. Baider, A., and R. C. Churchill, "Uniqueness and Non-Uniqueness of Normal Form Expansions for Vector Fields," *Proc. Roy. Soc.* (Edinburgh) **A108**, 27 (1988).
22. Kummer, M., "How to Avoid Secular Terms in Classical and Quantum Mechanics," *Nuov. Cim.* **1B**, 123 (1971).
23. Liu, J. C., "The Uniqueness of Normal Forms Via Lie Transformations and Its Applications to Hamiltonian Systems," *Celest. Mech.* **36**, 89 (1985).
24. Kahn, P. B. and Y. Zarmi, "Minimal Normal Forms in Harmonic Oscillations with Small Nonlinear Perturbations," *Physica D* **54**, 65 (1991).

2

BASIC CONCEPTS

2.1 DEFINITIONS

2.1.1 Fixed point

Consider a dynamical system, the equations of motion of which are

$$\frac{d\mathbf{x}}{dt} = \mathbf{F}(\mathbf{x};\boldsymbol{\mu}) \qquad \mathbf{x}(0) = \mathbf{x}_0 \qquad (2.1)$$

In Equation (2.1), **x** is an n-dimensional vector, **F** is an n-dimensional vector field, and **μ** is a set of parameters. A point $\mathbf{x}_0(\boldsymbol{\mu})$ is called a *fixed point* of the system, if

$$\left.\frac{d\mathbf{x}}{dt}\right|_{\mathbf{x}=\mathbf{x}_0} = \mathbf{F}(\mathbf{x}_0;\boldsymbol{\mu}) = 0 \qquad (2.2)$$

This means that

$$\mathbf{x}(t;\boldsymbol{\mu}) \equiv \mathbf{x}_0(\boldsymbol{\mu})$$

is a solution of Equation (2.1).

Example 2.1: The logistic equation. The system

$$\frac{dx}{dt} = \lambda x(K - x) \qquad (2.3)$$

has two fixed points, at $x=0$ and $x=K$, both yielding $dx/dt=0$.

Example 2.2: The Duffing oscillator. The two-dimensional system

$$\frac{dx}{dt} = y \qquad \frac{dy}{dt} = -x - \varepsilon x^3$$

has a fixed point at $(x,y)=(0,0)$. If $\varepsilon<0$, the system also has fixed points at $(\pm(-1/\varepsilon)^{1/2}, 0)$.

2.1.2 Hyperbolic fixed point

One is often interested in the behavior of a system near a fixed point $\mathbf{x}_0(\boldsymbol{\mu})$. If the vector field $\mathbf{F}(\mathbf{x},\boldsymbol{\mu})$ is sufficiently well behaved near $\mathbf{x}_0(\boldsymbol{\mu})$, e.g., all its first derivatives exist and are continuous in some neighborhood of $\mathbf{x}_0(\boldsymbol{\mu})$, then one can write

$$\frac{d\mathbf{x}}{dt} = \mathbf{A}(\boldsymbol{\mu})(\mathbf{x}-\mathbf{x}_0(\boldsymbol{\mu})) + \tilde{\mathbf{G}}(\mathbf{x}-\mathbf{x}_0(\boldsymbol{\mu});\boldsymbol{\mu}) \qquad \|\mathbf{x}-\mathbf{x}_0(\boldsymbol{\mu})\| < R \quad (2.4)$$

where the norm is appropriately defined and R defines a neighborhood of the fixed point within which Equation (2.4) holds. The function $\tilde{\mathbf{G}}(\mathbf{x}-\mathbf{x}_0(\boldsymbol{\mu});\boldsymbol{\mu})$ is nonlinear in $\mathbf{x}-\mathbf{x}_0(\boldsymbol{\mu})$. $\mathbf{A}$ is the matrix of the derivatives of $\mathbf{F}(\mathbf{x},\boldsymbol{\mu})$ at $\mathbf{x}=\mathbf{x}_0(\boldsymbol{\mu})$:

$$A_{ij} = \left.\frac{\partial F_i(\mathbf{x},\boldsymbol{\mu})}{\partial x_j}\right|_{\mathbf{x}=\mathbf{x}_0(\boldsymbol{\mu})}$$

Shifting the fixed point to the origin ($\mathbf{x}=0$), Equation (2.4) becomes

$$\frac{d\mathbf{x}}{dt} = \mathbf{A}(\boldsymbol{\mu})\mathbf{x} + \mathbf{G}(\mathbf{x};\boldsymbol{\mu}) \qquad \|\mathbf{x}\| < R \quad (2.5)$$

The $\boldsymbol{\mu}$ dependence of $\mathbf{x}_0(\boldsymbol{\mu})$ is absorbed in $\mathbf{G}(\mathbf{x};\boldsymbol{\mu})$.

Near the fixed point, the character of the solution depends on the eigenvalues of $\mathbf{A}$, $\lambda_1,\ldots,\lambda_n$. If all have nonzero real parts (i.e., excluding complex conjugate pairs, $\lambda_k=\pm i\omega$, and zero, $\lambda_l=0$), the fixed point is called *hyperbolic*.

Example 2.3. $\mathbf{x}=0$ is a hyperbolic fixed point of the system

$$\frac{d}{dt}\begin{pmatrix} x_1 \\ x_2 \\ x_3 \end{pmatrix} = \begin{pmatrix} 1+2i & & \\ & 1-2i & \\ & & -5 \end{pmatrix}\begin{pmatrix} x_1 \\ x_2 \\ x_3 \end{pmatrix} + \begin{pmatrix} 2x_1 x_2^3 \\ 9x_3^5 \\ -0.1x_1^2 + x_2^2 \end{pmatrix}$$

DEFINITIONS

If all the eigenvalues have negative real parts, there is a neighborhood of the origin such that when, at some time, the solution is within that neighborhood, it will be attracted to the origin (this will be clarified later on, when the Poincaré–Lyapunov theorem is discussed). Often, such a hyperbolic fixed point is called a *sink*. Similarly, if all the eigenvalues have positive real parts, then there is a neighborhood within which solutions are repelled away from the origin. Often, such a hyperbolic fixed point is called a *source*.

If some of the eigenvalues have positive real parts and some have negative real parts, than, along some directions the solution will be attracted toward the origin, whereas along others, it will be repelled away from it. Such a point is called a *saddle fixed point*.

Example 2.4. Consider the two-dimensional system

$$\dot{x} = x + \varepsilon x y \qquad \dot{y} = -y + \varepsilon y^2 \qquad (2.6)$$

The origin is a saddle fixed point. In its vicinity, the y component decays to zero, while the x component grows. Typical trajectories are shown in Figure 2.1.

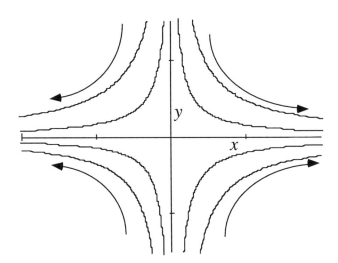

Figure 2.1 Solutions of Equations (2.6) for $\varepsilon=0.1$. Arrows indicate the direction of evolution of the solutions in time.

2.1.3 Center fixed point

A fixed point is called a *center*, if all the eigenvalues are pure imaginary and do not Vanish. In the two dimensional case, a center is characterized by a neighborhood within which an infinite number of closed orbits may exist.

Example 2.5: The pendulum. Consider the equation

$$\ddot{x} + \sin x = 0 \tag{2.7a}$$

Defining $y=(dx/dt)$, this second-order equation becomes a set of two first-order equations:

$$\dot{x} = y \qquad \dot{y} = -\sin x \tag{2.7b}$$

The fixed points of Equations (2.7b) are at $(x,y)=(n\pi,0)$. The linear stability analysis is simple. Near the fixed point we write

$$x = n\pi + \xi \qquad |\xi| \ll 1$$
$$|y| \ll 1$$

Expanding Equations (2.7b) in ξ and retaining the linear parts only, we find

$$\dot{\xi} = y \qquad \dot{y} = (-1)^{n+1}\xi \tag{2.7c}$$

The eigenvalues of the linear part are $\pm(-1)^{(n+1)/2}$. For *odd n*, they are ± 1 (a saddle). For *even n*, they are $\pm i$ (a center). We now concentrate on the first three fixed points: $(0,0)$ and $(\pm\pi,0)$.

The system of Equations (2.7) has a first integral:

$$\tfrac{1}{2}y^2 - \cos x = E - 1 \tag{2.8}$$

where E is the energy. We note that $E=0$ at the center $[(x, y)=(0,0)]$. For $0<E<2$, $y=0$ is a possible value for the velocity, occurring when

$$-1 < \cos x = 1 - E < 1$$

For $0<E<2$, the solution is a closed orbit, its period given by

$$T = 4\int_0^a \frac{dx}{\sqrt{2(E-1+\cos x)}} \qquad a = \cos^{-1}(1-E) \tag{2.9}$$

DEFINITIONS

For $E<2$, the period is finite. The only region in the interval of integration that may cause a divergence of the integral occurs near $x=a$. In that vicinity, we write

$$x = a - \xi$$

To leading powers in ξ, the integrand in equation (2.9) then becomes

$$\frac{1}{\sqrt{(E-1)\xi^2 + 2\sqrt{E(2-E)}\,\xi}}$$

For $0<E<2$, the linear term in the square root dominates as $\xi \to 0$. The integrand behaves like $\xi^{-1/2}$, so that the integral converges (in particular, as $E \to 0$, it yields $T \to 0$). However, for $E \to 2$, the quadratic term dominates, causing the integral to diverge as $\xi \to 0$.

The solution for $E=2$ is called the *separatrix*. It separates between the phase space domain ($E<2$) within which a continuum of periodic solutions (one for each value of E) exists, all oscillating around the center at (0,0). For $E>2$, the velocity, y, never Vanishes. It stays finite all the time, causing x to become unbounded over long times. Typical trajectories are shown in Figure 2.2.

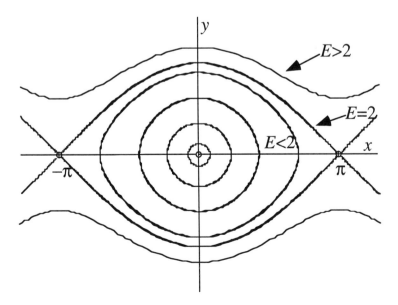

Figure 2.2 Sample of phase-plane trajectories of Equation (2.7).

2.1.4 Stability of fixed points

A fixed point is called *stable* if a neighborhood around it exists, so that when the solution is within that neighborhood, it is attracted toward the fixed point. The term *local stability* is used if that neighborhood covers only a part of phase space. The term *global stability* is used if the solution is attracted toward the fixed point from anywhere in phase space.

Example 2.6. The point $x=0$ is a locally stable fixed point of the equation

$$\dot{x} = -x(1-x)$$

The solution is attracted to $x=0$ only for initial conditions $x_0<1$.

Example 2.7. The point (0,0) is a globally stable fixed point of the system (harmonic oscillator with cubic damping)

$$\dot{x} = y \qquad \dot{y} = -x - y^3 \qquad (2.10)$$

A sample of phase-space trajectories of this system is shown in Figure 2.3. All trajectories are attracted toward the origin as $t\rightarrow\infty$.

If the fixed point does not obey the previous definition, it is called *unstable*. For example, a saddle fixed point is unstable (at least in one direction in phase space). If, in Equation (2.10), the sign of the cubic damping term is reversed, then the origin becomes unstable, a *repeller*.

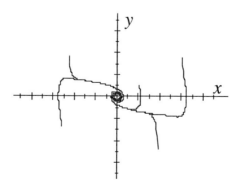

Figure 2.3 Sample of phase-space trajectories of Equation (2.10). All converge onto the origin.

DEFINITIONS

2.1.5 Stability of solutions

Often, one is interested in the stability of solutions of the dynamical equations. For example, if a perturbation method is employed, it is desirable to know for how long in time the approximation is close to the exact, but unknown, solution. In other problems, one is interested in finding whether the solution of a given problem does or does not approach some asymptotic form. If it does, the rate of approach may be of interest. Let $\mathbf{x}(t)$ be a solution of Equation (2.1).

Lyapunov stability. One is first interested in the of sensitivity of the solution to a slight change in the initial condition. Let $\varepsilon > 0$ be an allowed deviation in the solution, and let $\delta > 0$ be the allowable range of variation in the initial condition. If, for any arbitrarily small ε, there exists a δ such that for two initial conditions $\mathbf{x}_0$ and $\mathbf{x}'_0$ satisfying

$$\|\mathbf{x}_0 - \mathbf{x}'_0\| \leq \delta$$

the solution $\mathbf{x}'(t)$ emanating from the new initial condition satisfies

$$\|\mathbf{x}'(t) - \mathbf{x}(t)\| < \varepsilon$$

$\mathbf{x}(t)$ is called *stable in the Lyapunov sense.* See Figure 2.4.

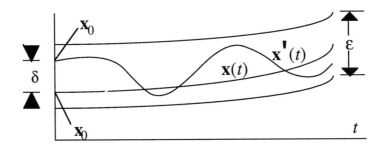

Figure 2.4 $\mathbf{x}(t)$ is stable in the Lyapunov sense.

Example 2.8. Consider a harmonic oscillator

$$\ddot{x} + x = 0$$

which can be transformed into two first-order equations:

$$\frac{d}{dt}\begin{pmatrix} x \\ y \end{pmatrix} = \begin{pmatrix} y \\ -x \end{pmatrix} = \begin{pmatrix} 0 & 1 \\ -1 & 0 \end{pmatrix}\begin{pmatrix} x \\ y \end{pmatrix}$$

Define

$$z \equiv (x, y)$$

then $z=(0,0)$ is clearly a solution. Now assume that the initial condition for the problem is z_0, satisfying

$$\|z_0\| = |x_0| + |y_0| \leq \delta \quad \Rightarrow \quad |x_0| \leq \delta, \quad |y_0| \leq \delta$$

This equation is solved by

$$z = \begin{pmatrix} x \\ y \end{pmatrix} = \exp\left(-\begin{pmatrix} 0 & 1 \\ -1 & 0 \end{pmatrix} t\right)\begin{pmatrix} x_0 \\ y_0 \end{pmatrix} = \begin{pmatrix} x_0 \cos t - y_0 \sin t \\ y_0 \cos t + x_0 \sin t \end{pmatrix}$$

which satisfies

$$\|z\| = |x| + |y| \leq 2(|x_0| + |y_0|) \leq 2\delta$$

Choose $\delta = \varepsilon/2$ to satisfy the Lyapunov stability condition. Thus, for an initial condition that is close to $z=0$, the solution always stays within the same circle. However, *it need not tend to zero*.

Example 2.9. Any solution of the equation

$$\dot{x} = 1 \qquad x(0) = x_0$$

is unbounded, but is stable in the Lyapunov sense.

Example 2.10. $x=0$ is a Lyapunov stable solution of the equation

$$\dot{x} = -kx \qquad x(0) = 1$$

Example 2.11. Any solution of the equation

$$\dot{x} = x \qquad x(0) = x_0$$

is unbounded and unstable in the Lyapunov sense.

DEFINITIONS

Positive attractor. The solution $\mathbf{x}(t)$ is a *positive attractor*, if for another solution, $\mathbf{x}'(t)$, that is sufficiently close to $\mathbf{x}(t)$, the two coincide asymptotically. Thus, for sufficiently small δ, one has

$$\|\mathbf{x}_0' - \mathbf{x}_0\| \leq \delta \implies \lim_{t \to \infty} |\mathbf{x}'(t) - \mathbf{x}(t)| = 0$$

Note that the two may be distant from each other *only* for a while. See Figure 2.5.

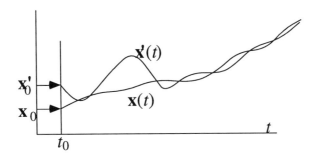

Figure 2.5 Example of positive attractor.

Asymptotic stability. $\mathbf{x}(t)$ is called *asymptotically stable* if it is a Lyapunov stable positive attractor. A solution $\mathbf{x}'(t)$ that starts close to $\mathbf{x}(t)$, at $t = 0$ (i.e., within a δ neighborhood) stays close to the latter for all times, and tends to the latter for $t \to \infty$. See Figure 2.6.

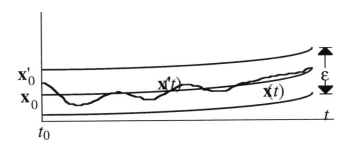

Figure 2.6 Example of asymptotic stability.

Example 2.12. For a spring with friction

$$\ddot{x} + \gamma \dot{x} + x = 0$$

the point $x=0$, $dx/dt=0$ is an asymptotically stable solution.

Example 2.13: Lyapunov stability in two dimensions. Initial conditions are within a circle of radius δ. The solutions remain within distance ε (but need not approach one other as $t \to \infty$). See Figure 2.7.

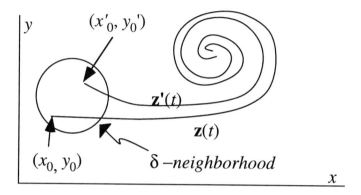

Figure 2.7 Lyapunov stability in two-dimensional phase plane.

Example 2.14: Positive attractor in two dimensions. The solutions start at a distance $\leq \delta$ and approach each other as $t \to \infty$, although they are not always within ε of each other. See Figure 2.8.

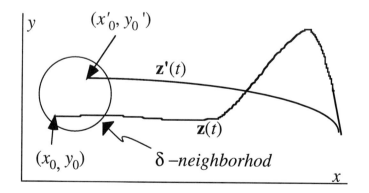

Figure 2.8 Positive attractor in two-dimensional phase plane.

Example 2.15: Asymptotic stability in two dimensions. Solutions are always close to each other and approach one another as $t \to \infty$. See Figure 2.9.

DEFINITIONS

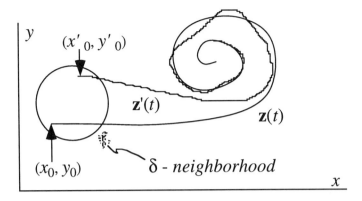

Figure 2.9 Asymptotic stability in two-dimensional phase plane.

2.1.6 Limit cycle

Interesting phenomena occur when dissipative terms are added to the equations describing a system that has a center fixed point. For example, consider the equation of a harmonic oscillator

$$\ddot{x} + x = 0$$

The point $(x, dx/dt) = (0,0)$ is a center. There is an infinite family of periodic closed-orbit solutions (circles) that depend continuously on the total energy and oscillate around the origin in the phase plane.

If a pure friction term is added to the equation, whether it is linear

$$\ddot{x} + x = \varepsilon \dot{x}$$

or nonlinear, e.g., cubic,

$$\ddot{x} + x = \varepsilon \dot{x}^3$$

then the fixed point at the origin ceases to be a center. The infinity of closed-orbit solutions is replaced by an infinite family of curves that spiral toward the origin for $\varepsilon<0$ (a *sink*) and away from the origin for $\varepsilon>0$ (a *source*). The solutions behave in that way because the friction terms modify the energy of the system in one direction only. For instance, in the linear case, multiplying the equation by dx/dt we find

$$\frac{dE}{dt} = \frac{d}{dt}\left(\tfrac{1}{2}\dot{x}^2 + \tfrac{1}{2}x^2\right) = \varepsilon \dot{x}^2$$

This corresponds energy dissipation (generation) of for ε<0 (ε>0).

We make ε x-dependent, so that it changes signs as one goes from small to large amplitudes. For instance, consider the Van der Pol equation, to be discussed at length later on in the book

$$\ddot{x} + x = \varepsilon \dot{x}\left(1 - x^2\right)$$

For ε>0, at small amplitudes the r.h.s. generates energy. Hence, if the system is near the origin, the solution spirals outward. At large amplitudes, the friction term is dissipative, causing a reduction of the energy: the solution spirals toward smaller amplitudes. It can be shown [1] that, in such a situation, there is a limiting closed, periodic curve, called a *stable limit cycle*, toward which the solution is attracted. (See Figure 2.10.) On the limit cycle, energy dissipation and generation occur over different parts of the phase plane in such a manner that the total change in the energy over one cycle Vanishes.

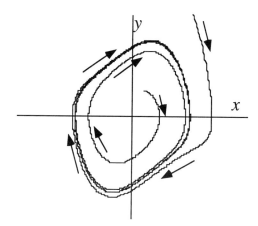

Figure 2.10 Example of a stable limit cycle. Arrows indicate direction of motion as time grows.

A limiting curve also exists for ε<0. However, the solution spirals away from it. Inside the limiting curve it decays toward the origin, and outside it diverges. The limit cycle is *unstable*. (See Figure 2.11.)

Another characteristic of the limit cycle is that there are no other closed-orbit periodic solutions in its vicinity. For small ε, the limit cycle is very close to a circle of a particular radius. As ε→0, it degenerates to that circle (its radius equal to 2 in the case of the Van der Pol equation). Thus, of the infinite family of circles allowed around the center fixed point, only one is selected by the peculiar friction term.

DEFINITIONS

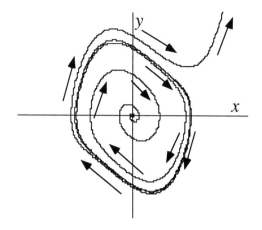

Figure 2.11 Example of an unstable limit cycle.

2.1.7 The O symbol

Often one wants to state the size of a quantity in relation to another one. In this book, we will address the length of time scales and the level of an approximation. For these two we use the O symbol.

Given a small parameter, ε, we say that a quantity ξ is $O(\varepsilon)$ if a positive constant K exists such that

$$|\xi| \leq K|\varepsilon| \qquad \varepsilon \to 0$$

Example 2.16

$$\varepsilon = O(\varepsilon) \qquad \varepsilon^2 = O(\varepsilon) \qquad \frac{2+\varepsilon}{1+\varepsilon} = 2 - \varepsilon + \varepsilon^2 + O(\varepsilon^3)$$

Within the context of a perturbation expansion, the range in time over which a certain approximation is valid, is expressed in terms of powers of $(1/\varepsilon)$. We say that $t = O(1/\varepsilon)$ if for some positive constant T

$$t \leq T/\varepsilon \qquad \varepsilon \to 0$$

Example 2.17 $\qquad T_1/\varepsilon^2 = O(1/\varepsilon)$

$$\sin\!\left((1 + \varepsilon\omega_1 + \varepsilon^2\omega_2)t\right) = \begin{cases} \sin t + O(\varepsilon) & t = O(1) \\ \sin\!\left((1 + \varepsilon\omega_1)t\right) + O(\varepsilon) & t = O(1/\varepsilon) \end{cases}$$

2.1.8 Asymptotic series

In perturbation theory, one generates the solution as a power series in the small perturbation parameter. The series is known to converge only in very few cases. This situation is related to the arbitrary manner in which we impose the expansion in powers of the small parameter, ignoring the fact that the full solution may not be describable in terms of such an expansion. (See the discussion following the presentation of the Hartman–Grobman theorem in Section 2.2.2.) We do so owing to the lack of better tools. However, in many problems, although the infinite series diverges, a finite number of terms in the sum provides an approximation with a specified error to the full (unknown) solution. One then says that the series is an *asymptotic* one.

The concept of asymptotic series is discussed thoroughly in the literature, and we refer the reader to standard texts that cover this topic [2–4]. It will be easiest to understand the idea, starting from the familiar notion of a convergent series. The series

$$\sum_{n=0}^{\infty} a_n \varepsilon^n \qquad (2.11)$$

is convergent for $|\varepsilon| \leq \varepsilon_0$, if the remainder of the series obeys

$$R_N(\varepsilon) = \sum_{N+1}^{\infty} a_n \varepsilon^n \to 0, \qquad N \to \infty; \quad \varepsilon \text{ fixed} \qquad (2.12)$$

Equation (2.12) implies, in particular, that a positive constant K exists such that

$$|R_N| \ll K |\varepsilon|^N, \qquad \varepsilon \to 0; \quad N \text{ fixed} \qquad (2.13)$$

Equation (2.13) defines an *asymptotic* series. The remainder goes to zero faster than $|\varepsilon|^N$ as $\varepsilon \to 0$, but need not tend to zero for fixed ε, as $N \to \infty$. Equation (2.13) is a prerequisite for convergence, but does not guarantee the latter. Convergence imposes a more stringent requirement on the remainder.

An infinite series is called *asymptotic to a function* $f(\varepsilon)$ as $\varepsilon \to 0$, if for *every* N, one has

$$\left| f(\varepsilon) - \sum_{n=0}^{N} a_n \varepsilon^n \right| \ll K |\varepsilon|^n, \qquad \varepsilon \to 0 \qquad (2.14)$$

DEFINITIONS

Again, the remainder has to Vanish as $\varepsilon \to 0$, faster than the last tern retained in the sum. Notice that the series need not be convergent. The symbol for a series that is asymptotic to a function is

$$f(\varepsilon) \sim \sum_{n=0}^{\infty} a_n \varepsilon^n, \qquad \varepsilon \to 0$$

Example 2.18. We wish to study the large x behavior of the function, defined for all $x>0$:

$$f(x) = \int_0^{\infty} \frac{e^{-t}}{x+t} dt$$

Repeated integration by parts yields the following after N iterations:

$$f(x) = \sum_{n=0}^{N} (-1)^n \frac{n!}{x^{n+1}} + R_N$$

where

$$R_N = (-1)^{N+1} (N+1)! \int_0^{\infty} \frac{e^{-t}}{(x+t)^{N+2}} dt$$

yielding

$$|R_N| \ll \frac{(N+1)!}{x^{N+2}}$$

The requirement for the series to be asymptotic to $f(x)$, is obeyed:

$$f(x) \sim \sum_{n=0}^{\infty} (-1)^n \frac{n!}{x^{n+1}}$$

However, the series diverges for any $x>0$. A truncated version of it provides a good approximation for $f(x)$ for fixed N as $x \to \infty$. The larger is x [hence, the smaller is $(1/x)$], the smaller is the number of terms required to provide a good description of $f(x)$.

2.1.9 Near-identity transformation

In performing a perturbation analysis, we will often use the term *near-identity transformation*. Consider the system of equations

$$\frac{d\mathbf{x}}{dt} = \mathbf{A}\mathbf{x} + \varepsilon \mathbf{G}(\mathbf{x};\varepsilon) \quad \|\mathbf{x}\| < R \quad \quad |\varepsilon| \ll 1 \quad \quad (2.15)$$

where $\mathbf{x}$ is an N-dimensional vector. One now postulates that $\mathbf{x}$ can be expanded in powers of ε. The relation between $\mathbf{x}(t)$ and its zero-order approximation $\mathbf{y}(t)$

$$\mathbf{x} = \mathbf{y} + \sum_{n \geq 1} \varepsilon^n \mathbf{T}_n(t;\mathbf{y}) \quad \quad (2.16)$$

is called the *near-identity transformation*. Formally, it can be viewed as a transformation from the N-dimensional space into itself. As pointed out earlier, Equation (2.16) is imposed with no guarantee that the infinite series converges. The near-identity transformation may then be just an asymptotic expansion of the solution of Equation (2.15).

2.2 FUNDAMENTAL THEOREMS

2.2.1 Existence and uniqueness of solutions: Lipschitz condition

Although we do not delve into the theory of ordinary differential equations in this book, we will deal only with systems for which the fundamental theorem for existence and uniqueness of the solutions is obeyed. There are various conditions that guarantee that the theorem holds. Of these, the most common one is the *Lipschitz condition*.

Let $\mathbf{f}(\mathbf{x},t;\varepsilon)$ be an N-dimensional vector function of an N-dimensional vector $\mathbf{x}$, of time, and of a set of parameters ε. If in a domain $\mathbf{x} \in D \subset \mathbf{R}^N$, for some time interval $t_0 \leq t \leq t_1$, and some domain in the parameter space $\|\varepsilon\| \leq \varepsilon_0$, there exists a positive constant L so that

$$\|\mathbf{f}(\mathbf{x}_1,t;\varepsilon) - \mathbf{f}(\mathbf{x}_2,t;\varepsilon)\| \leq L\|\mathbf{x}_1 - \mathbf{x}_2\| \quad \mathbf{x}_1,\mathbf{x}_2 \in D$$

then we say that $\mathbf{f}$ satisfies a Lipschitz condition in $\mathbf{x}$ with Lipschitz constant L. If $\mathbf{f}(\mathbf{x}, t; \varepsilon)$ is continuous with respect to $\mathbf{x},t,\varepsilon$, then there is a domain within which the initial value problem

FUNDAMENTAL THEOREMS

$$\dot{\mathbf{x}} = \mathbf{f}(\mathbf{x},t;\varepsilon) \quad \mathbf{x}(0) = \mathbf{x}_0$$

has a unique solution.

Example 2.19. Consider the differential equation

$$\dot{x} = x^\alpha, \qquad x(0) = 0 \qquad (2.17)$$

In a naive, straightforward manner, we find that the general solution of Equation (2.17) is

$$x = \left(x_0^{1-\alpha} + (1-\alpha)(t-t_0)\right)^{1/(1-\alpha)}$$

However, we have to distinguish between $\alpha > 1$ and $\alpha < 1$. For $\alpha > 1$, $1-\alpha$ is negative, and it is better to write the general solution as

$$x = \frac{x_0}{\left(1 - (\alpha-1)x_0^{\alpha-1}(t-t_0)\right)^{1/(\alpha-1)}}$$

The only way to obtain a solution that obeys the initial condition of Equation (2.17) is to set $x_0 = t_0 = 0$, yielding $x(t) \equiv 0$ as the only solution. On the other hand, for $\alpha < 1$, the one parameter family

$$x(t) = \begin{cases} 0 & t \leq t_0 \\ \left((1-\alpha)(t-t_0)\right)^{1/(1-\alpha)} & t > t_0 \end{cases}$$

solves the equation. At $t = t_0$, the function and its derivative on either side of the point join smoothly.

For $\alpha > 1$ the r.h.s. of Equation (2.17) satisfies a Lipschitz condition, whereas for $\alpha < 1$ it does not.

2.2.2 Hartman–Grobman theorem

In this book we concentrate on series expansions of solutions of the equations describing dynamical systems. Very often these expansions are limited in validity: They may either provide a given level of approximation only for a limited range in time, or may break down at a finite order. Often, the series is divergent and is only an asymptotic one. This is an intrinsic aspect of perturbation theory, which is intimately related to the fundamental theorem of Hartman and

Grobman. The theorem is proved in many texts, and we refer the reader to [5–7].

Theorem. An N-dimensional nonlinear dynamical system is given by

$$\dot{\mathbf{x}} = \mathbf{A}\mathbf{x} + \mathbf{F}(\mathbf{x};\varepsilon) \qquad \mathbf{x} \in \mathbf{R}^N$$

where $\mathbf{F}(\mathbf{x};\varepsilon)$ is nonlinear (i.e., it is at least of second order in the components of $\mathbf{x}$). The eigenvalues of the matrix $\mathbf{A}$ all have non-Vanishing real parts. (The origin is a hyperbolic fixed point.) There exists a one-to-one mapping (hence invertible) ψ of $\mathbf{R}^N$ onto $\mathbf{R}^N$ (such a mapping is called a *homeomorphism*)

$$\mathbf{x} = \psi(\mathbf{y}) \qquad \mathbf{y} = \psi^{-1}(\mathbf{x})$$

and there exists a neighborhood U of the origin such that, within that neighborhood

$$\dot{\mathbf{y}} = \mathbf{A}\mathbf{y}$$

The procedure of finding a mapping that transforms the original nonlinear problem to its linear, unperturbed part, is also called a *linearization* of the differential equation.

We do not present the proof of the theorem, but point out that it depends on the hyperbolic nature of the fixed point at the origin. More specifically, we note that, if all eigenvalues of $\mathbf{A}$ have nonzero real parts, we can assume that it is of the form

$$\mathbf{A} = \begin{pmatrix} \mathbf{P} & 0 \\ 0 & \mathbf{Q} \end{pmatrix}$$

where the eigenvalues of the submatrix $\mathbf{P}$ are those with positive real parts and those of $\mathbf{Q}$, with negative real parts. This implies that

$$\|e^{\mathbf{P}t}\| > 1 \qquad \|e^{\mathbf{Q}t}\| < 1 \qquad (2.18)$$

The existence of the mapping ψ is proved by a sequence of successive approximations. The proof that the sequence converges exploits Equation (2.18). A geometrical interpretation of this result will be given further on.

The Hartman–Grobman theorem stipulates that the motion of a

FUNDAMENTAL THEOREMS

system close to a hyperbolic fixed point can be transformed to the motion of the solution of the linear unperturbed system.

Note that the theorem does not state that the mapping ψ is also differentiable. (If it had k continuous derivatives, it would have been called a *C^k diffeomorphism*. As it is, one also denotes it as a C^0 mapping.) For some systems, the mapping might be differentiable, or even analytic. In general, it is not. This is the essence of the difficulties encountered in perturbation theory. We expand the solution **x**, in a near-identity transformation, relating it to the zero-order term, **y**. For the series to converge, the transformation must satisfy very stringent requirements (e.g., being infinitely differentiable with respect to **y**). The Hartman–Grobman theorem says that, in general, such analyticity properties cannot be guaranteed. In particular, the transformation may not be analytic in ε, so that the expansion may not provide a convergent series, but just an asymptotic one.

Example 2.20. The question of the requirements that a differential equation must satisfy so that its linearization mapping in the vicinity of a hyperbolic fixed point is not just C^0 but, say, C^k, has been studied [8]. We analyze here a version of an example discussed in [9–11]:

$$\dot{x} = \lambda x + k y^n \qquad \dot{y} = \mu y \qquad (2.19)$$

The linearized system is

$$\dot{u} = \lambda u \qquad \dot{v} = \mu v \qquad (2.20)$$

Without loss of generality, assume the same initial conditions for the full system and for the linearized one. Equations (2.20) are solved by

$$u = x_0 \exp(\lambda t) \qquad v = y_0 \exp(\mu t)$$

$$\Rightarrow u = x_0 \left(\frac{v}{y_0}\right)^{\lambda/\mu} \qquad (2.21)$$

The properties of the curve $u=u(v)$ depend on the ratio λ/μ.

I. (λ/μ) not an integer. Let $k=[\lambda/\mu]$, be the integral part λ/μ. Only the first k derivatives of $u=u(v)$ with respect to v are defined and continuous near the origin. Hence the curve is C^k.

II. (λ/μ) an integer. The curve $u=u(v)$ is infinitely differentiable near the origin. The curve is C^∞.

In Equations (2.18), y is solved by

$$y = v = y_0 \exp(\mu t)$$

For the solution of x, we have, again, to distinguish two cases.

I. $(\lambda/\mu) \neq n$

$$x = A\exp(\lambda t) + k\frac{y_0^n}{(n\mu - \lambda)}\exp(n\mu t)$$

$$\Rightarrow x = \left(\frac{A}{x_0}\right)u + k\frac{v^n}{(n\mu - \lambda)} = \left(\frac{A}{x_0}\right)\left(\frac{y}{y_0}\right)^{\lambda/\mu} + k\frac{y^n}{(n\mu - \lambda)}$$

$$\left(A = x_0 - k\frac{y_0^n}{(n\mu - \lambda)}\right)$$

The function $x=x(u,v)$ is C^∞. Hence, the linearizing transformation $(x,y) \Rightarrow (u,v)$ is C^∞. Notice that the curve $x=x(y)$ is also C^∞. Thus, as expected, so is the curve $u=u(v)$.

II. $(\lambda/\mu) = n$ *(resonance)*

$$x = x_0 \exp(\lambda t) + k y_0^n\, t \exp(\lambda t)$$

$$\Rightarrow x = u + k v^n \ln\left(\frac{v}{y_0}\right) = x_0 \left(\frac{y}{y_0}\right)^{\lambda/\mu} + k y^n \ln\left(\frac{y}{y_0}\right)$$

Of the two terms in the expression for the curve $x=x(y)$, the first is C^∞, but the second is only C^1 near the origin. So is the function $x=x(u,v)$. Hence, the linearizing transformation $(x,y) \Rightarrow (u,v)$ is C^1. Thus, although the curve $u=u(v)$ is C^∞, the curve $x=x(y)$ is C^1 like the linearizing transformation.

2.2.3 Poincaré–Lyapunov theorem

The Poincaré–Lyapunov theorem is an important tool in the analysis of the stability of nonlinear systems. Consider the system

$$\dot{\mathbf{x}} = (\mathbf{A} + \mathbf{B}(t))\mathbf{x} + \mathbf{g}(t,\mathbf{x}) \qquad \mathbf{x}(t_0) = \mathbf{x}_0 \qquad \mathbf{x} \in \mathbf{R}^N \qquad (2.22)$$

FUNDAMENTAL THEOREMS

where **A** is an $N \times N$ constant matrix. All the eigenvalues of **A** have *negative real parts*. The $N \times N$ matrix $\mathbf{B}(t)$ is continuous in t and, in the norm used in the space, it satisfies

$$\lim_{t \to \infty} \|\mathbf{B}(t)\| = 0 \tag{2.23}$$

The vector field $\mathbf{g}(t, \mathbf{x})$ is continuous in t and in $\mathbf{x}$, and has a continuous derivative in $\mathbf{x}$. It is a nonlinear function of $\mathbf{x}$, satisfying

$$\frac{\|\mathbf{g}(t,\mathbf{x})\|}{\|\mathbf{x}\|} \xrightarrow[\|\mathbf{x}\| \to 0]{} 0 \tag{2.24}$$

uniformly in t (namely, when $\|\mathbf{x}\| \to 0$, $\|\mathbf{g}(t, \mathbf{x})\|/\|\mathbf{x}\| \to 0$ at a rate that is bonded independently of t).

Under these conditions, three constants C, μ, δ exist such that

$$\|\mathbf{x}_0\| < \delta \implies \|\mathbf{x}(t)\| < C\|\mathbf{x}_0\|\exp(-\mu(t - t_0)) \tag{2.25}$$

The proof is give in many texts. (See, e.g., Refs. 2 and 12).

Significance of the theorem. If we omit the matrix $\mathbf{B}(t)$ and the nonlinear perturbation, $\mathbf{g}(t,\mathbf{x})$, in Equation (2.22), then the solution is a decaying exponential. That is, the origin is a stable fixed point. For sufficiently long times, Equation (2.23) says that the effect of the matrix $\mathbf{B}(t)$ becomes Vanishingly small. For sufficiently small $\mathbf{x}$, Equation (2.24) says that $\mathbf{g}(t,\mathbf{x})$ is small compared to $\mathbf{x}$ when the latter tends to zero. Now return $\mathbf{B}(t)$ and $\mathbf{g}(t,\mathbf{x})$ into the equation. The theorem states that sufficiently close to the origin and for sufficiently long time, the effect of both $\mathbf{B}(t)$ and $\mathbf{g}(t,\mathbf{x})$ cannot destroy the asymptotic stability of the origin.

Comments. The range $\|\mathbf{x}_0\| \leq \delta$, within which attraction of the solution to zero is exponential, is called the *Poincaré–Lyapunov domain* of the equation.

The existence and uniqueness theorem, stated in Section 2.2.1, guarantees that a unique solution exists for some time interval $[t_0, t_1]$. The Poincaré–Lyapunov theorem states that the solution exists for all t and tends to zero exponentially when $t \to \infty$.

Why are we interested in the vicinity of zero? Consider a general equation of the form

$$\dot{\mathbf{y}} = \mathbf{F}(\mathbf{y},t)$$

Denote its solution by $\mathbf{y} = \boldsymbol{\phi}(t)$. The asymptotic stability or instability of $\boldsymbol{\phi}(t)$ can be analyzed by studying the behavior around zero of

$$\mathbf{x} = \mathbf{y} - \boldsymbol{\phi}(t)$$

if the equation for $\mathbf{x}$ can be cast in the form of Equation (2.22).

Simple result. Let $\mathbf{x}_1(t)$ and $\mathbf{x}_2(t)$ be two different solutions of Equation (2.22). From the Poincaré–Lyapunov theorem it follows that

$$\|\mathbf{x}_1(t) - \mathbf{x}_2(t)\| \le C\|\mathbf{x}_1(t_0) - \mathbf{x}_2(t_0)\|\exp(-\mu(t - t_0))$$

That is, although the two solutions start at different points, both converge asymptotically to zero. So does the distance between them.

2.2.4 Difference between nonlinear and linear problems

The Poincaré–Lyapunov theorem indicates a basic difference between linear and nonlinear systems. In nonlinear systems, asymptotic stability of the solution of its linear part, guarantees asymptotic stability in the full system, if the added nonlinearity is small. In linear systems, the addition of a small *linear* perturbation to an equation that originally had an asymptotically stable solution, may destroy the stability. Consider a linear two-dimensional systlem, in which a small ($O(\varepsilon^2)$] *linear* perturbation is added [13]:

$$\dot{\mathbf{x}} = \mathbf{A}\mathbf{x} + \mathbf{g}\mathbf{x}$$

$$\mathbf{x} = \begin{pmatrix} x_1 \\ x_2 \end{pmatrix} \quad \mathbf{A} = \begin{pmatrix} -\varepsilon & 1 \\ 0 & -\varepsilon \end{pmatrix} \quad \mathbf{g} = \begin{pmatrix} 0 & 0 \\ a^2\varepsilon^2 & 0 \end{pmatrix}$$
(2.26)

$$\Rightarrow \|\mathbf{A}\| = 1 + 2\varepsilon \qquad \|\mathbf{g}\| = a^2\varepsilon^2$$

$$\dot{\mathbf{x}} = \begin{pmatrix} -\varepsilon & 1 \\ a^2\varepsilon^2 & -\varepsilon \end{pmatrix}\mathbf{x}$$

(The norm is defined here as the sum of the absolute values of the elements.) $\mathbf{A}$ has one (double) eigenvalue, $-\varepsilon$, whereas the

FUNDAMENTAL THEOREMS

eigenvalues of the matrix of the full equation are $-\varepsilon \pm a \cdot \varepsilon$. Thus, the stability properties of the problem change completely when a small perturbation is added: It is stable for a<1 and unstable for a>1. The reason is that although the perturbation is of higher order in the small parameter, ε, it does not Vanish faster than x when $\|x\| \to 0$.

2.2.5 One dimensional examples of Poincaré–Lyapunov theorem

Example 2.21. The equation

$$\frac{dx}{dt} = -\frac{x}{1+x} \qquad (2.27)$$

is solved by

$$\frac{x}{x_0}\exp(x - x_0) = \exp(-(t - t_0))$$

Clearly, for any initial condition x_0, we have $x \xrightarrow[t\to\infty]{} 0$.

Let us study this problem for $|x| \ll 1$. The r.h.s. of Equation (2.27) can be written as

$$\frac{dx}{dt} = -x + \frac{x^2}{1+x} \qquad (2.28)$$

The nonlinear perturbation Vanishes faster than the linear part, as $x \to 0$. As a result, the conditions of the Poincaré–Lyapunov theorem are satisfied, and the stability of the solution of the linear problem

$$\frac{dx}{dt} = -x \Rightarrow x = x_0 \exp(-(t - t_0))$$

is carried over to the nonlinear problem: $x=0$ is an asymptotically stable attractor also of the nonlinear problem, as $t \to \infty$.

Example 2.22. $x=0$ is a solution of the equation

$$\dot{x} = -1 + \exp(-x) \qquad (2.29)$$

We isolate the linear part in Equation (2.29) and obtain:

$$\frac{dx}{dt} - x + \underbrace{\{x + \exp(-x) - 1\}}_{\text{nonlinear perturbation}} \tag{2.30}$$

In its present form, the equation satisfies the conditions of the theorem, so that, for |x|<1, x=0 is an attractor.

2.2.6 Gronwall's lemma

The inequality due to Gronwall [14] is the standard tool for estimating the error in perturbation theory. Error estimates obtained in this book will be based mainly on Gronwall-type arguments.

Lemma. Let $a(t) \geq 0$, $b(t) \geq 0$, $c(t) \geq 0$, and $Z(t) \geq 0$, be continuous functions in the interval $[t_0, t_0+T]$ satisfying the inequality

$$Z(t) \leq a(t) \int_{t_0}^{t} b(s) Z(s) ds + c(t) \qquad t_0 \leq t \leq t_0 + T \tag{2.31}$$

Then $Z(t)$ satisfies

$$Z(t) \leq a(t) \int_{t_0}^{t} b(s) c(s) \exp\left(\int_{s}^{t} a(s') b(s') ds' \right) ds + c(t) \tag{2.32}$$

The proof can be found in many texts. (See, e.g., Refs. 14, 2, 12, and 15.)

Example 2.23. Consider the Duffing equation

$$\ddot{x} + x + \varepsilon x^3 = 0 \qquad x(0) = a, \; \dot{x}(0) = 0$$

In complex notation ($z = x + i\,\dot{x}$) this becomes

$$\dot{z} = -i\,z - i\,\varepsilon \tfrac{1}{8}(z + z^*)^3 \qquad z(0) = a + i\,0$$

We approximate the solution by $x \cong x_0 = a \cdot \cos t$ ($z \cong u = a \cdot \exp(-it)$], and write the error as

$$z - u = a\eta \exp(-it)$$

FUNDAMENTAL THEOREMS

We find that η satisfies the following equation:

$$\dot{\eta} = \left(\frac{1}{8}i\frac{\varepsilon}{a}\right)\exp(i\,t)\left\{\begin{array}{l}(u+u^*)^3 + 3(u+u^*)^2(\eta+\eta^*) \\ + 3(u+u^*)(\eta+\eta^*)^2 + (\eta+\eta^*)^3\end{array}\right\} \quad (2.33)$$

The first term on the r.h.s. of Equation (2.33) can be integrated, yielding

$$\eta = \tfrac{1}{8}i\,\varepsilon a^2 \left\{\begin{array}{l}\dfrac{\exp(-2i\,t)-1}{-2i} + 3t \\ + \dfrac{\exp(2i\,t)-1}{2i} + \dfrac{\exp(4i\,t)-1}{4i}\end{array}\right\}$$

$$+\left(\tfrac{1}{8}i\,\dfrac{\varepsilon}{a}\right)\int_0^t \exp(i\,s)\left\{\begin{array}{l}+3(u+u^*)^2(\eta+\eta^*) \\ +3(u+u^*)(\eta+\eta^*)^2 + (\eta+\eta^*)^3\end{array}\right\}ds \quad (2.34)$$

We wish to find the length of time over which η remains a small correction. To this end, we note that $|u|=a$ and that, as long as $|\eta|$ is small, $|\eta|^2$ and $|\eta|^3$ can be replaced by $|\eta|$. This yields a Gronwall-type integral inequality:

$$|\eta| \leq 7\varepsilon a^2 \int_0^t |\eta|\,ds + \tfrac{1}{8}\varepsilon a^2\left(\tfrac{9}{2}+3t\right) \quad (2.35)$$

Using Equation (2.31), Equation (2.34) yields

$$|\eta| \leq \varepsilon a^2\left(At\exp(7\varepsilon a^2 t) + B\exp(7\varepsilon a^2 t) + Ct + D\right) \quad (2.36)$$

Here A, B, C, and D are known coefficients. We see that, for $t=O(1)$, $|\eta|=O(\varepsilon)$, while for $t=O(1/\varepsilon)$, $|\eta|=O(1)$. Thus, a Gronwall-type inequality provides a simple tool for estimating the error.

If, on the other hand, we choose z_0 to have a modified frequency

$$z_0 = a\exp(-i\,(1+\varepsilon\omega_1)t)$$

the resulting equations are modified. Equation (2.34) now changes to

$$\eta = \underline{(\exp(-i\,\varepsilon\omega_1 t)-1)}$$

$$+ \tfrac{1}{8} i\, \varepsilon a^2 \left\{ \begin{array}{l} \dfrac{\exp(-2it)-1}{-2i} + 3\,\underline{\dfrac{\exp(-i\,\varepsilon\omega_1 t)-1}{-i\,\varepsilon\omega_1}} \\[2mm] + \dfrac{\exp(2it)-1}{2i} + \dfrac{\exp(4it)-1}{4i} + O(\varepsilon) \end{array} \right\}$$

$$\left(\tfrac{1}{8} i\, \dfrac{\varepsilon}{a}\right)\int_0^t \exp(is)\left\{\begin{array}{l}+3(u+u^*)^2(\eta+\eta^*)\\ +3(u+u^*)(\eta+\eta^*)^2 + (\eta+\eta^*)^3\end{array}\right\}ds \qquad (2.37)$$

Note that the two underlined terms in Equation (2.37) cancel exactly if

$$\omega_1 = \tfrac{3}{8} a^2$$

thereby eliminating the potentially dangerous $(1/\varepsilon)$ term [the "reincarnation" of the term proportional to t in Equation (2.34)]. The error estimate now becomes

$$|\eta| \leq \varepsilon a^2 \left((E + O(\varepsilon))\exp(7\varepsilon a^2 t) + F + O(\varepsilon) \right)$$

Hence, the validity of the approximation has been extended over one additional timescale—the error η is $O(\varepsilon)$ for $t=O(1/\varepsilon)$.

Note that, owing to the appearance of the product $(\varepsilon \cdot t)$ in the exponential, *the error estimate cannot be extended timescales longer than $O(1/\varepsilon)$*.

2.3 STABLE, UNSTABLE, AND CENTER MANIFOLDS

The topological structure of the solutions of higher-dimensional systems is a topic of intensive study in the literature. It is beyond the scope of this book. Of this rich subject we briefly mention the concepts of the *stable, unstable,* and *center manifolds* of a dynamical system. We develop these notions through a sequence of examples.

2.3.1 Stable and unstable manifolds

Consider the linear two-dimensional system

STABLE, UNSTABLE, AND CENTER MANIFOLDS

$$\frac{d}{dt}\begin{pmatrix} x \\ y \end{pmatrix} = \begin{pmatrix} 1 & 0 \\ 0 & -1 \end{pmatrix}\begin{pmatrix} x \\ y \end{pmatrix} \qquad (2.38)$$

for which the origin is a hyperbolic fixed point.

The lines $x=0$ and $y=0$ are both *invariant manifolds*; namely, a trajectory emanating from an initial condition $(0,y_0)$ on the line $x=0$ remains on that line. In a similar manner, the trajectory evolving from an initial condition $(x_0, 0)$ on the line $y=0$, remains on the latter.

The line $x=0$ is the *stable manifold*, E^s, passing through the fixed point at the origin. Any trajectory on that line tends exponentially in time toward the origin as $t \to \infty$. The line $y=0$ is the *unstable manifold*, E^u, passing through the fixed point. Any trajectory on that line diverges exponentially in time away from the origin as $t \to \infty$. Typical trajectories of solutions of Equation (2.38) are shown in Figure 2.12. Note that E^s is the subspace spanned by the eigenvector of the matrix in Equation (2.38) that has eigenvalue -1. Similarly, E^u is spanned by the eigenvector with eigenvalue $+1$.

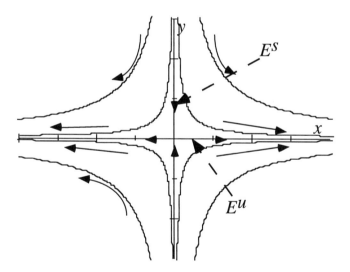

Figure 2.12 Trajectories of solutions of Equations (2.38). Arrows indicate direction of evolution in time.

Comment. The two invariant manifolds need not be along the axes. For example, in the system

$$\frac{d}{dt}\begin{pmatrix} x \\ y \end{pmatrix} = \begin{pmatrix} \frac{1}{\sqrt{2}} & \frac{1}{\sqrt{2}} \\ \frac{1}{\sqrt{2}} & -\frac{1}{\sqrt{2}} \end{pmatrix} \begin{pmatrix} x \\ y \end{pmatrix}$$

the invariant manifolds are straight lines subtending the angles ($\pi/8$) and ($9\pi/8$) to the x axis.

We now add nonlinear terms to Equations (2.38), for example

$$\frac{d}{dt}\begin{pmatrix} x \\ y \end{pmatrix} = \begin{pmatrix} 1 & 0 \\ 0 & -1 \end{pmatrix} \begin{pmatrix} x \\ y \end{pmatrix} + \frac{1}{2}\begin{pmatrix} x^2 + y^2 \\ x^2 + y^2 \end{pmatrix} \qquad (2.39)$$

This system has two fixed points, at $(x,y)=(0,0)$ and $(-1,1)$. The first point is a saddle, as in Equation (2.38), whereas the second one is a center. Rather than performing a detailed analysis of the structure of the solutions, we show typical ones in Figure 2.13.

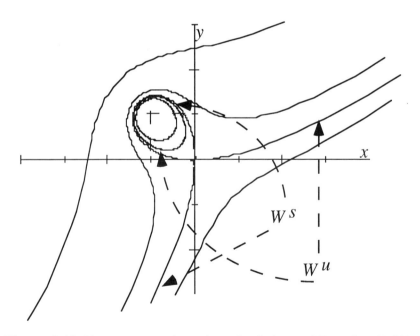

Figure 2.13 Phase-space trajectories of solutions of Equation (2.39).

Very close to the origin, the linear terms in Equation (2.39) dominate. Hence, the stable and unstable manifolds (denoted here by W^s and W^u) start off identical to the corresponding manifolds, E^s and

STABLE, UNSTABLE, AND CENTER MANIFOLDS

E^u, of the linear system, Equation (2.38). (More precisely, each manifold is tangent to the corresponding unperturbed one at the origin, so that the linear approximations for the equations of the unperturbed and the full manifolds are the same.)

For large x and y, the lines $x=0$ and $y=0$ are no longer invariant, and are no longer parts of W^s or W^u. The latter are now distorted lines. Note, also, that, in this particular example, the upper branch of W^s and the left branch of W^u join smoothly in one curve. They constitute the *separatrix* enclosing the domain around the center at (-1,1), where there is an infinite family of periodic orbits.

Thus, the invariant stable and unstable manifolds exist in the nonlinear case as distorted versions of the unperturbed ones. Intersection or joining of manifolds is a common occurrence in nonlinear systems. The extension of the concepts discussed here to higher dimensional systems is obvious. In a system described by

$$\dot{\mathbf{x}} = \mathbf{A}\mathbf{x} + \mathbf{g}(\mathbf{x})$$

where $\mathbf{A}$ is an n×n matrix and $\mathbf{g}$ is a nonlinear function of the coordinates, E^s (E^u), the stable (unstable) manifold of the linearized system, is spanned by those coordinates that correspond to eigenvalues with negative (positive) real parts. These become the distorted manifolds W^s (W^u) in the nonlinear system. The topological structure of these manifolds provides ample information about the characteristics of the solutions. In particular, it can provide indications of the onset of chaotic behavior.

2.3.2 Center manifold

As long as the origin is a hyperbolic fixed point, it has one or both of the stable and unstable manifolds. When the linearized system has eigenvalues with zero real part, a third invariant manifold emerges, the *center manifold*. Consider, for example, a linear unperturbed system:

$$\dot{x} = y \qquad \dot{y} = -x \qquad \dot{w} = -w \qquad (2.40)$$

The motion is a spiral in which w decays to zero and x and y execute circular motion in the x–y plane. There is a stable manifold, E^s: the w axis. The x–y plane ($w=0$) is the center manifold, E^c, of the unperturbed problem; it is invariant, in the sense that motion that starts in that plane remains there. The motion is periodic, around the center, (0,0), and there is an infinite family of allowed circles. An example is shown in Figure 2.14.

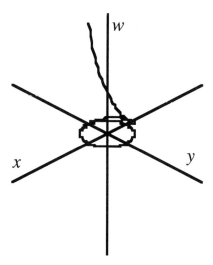

Figure 2.14 Typical trajectory of solution of Equations (2.40).

When nonlinear terms are added to the equations, the manifolds are distorted, as expected, and, in the lowest approximation, coincide with the unperturbed ones (i.e., they are tangent to them at the origin). For instance, consider the nonlinear system of equations

$$\dot{x} = y$$
$$\dot{y} = -x + 0.1y(1 - x^2 w) \qquad (2.41)$$
$$\dot{w} = -w + 0.1(x^2 + y^2)$$

If not for the nonlinear coupling, w would have decayed to zero. It deviates from zero as a consequence of the dependence of the nonlinear coupling on x and y. The resulting center manifold, W^c, is a surface tangent to the unperturbed center manifold (the x–y plane), E^c, but is *not* identical to it. On the center manifold, asymptotically in time, w is not an independent variable, but a *trailing* coordinate, dragged along by the *dominant* coordinates, x and y,

$$w = h(x, y)$$

Near the origin $h(x,y)$ is at least quadratic in the two variables. A typical solution of Equation (2.41) is shown in Figure 2.15.

In a general multidimensional system, one distinguishes between the dominant coordinates, that have eigenvalues with zero real part, and the trailing ones with eigenvalues that have nonzero real parts. One writes the equations as

STABLE, UNSTABLE, AND CENTER MANIFOLDS

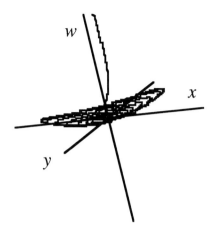

Figure 2.15 Typical trajectory of solution of Equation (2.41).

$$\dot{x} = Ax + M(x,y) \qquad \dot{y} = By + N(x,y) \qquad (2.42)$$

The eigenvalues of the $n_c \times n_c$ matrix A are pure imaginary. The eigenvalues of B have nonzero real parts. y denotes the trailing coordinates and x, the dominant ones. M and N are nonlinear vector functions of x and y. As in the case of Equation (2.41), on the center manifold, asymptotically one has:

$$y = h(x)$$

Near the origin, $h(x)$ is at least quadratic in the components of x.

Extension of Hartman–Grobman theorem. An extension of the theorem exists in this case, stating that a homeomorphism (i.e., a continuous, one-to-one, hence invertible, mapping) exists that transforms the full motion of solutions for x and y of Equations (2.42) into the motion of a reduced system of variables:

$$\dot{u} = Au + M(u, h(u)) \qquad \dot{v} = Bv \qquad (2.43)$$

In Equation (2.43), the motion of the "trailing coordinates" is replaced by an exponential decay along the unperturbed stable manifold, E^s, and/or divergence along the unperturbed unstable manifold, E^u. The motion of the dominant x variable is replaced by its asymptotic behavior on the *full* center manifold, W^c. The theorem is by Pliss [16] and Shoshitaishvili [17,18] and is reviewed by Crawford [19].

REFERENCES

1. Andronov, A. A, E. A. Leontovich, I. I. Gordon, and A. G. Maier, *Theory of Bifurcations of Dynamic Systems on a Plane*, Israel Program of Scientific Translations, Jerusalem (1971).
2. Coddington, E. A., and N. Levinson, *Theory of Ordinary Differential Equations,* Krieger, Malabar, FL (1985).
3. Wasow, A, *Asymptotic Expansions of Ordinary Differential Equations,* Dover, New York (1987).
4. Bender, C. M., and S. A. Orszag, *AdVanced Numerical Methods for Scientists and Engineers*, McGraw-Hill, New York (1978).
5. Hartman, P., *Ordinary Differential Equations,* Birkhauser, Boston (1982).
6. Perko, L., *Differential Equations and Dynamical Systems,* Springer-Verlag, New York (1991).
7. Robinson, C., *Dynamical Systems: Stability, Symbolic Dynamics and Chaos,* CRC Press, Boca Raton, FL (1995).
8. Sternberg, S., "On the Structure of Local Homeomorphisms of Euclidean n–space II," *Am. J. Math.* **80**, 623 (1958).
9. Guckenheimer, J., and P. Holmes, *Nonlinear Oscillations, Dynamical Systems and Bifurcations of Vector Fields,* Springer-Verlag, New York (1983).
10. Meyer, K. R., "Counter Examples in Dynamical Systems via Normal form Theory," *SIAM Rev.* **28**, 41 (1986).
11. Wiggins, S., *Introduction to Applied Nonlinear Dynamical Systems and Chaos*, Springer-Verlag, New York (1990).
12. Sanders, J. A., and F. Verhulst, *Averaging Method in Nonlinear Dynamical Systems*, Springer-Verlag, New York (1985).
13. Robinson, C., "Stability of Periodic Solutions from Asymptotic Expansions," in *Classical Mechanics and Dynamical Systems: Lecture Notes in Pure and Applied Mathematics,* Vol. 70, pp. 173–185, R. DeVanay and Z. Nitecki, eds., Marcel Dekker (1981).
14. Gronwall, T. H., "Note on the Derivative with Respect to a Parameter of the Solutions of a System of Differential Equations," *Annals Math.* **20**, 292–296 (1919).
15. Smith, D. R., *Singular Perturbation Theory: An Introduction with Applications*, Cambridge University Press, London (1985).
16. Pliss, V. A., "A Reduction Principle in Stability Theory of Motions," *Izv. Akad. Nauk SSSR, Ser Mat.* **28**, 1297 (1964).
17. Shoshitaishvili, A. N., "Bifurcations of Topological Type at Singular Points of Parametrized Vector Fields," *Funct. Anal. Appl.* **6**, 169 (1972).
18. Shoshitaishvili, A. N., "Bifurcation of the Topological Type of Singular Points of Vector fields that Depend on Parameters," *Trudy Seminara I. G. Petrovskogo* **1**, 279 (1975).
19. Crawford, J. D., "Introduction to Bifurcation Theory," *Rev. Mod. Phys.* **63**, 991–1037 (1991).

3

NAIVE PERTURBATION THEORY (NPT)

In this chapter, we introduce the basic aspects of *naive perturbation theory* (NPT), a natural procedure that works well in many situations. However, for a broad class of problems, it is not an effective expansion method, as it can lead to *secular* terms that limit the validity of the approximations to short times. Often, these terms are not a property of the full solution, but are generated in perturbation expansions for systems that evolve in periodic orbits, as a direct consequence of the expansion procedure. In such cases, one denotes them as spurious secular terms. Secular terms are in the form of a product of a periodic function and a polynomial in time, for example

$$t \sin t \quad \text{or} \quad t^2 \cos t$$

Note that the multiplicative factors of t destroy the boundedness of the pure trigonometric functions. To be clear, note that terms of the form

$$t \exp(-\lambda t) \quad \text{or} \quad t^2 \exp(\lambda t) \quad (\text{Re}\,\lambda \neq 0)$$

are not secular.

Our exposition will be developed through the analysis of a sequence of simple examples. We begin with an introduction to naive perturbation theory and illustrate a situation in which it is an effective procedure. We then discuss problems in which the theory leads to spurious secular terms in the expansion. The latter usually arise when the unperturbed linear part is a simple harmonic oscillator of frequency ω corresponding to pure imaginary eigenvalues, $\pm i\omega$.

It is important to realize how plausible and natural naive perturbation theory is. As will become clear in the following, NPT emerges as the obvious way to analyze systems that are amenable to successive approximations in terms of a small parameter and there are many ways to "fix" it so that it works.

NAIVE PERTURBATION THEORY (NPT)
3.1 DEFINITION AND BASIC ASPECTS OF NPT

When we develop an approximate solution to a differential equation in a perturbation expansion, we prefer the individual terms in the expansion to reflect the true character of the full solution. Thus, for example, if the full solution is periodic, we would like each term in the approximation to also be periodic. A possible procedure to find a perturbative solution to our fundamental equation

$$\frac{d\mathbf{x}}{dt} = \mathbf{A}\mathbf{x} + \varepsilon \mathbf{F}(\mathbf{x}; \varepsilon) \qquad (3.1)$$

in the neighborhood of its fixed points, $dx/dt=0$, might be to use the known solutions, $\mathbf{x}_0(t)$, of the unperturbed equation, $d\mathbf{x}_0/dt = \mathbf{A}\mathbf{x}_0$, as the starting point. This is often called a *straightforward* or *naive* expansion, since it uses the unperturbed solution to generate the perturbation series. One writes

$$\mathbf{x}(t) = \mathbf{x}_0(t) + \sum_{n \geq 1} \varepsilon^n \mathbf{x}_n(t; \varepsilon) \qquad (3.2)$$

where $\mathbf{x}_n(t)$ are correction terms. The perturbation expansion is developed by rearranging Equation (3.1) to the form

$$\frac{d}{dt}(\exp(-\mathbf{A}t)\mathbf{x}) = \varepsilon \exp(-\mathbf{A}t) \mathbf{F}(\mathbf{x}; \varepsilon) \qquad (3.3)$$

Integrating, we obtain

$$\mathbf{x}(t) = \mathbf{x}(0)\exp(\mathbf{A}t) + \varepsilon \int_0^t \exp\mathbf{A}(t-t') \mathbf{F}[\mathbf{x}(t'); \varepsilon] dt' \qquad (3.4a)$$

Within the integral, we substitute Equation (3.4a) for $\mathbf{x}(t')$ to obtain

$$\mathbf{x}(t) = \exp(\mathbf{A}t)\mathbf{x}(0)$$

$$+ \varepsilon \int_0^t \exp\mathbf{A}(t-t') \mathbf{F} \begin{bmatrix} \mathbf{x}(0)\exp(\mathbf{A}t') \\ + \varepsilon \int_0^{t'} \exp\mathbf{A}(t'-t'') \mathbf{F}[\mathbf{x}(t''); \varepsilon] dt'' \end{bmatrix} dt' \qquad (3.4b)$$

DEFINITION AND BASIC ASPECTS OF NPT

Repetition of the substitution [now of $\mathbf{x}(t'')$ by the full expression of Equation (3.4a)] provides the pattern of the NPT expansion. Thus, in the formalism suggested by Equation (3.4b), we are using the solution of the unperturbed equation as the generator of the expansion. The linear unperturbed problem

$$\frac{d\mathbf{x}_0}{dt} = \mathbf{A}\mathbf{x}_0$$

is solved by

$$\mathbf{x}_0(t) = \exp(\mathbf{A}t)\mathbf{x}(0)$$

A first-order approximation for $\mathbf{x}(t)$ is obtained by truncating the expansion and writing in Equation (3.4b):

$$\mathbf{x}(t) \approx \exp(\mathbf{A}t)\mathbf{x}(0) + \varepsilon \int_0^t \exp\mathbf{A}(t-t')\mathbf{F}\big[\mathbf{x}(0)\exp(\mathbf{A}t';\varepsilon)\big]dt' \quad (3.4c)$$

In deriving this result, our basic assumption is that, for the time under consideration, $\mathbf{x}(t)$ is well characterized by its unperturbed approximation. Higher-order terms are obtained by repeated substitution of iterates of the unperturbed solution. This procedure is plausible when there is no eigenvalue update. The latter is a central issue in the construction of alternatives to NPT.

The eigenvalues λ_i of the matrix $\mathbf{A}$ determine the timescales $(1/\lambda_i)$ characterizing the time dependence of the solution of the unperturbed problem. In the naive expansion, the unperturbed solution is used as the generator of the expansion. This is satisfactory if the timescales characterizing the behavior of the full solution are unaffected by the perturbation and remain the same as those of the unperturbed problem. (The perturbation does not cause "updating" of these scales.)

However, often, the characteristic timescales of the full solution are updated and are different from the unperturbed ones. (This may happen, for example, when the unperturbed problem is periodic, so that the linear part of the equation has pure imaginary eigenvalues.) Then, a clock that uses the unperturbed scale (as is done in the naive expansion) will get out of step with the true solution and exhibit secular behavior. In particular, the naive expansion may exhibit such behavior in each order; only when the entire expansion is included does updating of the oscillation period emerge. The structure of the approximation, obtained by the truncation of the perturbation series in any finite order, is fundamentally different from that of the full solution. By contrast, we will see that the method of normal forms retains the character of the solution in each order of the expansion.

Comments

1. Writing x_0, the fundamental, zero-order term, as $\rho \cdot \exp(-i\varphi)$, we observe that, in conservative one-dimensional systems, the radius ρ is constant and the phase factor $\exp(-i\varphi)$ lies always on the unit circle, regardless of the size of the perturbation. In NPT, one generates the approximation by expanding the exponential phase factor in a power series. The resulting correction terms do not lie on the unit circle. In contrast to NPT, the normal form expansion develops the approximation in terms of corrections to the frequency, ω. Therefore, an approximation to $\exp(-i\varphi)$, obtained by termination of the expansion for ω at any finite order, remains on the unit circle.
2. Equation (3.4a) is an integral equation: a *nonlinear Volterra equation*. [1]. It is usually solved by an infinite series obtained through the iteration procedure indicated by Equation (3.4b). When the integrand on the r.h.s. of Equation (3.4a) is bounded appropriately, the resulting infinite sum converges. This is the case in problems where the eigenvalues of the matrix **A** are pure imaginary and the full solution is periodic. Thus, the infinite series in a naive perturbation expansion may converge. However, if secular terms are generated, individual terms in the series blow up in time, as the expansion is in powers of εt rather than ε. Consequently, any approximation of the solution by a finite number of terms will also blow up in time.
3. This unpleasant scenario is very common and occurs in far less complicated situations. Consider, for example, the function $\exp(i\varepsilon t)$. Its Taylor expansion in powers of (εt) converges for all ε and all t. If one approximates the function by a finite part of the Taylor expansion *within a prescribed accuracy*, the number of terms that have to be included becomes progressively larger when t increases. However, if one insists on approximating the function by a finite number of terms in the sum *regardless* of the value of (εt), then, for $t \geq O(1/\varepsilon)$, the finite sum becomes a poor approximation.

With these observations made, we divide our development into two parts according to the spectrum of the matrix **A**:

I. For the class of problems described by Equation (3.1), if the eigenvalues of the matrix **A** have only negative (or only positive) real parts, the terms in a perturbation expansion, given by the development

DEFINITION AND BASIC ASPECTS OF NPT 59

that begins with Equation (3.4c), have the same character as the full solution and secular terms do not arise. For these situations, NPT is a satisfactory method and yields an expansion that is uniformly valid in time. We will see that it is *precisely* equivalent to the expansion obtained by the method of normal forms. (The perturbation schemes simply give a reorganization of the expansion.) In Chapter 5, we will learn how that this follows from the fact that the solutions of the unperturbed system are an appropriate choice as generators of the expansion. In such situations, we say that the eigenvalues are not updated.

II. For problems in which the eigenvalues of the matrix **A** are updated, as usually happens when they lie on the imaginary axis, the naive perturbation expansion, in its truncated form, has spurious secular terms that are an artifact of the truncation procedure and are not present in the full solution. (This happens, for example, when the full solution is periodic, while individual terms in the expansion are not periodic.) One can think of the source of spurious secular terms as originating from the use of an incorrect "clock" to pace the development of the expansion. To see this, refer to Equation (3.2) and observe that we write the argument everywhere as "t," indicating that the clocks that measure time associated with the unperturbed and the perturbed problems are the same. This assumption leads to the appearance of the spurious secular terms in the expansion, when the clocks become different due to an eigenvalue update.

Finally, one has to be careful in reaching general conclusions about the validity or lack thereof of NPT, based solely on the location of the eigenvalues of the matrix **A**. In Example (3.6), we discuss a system that has pure imaginary eigenvalues without an eigenvalue update. In this situation, NPT yields a uniformly valid expansion, i.e., one without secular terms.

3.1.1 Eigenvalues with negative real part

In this section we consider Equation (3.1) for a simple situation that is sufficient to illustrate the basic aspects of the method. (We reconsider problems of this class in Chapter 5, where we discuss more general situations.) The matrix **A** is taken to be 2×2 with real, negative, distinct eigenvalues and the perturbation term is a polynomial of the vector **x**(t). Furthermore, we limit ourselves to a first-order perturbation expansion, since for sufficiently small amplitudes, we expect the components of the vector $x(t)$ to decay to the origin. A first-order approximation should be sufficient to capture the structure of the solution. (Keep in mind that, to simplify the algebra further, we always assume that the matrix **A** is in diagonal form.) Assuming for the perturbation **F**(**x**;ε) a polynomial of degree ≥2 in the vector **x**, we turn to Equation (3.4c) and write expressions for the components $x_1(t)$ and $x_2(t)$:

$$x_1(t) \approx x_1(0)\exp(\lambda t)$$
$$+ \varepsilon \int_0^t \exp\lambda(t-t') \sum_{\substack{p,q \geq 0 \\ p+q \geq 2}} a_{p,q} x_1(0)^p x_2(0)^q \exp((p\lambda + q\mu)t') dt' \quad (3.5a)$$

$$x_2(t) \approx x_2(0)\exp(\mu t)$$
$$+ \varepsilon \int_0^t \exp\mu(t-t') \sum_{\substack{p,q \geq 0 \\ p+q \geq 2}} b_{p,q} x_1(0)^p x_2(0)^q \exp((p\lambda + q\mu)t') dt' \quad (3.5b)$$

Here $a_{p,q}$ and $b_{p,q}$ are the coefficients in the expansion of the components of **F**(**x**;ε) at ε=0. We order the eigenvalues ($|\mu| > |\lambda|$) and discuss the components of x(t) separately.

The first component. Evaluating the integral in Equation (3.5a) formally, we obtain the following for the first-order correction:

$$x_1(t) \approx x_1(0)\exp(\lambda t)$$
$$+ \varepsilon \sum_{p,q} \frac{a_{p,q}}{[(p-1)\lambda + q\mu]} x_1(0)^p x_2(0)^q \exp[((p-1)\lambda + q\mu)t] \quad (3.6a)$$

A problem may arise in Equation (3.6a) if the integers p and q obey

$$(p-1)\lambda + q\mu = 0$$

However, since we have assumed that $|\mu| > |\lambda|$ and $p+q \geq 2$, *no* such integer pairs exist! Thus, Equation (3.6a) always gives x_1 correctly.

The second component. Consider the integral in Equation (3.5b). The formal expression for the integral will be similar in form to Equation (3.6a). However, now the denominator of Equation (3.6a) vanishes for p and q (both ≥0), that satisfy

$$p\lambda + (q-1)\mu = 0$$

DEFINITION AND BASIC ASPECTS OF NPT

When this occurs, we say that a *resonance* is present. Integers that yield a resonance may or may not exist, depending on the values of the two eigenvalues. For example, for $\lambda=-2$ and $\mu=-6$, resonance occurs only for $(p,q)=(3,0)$ and $(0,1)$. With these comments made, the integral in Equation (3.5b) is evaluated leading to one of two possible results

Case 1. $p\lambda+(q-1)\mu \neq 0$ for all $p,q \geq 0$, then

$$x_2(t) \approx x_2(0)\exp(\mu t)$$

$$+ \varepsilon \sum_{p,q} \frac{b_{p,q}}{[p\lambda+(q-1)\mu]} x_1(0)^p x_2(0)^q \exp[(p\lambda+(q-1)\mu)t]$$

(3.6b)

Case 2. There are values $p=p'$, $q=q'$ such that $p'\lambda+(q'-1)\mu=0$; such a term is excluded from the sum and we obtain

$$x_2(t) \approx x_2(0)\exp(\mu t)$$

$$+ \varepsilon \sum_{\substack{p,q \\ p \neq p', q \neq q'}} \frac{b_{p,q}}{[p\lambda+(q-1)\mu]} x_1(0)^p x_2(0)^q \exp[(p\lambda+(q-1)\mu)t]$$

$$+ \varepsilon x_1(0)^{p'} x_2(0)^{q'} t \exp(\mu t) b_{p',q'}$$

(3.6c)

Clearly, if more than one combination of p' and q' occurs, than all such terms should be treated separately as in Equation (3.6c). It is important to note that the term linear in t is bounded by the exponential factor since $\text{Re}(\mu)<0$.

Conclusions. The higher-order terms in the expansion are readily generated by repeated substitution of the unperturbed solutions. In the development of the expansion, one must remember that convergence is not guaranteed, except, possibly, for small initial amplitudes.

For problems of the class considered, it that NPT appears to be an effective procedure in that the individual terms in the expansion capture the correct behavior of the full solution. It is be possible to pursue this development through a more detailed discussion, but this would take us away from the primary objective of this chapter, which is to illustrate situations in which NPT is not a satisfactory method.

3.1.2 Pure imaginary eigenvalues

In this section we discuss a sequence of simple examples that illustrate how NPT can lead to the presence of spurious secular terms. We begin by the analysis of a linear problem.

Example 3.1. Perturbed linear oscillator. Consider

$$\frac{d^2x}{dt^2} + (1+\varepsilon)x = 0; \qquad x(0) = A; \qquad \dot{x}(0) = 0 \qquad (3.7)$$

This equation represents a simple harmonic oscillator with a slightly shifted spring constant that modifies its frequency. The full solution

$$x(t) = A\cos\{(1+\varepsilon)^{1/2}t\}$$

is periodic. Since ε is a small parameter, we can expand the argument of the cosine and write an approximate solution as

$$x(t) = A\cos\{[1 + \tfrac{1}{2}\varepsilon + O(\varepsilon^2)]t\}$$

Observe that the error term is *within* the argument of the cosine, so that the approximation has the structure we desire. It retains the periodic nature of the full solution, by incorporating the slight frequency shift due to the perturbation. However, while the full solution and the approximation are both bounded by $\pm A$, they get out of phase as time progresses. The approximation is close to the full solution within an error of $O(\varepsilon^2)$ for timescales of $O(1)$ and an error of $O(\varepsilon)$ for times of $O(1/\varepsilon)$. In our analysis, we would like to develop perturbation methods that have this property. For the present problem, naive perturbation theory does not yield such an expansion. Rather, the approximation one obtains is

$$x(t) = A\cos t - \tfrac{1}{2}\varepsilon t A\sin t + O(\varepsilon^2 t)$$

The ε-correction term is called *secular* and masks the periodic nature of the full solution. To see how this happens in naive perturbation theory, we write the cosine function in its expanded form:

$$\cos\{[1 + \tfrac{1}{2}\varepsilon]t\} = \cos t \cos(\tfrac{1}{2}\varepsilon t) - \sin t \sin(\tfrac{1}{2}\varepsilon t)$$

DEFINITION AND BASIC ASPECTS OF NPT

When expanded for small values of $\varepsilon \cdot t$, the $\sin(\tfrac{1}{2}\varepsilon t)$ term yields the aperiodic secular contribution.

Let us see how this occurs by deriving an approximate solution to Equation (3.7). First we write the equation in the form

$$\frac{d^2 x}{dt^2} + x = -\varepsilon x \tag{3.8a}$$

We write the exact solution in the form of an integral equation:

$$x(t) = A\cos t - \varepsilon \int_0^t \sin(t-t')x(t')\,dt' \tag{3.8b}$$

Substituting Equation (3.8b) for $x(t')$ within the integral yields

$$x(t) = A\cos t$$

$$- \varepsilon \int_0^t \sin(t-t')\left[A\cos t' - \varepsilon \int_0^{t'} \sin(t'-t'')x(t'')\,dt''\right]dt' \tag{3.8c}$$

Expanding, one has

$$x(t) = A\cos t - \varepsilon \int_0^t \sin(t-t')A\cos t'\,dt'$$

$$+ \varepsilon^2 \int_0^t dt' \int_0^{t'} \sin(t'-t'')x(t'')\,dt'' \tag{3.8d}$$

Equation (3.8d) is exact. It looks like an expansion in powers of ε, tempting one to truncate the expansion after the first term and write

$$x(t) = A\cos t - \varepsilon \int_0^t \sin(t-t')A\cos t'\,dt' + O(\varepsilon^2 t) \tag{3.8e}$$

We are really making a *valid* assumption that $x(t)$ is close to $x_0(t)$ for a finite range of time, $0 \le t \le T = O(1)$. Evaluating the integral, we obtain

$$x(t) = A\cos t - \tfrac{1}{2}\varepsilon t A\sin t + O(\varepsilon^2 t) \tag{3.8f}$$

so that we have a spurious secular term: $\{-\frac{1}{2}\varepsilon t A \sin(\varepsilon t)\}$. Thus, although the true solution is periodic, we have, while proceeding *correctly*, obtained an approximation that has an aperiodic structure.

Discussion. Generally, the approximations that naive perturbation theory provides for perturbed simple harmonic motion contain spurious secular terms. This follows directly from the choice of the unperturbed solution as the zero-order approximation. (One says that one uses the unperturbed functions, $A \cdot \cos t$ and $A \cdot \sin t$, as the generators of the expansion.) In the case of Equation (3.7), this means writing $x_0(t) = A \cdot \cos t$. If one thinks about it, there is ample freedom in the choice of the zero-order term. For instance, one could write

$$x_0(t) = A \cos \omega t$$

with $\omega \to 1$ as $\varepsilon \to 0$. If one writes

$$\omega = 1 + \sum_n \varepsilon^n \alpha_n$$

the coefficients α_n can be chosen to kill all the secular generating terms. This is precisely the method of Poincaré and Lindstedt [2,3]

3.1.3 An alternative approach

Another way to introduce a naive perturbation expansion is to expand the solution of Equation (3.7) by direct substitution of Equation (3.2). A hierarchy of equations is obtained by equating powers of ε:

$$\ddot{x}_0 + x_0 = 0 \qquad (3.9a)$$

$$\ddot{x}_1 + x_1 = -x_0 \qquad (3.9b)$$
$$\vdots$$

Note that Equation (3.9a) naturally yields

$$x_0(t) = A_0 \cos t$$

Substituting this result in the r.h.s. of Equation (3.9b) and solving the latter, we find

$$x_1(t) = B \cos t - \frac{1}{2}[t A_0 \sin t] \qquad (3.10a)$$

DEFINITION AND BASIC ASPECTS OF NPT

which yields the following first-order approxmation for $x(t)$:

$$x(t) \approx (A_0 + \varepsilon B)\cos t - \varepsilon A_0 (\tfrac{1}{2}t)\sin t \qquad (3.10b)$$

Thus, we obtain an approximation for $x(t)$, which, except for the error term, is precisely Equation (3.8f), with A replaced by $A_0 + \varepsilon B$. We observe that, within the framework of naive perturbation theory, the spurious secular term in the approximate solution is a direct consequence of the choice of a zero-order term that is unaffected by the perturbation.

Example 3.2. Often, we find it convenient to convert our second order differential equation to a pair of first order equations and then bring the matrix of the unperturbed problem to diagonal form. In the case of Example 3.1, we write

$$\frac{dx}{dt} = y \qquad \frac{dy}{dt} = -x - \varepsilon x$$

Introducing $z = x + i y$, and $z^* = x - i y$, the diagonalized equations are

$$\dot{z} = -i z - i \tfrac{1}{2}\varepsilon(z + z^*)$$

$$\dot{z}^* = i z^* + i \tfrac{1}{2}\varepsilon(z + z^*) \qquad (3.11)$$

We now expand z (and similarly, z^*) in a power series as in Equation (3.2):

$$z = z_0 + \varepsilon z_1 + \varepsilon^2 z_2 + \cdots \qquad (3.12)$$

We need only consider the equation for z, since that of z^* is obtained by complex conjugation. Substituting Equation (3.12) in Equation (3.11) and retaining terms through $O(\varepsilon)$, we have

$$\dot{z}_0 + \varepsilon \dot{z}_1 = -i z_0 - i \varepsilon z_1 - i \tfrac{1}{2}\varepsilon(z_0 + z_0^*) \qquad (3.13)$$

We implement the naive perturbation expansion by writing

$$\dot{z}_0 = -i z_0$$

yielding

$$z_0 = A_0 \exp(-i\, t)$$

The equation for z_1 becomes

$$\frac{dz_1}{dt} = -i\, z_1 - \frac{i}{2} A_0 \left[\exp(-i\, t) + \exp(+i\, t)\right]$$

This is solved by

$$z_1 = B\exp(-i\, t) - \tfrac{1}{2} i\, A_0 t \exp(-i\, t) - \tfrac{1}{4} A_0 \exp(+i\, t) \quad (3.14)$$

Note the secular term, $-\tfrac{1}{2} i A_0 t \cdot exp(-it)$. The coefficient B is arbitrary. Equation (3.8f) is recovered by writing $x = (z + z^*)/2$. The initial conditions are satisfied by writing $A = A_0 + \varepsilon B$.

Example 3.3. The secular behavior encountered above is not an artifact of this particular problem. It is quite general for problems in which the unperturbed eigenvalues are $\pm i$. For example, consider

$$\frac{d^2 x}{dt^2} + x = \varepsilon f(x)$$

$$0 < |\varepsilon| << 1 \quad x(0) = A \quad \frac{dx}{dt}(0) = B$$

(3.15a)

As the system is conservative, the phase-plane orbit is closed; a distorted circle for sufficiently small perturbations and initial conditions. We write the solution as an integral equation:

$$x(t) = A\cos t + B\sin t + \varepsilon \int_0^t f[x(t')] \sin(t - t')\, dt' \quad (3.15b)$$

An essential feature of NPT is that the unperturbed solutions, $\sin(t)$ and $\cos(t)$, are used as the generators of the expansion. The perturbation $f(x)$ is assumed to be a polynomial, or to have a Taylor expansion, in x. Imagine that we know $x(t)$. Inserting the solution in the expression for $f(x)$, we expect $f(x)$ to be of the general form

$$f(x) = \sum_n b_n \cos nt + c_n \sin nt + O(\varepsilon).$$

The coefficients b_n and c_n are expressible as expansions in powers of ε. The additional $O(\varepsilon)$ represents the cumulative effect of the

DEFINITION AND BASIC ASPECTS OF NPT

perturbation (e.g., frequency updating) on the form of $f(x)$.

Insisting on writing the solution in this form, we see that secular terms originate only from terms with n=1 in the sum [or in the $O(\varepsilon)$ correction if they occur there]. When multiplied by $\sin(t-t')$, they are the only ones that can yield constant terms in the integrand of Equation (3.15b) and hence yield terms in the expansion that are proportional to "t multiplied by a $\sin(t)$ or $\cos(t)$."

It is essential to keep in mind that, in contrast to linear equations, where one has final determination of the coefficients, here the coefficients are altered or updated as one includes higher-order terms in the expansion in powers of the small parameter.

Example 3.4: The Duffing oscillator. Let $f(x) = -x^3$. This system is called the *Duffing oscillator* which will serve, throughout the text, as a prototype of a conservative planar system. It is discussed in considerable detail in Chapter 6. Equation (3.15a) becomes

$$\frac{d^2x}{dt^2} + x + \varepsilon x^3 = 0; \quad x(0) = A, \quad \dot{x}(0) = 0 \quad (3.16)$$

Energy is conserved and the motion is periodic. The Hamiltonian for the system is

$$H = \tfrac{1}{2}(x^2 + \dot{x}^2) + \varepsilon \tfrac{1}{4} x^4 = \tfrac{1}{2} A^2 + \varepsilon \tfrac{1}{4} A^4 \quad (3.17)$$

To obtain an expansion by introducing the integral formulation given by Equation (3.4b), we *approximate* $x(t')^3$ by the unperturbed solution $\{A \cos(t)\}^3$ and only retain corrections of $O(\varepsilon)$. We get

$$x(t) = A\cos t - \varepsilon \int_0^t [A\cos t']^3 \sin(t - t')\, dt' + O(\varepsilon^2 t) \quad (3.18)$$

The solution is

$$x(t) = A\cos t - A^3\{\tfrac{3}{8} t \sin t + \tfrac{1}{32}[\cos t - \cos 3t\,]\} + O(\varepsilon^2 t) \quad (3.19)$$

Although the full solution is periodic, Equation (3.19) contains the *secular* aperiodic term $t \cdot \sin(t)$. This artifact of the approximation method is *not* a characteristic of the solution, but the "signature" of naive perturbation theory. The secular term limits the validity of the approximation to $t \approx O(1)$, as the first order correction ceases to be small [$O(\varepsilon)$] for $t \approx O(1/\varepsilon)$. Secular behavior is directly traceable to the presence in the integral of the unperturbed solution, acting as a forcing function. Thus, this perturbation method yields an expansion that does not retain the character of the full solution in each order. Because of this feature of naive perturbation theory, we seek a better method.

Comment. Often, Equation (3.16) appears in a form where the unperturbed frequency is ω rather then 1. In this case one simply rescales the time by introducing $\tau=\omega\cdot t$. This makes sense since in this way one is simply readjusting the rate at which one counts from units measured by the variable t to ones measured by the variable τ.

Example 3.5. We return to the Duffing oscillator and use the alternative analysis introduced in Section 3.2.1 to see other aspects of the naive perturbation expansion. We write the basic equation in the form

$$\ddot{x} + x + \varepsilon x^3 = 0; \qquad x(0) = A, \qquad \dot{x}(0) = 0 \qquad (3.20)$$

Repeating the procedure, described in Example 3.4, in a slightly different manner, substitute Equation (3.2) in Equation (3.20), and a hierarchy of dynamical equations is obtained. Through $O(\varepsilon)$ it yields

$$\ddot{x}_0 + x_0 = 0 \qquad (3.21a)$$

$$\ddot{x}_1 + x_1 = -x_0^3 \qquad (3.21b)$$

Equation (3.21a) is solved by

$$x_0 = A\cos t \qquad (3.22)$$

which, when substituted in Equation (3.17b), yields

$$\ddot{x}_1 + x_1 = -A^3\left\{\tfrac{3}{4}\cos t + \tfrac{1}{4}\cos 3t\right\} \qquad (3.23)$$

Equation (3.23) is solved by

$$x_1 = \underbrace{A^3\left\{-\tfrac{3}{8}t\sin t + \tfrac{1}{32}\cos 3t\right\}}_{\text{Particular solution of inhomogeneous equation}} + \underbrace{B\cos t}_{\text{General solution of homogeneous equation}} \qquad (3.24)$$

This is precisely the form of Equation (3.19) if we write $B = -\tfrac{1}{32}A^3$.

Technically, the l.h.s. of Equation (3.23) for x_1 describes a harmonic oscillator with unit angular frequency that is forced by the

DEFINITION AND BASIC ASPECTS OF NPT

r.h.s. of the equation. The latter includes a $\cos t$ term that resonates with the natural frequency of oscillation, generating, unbounded oscillations. As alluded to earlier, the difficulty in this form of naive perturbation theory (or origin of secular behavior) arises because, in the equation for x_0 [Equation (3.21a)], only the effect of the unperturbed part of the dynamical equation is included, whereas the time dependence of the full solution is affected by the perturbation. For instance, in the case of the Duffing equation, the solution of Equation (3.17a) given by x_0, has a period of oscillation equal to the unperturbed period, 2π, while the period of the full solution is modified by the perturbation and depends on both ε and on the oscillation amplitude, A. The correct period is given by

$$T = \frac{2\pi}{\omega(\varepsilon, A)}$$

with $\omega(\varepsilon,A) \neq 1$. There is a mismatch between the natural time scale of the selected x_0 and that of the full solution.

3.2 DISCUSSION

Many attempts have been made over more than a century to develop expansion schemes that overcome this handicap of NPT. One looks for an expansion for $x(t)$ around a zero order term that has the *same frequency of oscillations* as the full solution and assumes that the frequency can be expanded as

$$\omega(\varepsilon, A) = 1 + \varepsilon \omega_1 + \cdots \qquad (3.25)$$

and writes $x_0(t)$ in the form

$$x_0(t) = A \cos \omega t \qquad (3.26)$$

Using Equation (3.26) and the expansion, Equation (3.2), yields

$$-(1 + 2\varepsilon \omega_1) x_0(t) + x_0(t) + \varepsilon \ddot{x}_1(t) + \varepsilon x_1(t) + \varepsilon x_0^3(t) \\ = 0 + O(\varepsilon^2) \qquad (3.27)$$

The zero-order equation is trivially obeyed. The $O(\varepsilon)$ terms yield

$$\ddot{x}_1 + x_1 = -A^3 \left\{ \tfrac{3}{4} \cos \omega t + \tfrac{1}{4} \cos 3\omega t \right\} + 2\omega_1 A \cos \omega t \qquad (3.28)$$

The solution of Equation (3.28) is as follows:

$$x_1(t) = \frac{1}{4(9\omega^2 - 1)}\cos 3\omega t + \frac{2\omega_1 A - \left(\frac{3}{4}\right)A^3}{1 - \omega^2}\cos\omega t + B\cos t \qquad (3.29)$$

where B is an arbitrary constant. The potentially dangerous term is the $\cos\omega t$ term. The appearance of a secular term has been avoided by the replacement of the unperturbed frequency ($\omega=1$) by the updated one, $\omega(\varepsilon,A)$, but at a price. Using Equation (3.25), the denominator of the $\cos\omega t$ term obtains the form

$$1 - \omega^2 = -2\omega_1\varepsilon + O(\varepsilon^2)$$

As a result, x_1 has a term that is $O(1/\varepsilon)$ and, in the approximation

$$x(t) \approx x_0(t) + \varepsilon x_1(t)$$

that term becomes $O(1)$, spoiling the validity of the expansion. The problem can be resolved if the $\cos\omega t$ term is eliminated from Equation (3.29) by fixing ω_1, the update of the oscillation frequency, at

$$\omega_1 = \tfrac{3}{8}A^2$$

In summary, inclusion of the correctly updated frequency in the time dependence of x_0 prevents the appearance of a secular term, in this order of the expansion and guarantees that, in that order, the expansion is valid for $t=O(1/\varepsilon)$. This simple procedure is the basis of the method developed by Poincaré [2] and Lindstedt [3] almost a century ago.

Example 3.6. The model presented below serves to emphasize that the failure of NPT to generate a uniform expansion depends on whether an eigenvalue update is required. Consider

$$\dot{x} = -y + 2\varepsilon xy; \qquad \dot{y} = x + 2\varepsilon y^2$$

The model is exactly solvable, yielding

$$x = A\frac{\cos t}{[1 - 2\varepsilon A(1 - \cos t)]} \qquad y = A\frac{\sin t}{[1 - 2\varepsilon A(1 - \cos t)]}$$

Expanding the denominator in powers of ε, yields the naive expansion.

DISCUSSION

As there is no frequency update, no spurious *secular terms* appear. Hence, in this example, NPT yields a perfectly satisfactory expansion.

Example 3.7. A nonautonomous system. Genuine secular behavior may occur in nonautonomous problems. Consider

$$\frac{d^2x}{dt^2} + x = \varepsilon \cos t$$

The solution is

$$x(t) = A\cos t + B\sin t + \tfrac{1}{2}\varepsilon t \sin t$$

The solution has genuine secular behavior and the "frequency" of the full solution is the same as that of the unperturbed problem. (We put "frequency" in quotation marks since the motion is not periodic.)

Final comments. NPT uses the unperturbed solutions as generators. Therefore, it is incapable of accounting for an eigenvalue update associated with the perturbation. In the expansion, Equation (3.2), of the general problem, Equation (3.1), terms with a structure that differs from that of the full solution may appear. The truncated series may then misrepresent the long time characteristics of the full solution. In particular, in problems that have periodic solutions, NPT may generate spurious secular terms. In problems with aperiodic solutions, there may be true secular terms. (See Example 3.7.)

Various methods have been developed to obtain an expansion that is free from spurious secular terms and in which the structure of the full solution is reflected in a truncated expansion. Throughout the rest of this book, we will employ one method that overcomes the problem of secular terms in a systematic manner: the *method of normal forms*.

REFERENCES

1. Tricomi, F. G., *Integral Equations*, Dover, New York (1985).
2. Poincaré, H., *New Methods of Celestial Mechanics* [originally published as *Les Méthodes Nouvelles de la Méchanique Céleste*, Gauthier-Villars, Paris (1892)], American Institute of Physics (1993).
3. Lindstedt, A., "Über die Integration einer für die störungstheorie wichtigen Differentialgleichung." *Astron. Nach.* **103**, 211 (1882).

4

FORMALISM OF PERTURBATION EXPANSION

4.1 FREEDOM IN PERTURBATION EXPANSION OF SOLUTIONS OF DYNAMICAL SYSTEMS

There is much freedom in the perturbative expansions of solutions of dynamical systems. It enables one to simplify the form of the approximation generated in the expansion. This freedom is a central theme in the present book and the subject matter of this section.

Consider, for instance, the analysis of the Duffing equation in Example 3.5. The freedom is manifest already in first order. The solution for x_1, given by Equation (3.24), includes two terms: a particular solution of the inhomogeneous equation and a general solution of the homogeneous one. The former is clearly a functional of $\mathbf{x}_0$, while the latter introduces an explicit time dependence.

A similar pattern emerges in higher orders. Free terms may show up in each order. Their existence and the constraints one imposes on them affect the structure of the higher orders and, as a result, they also affect the form as well as the quantitative nature of the zero-order term. Depending on the way the free terms are structured, $\mathbf{x}_n$ may depend on $\mathbf{x}_0$ and may also depend explicitly on time.

While the example just cited has been analyzed within the framework of the naive perturbation theory, the freedom in the expansion procedure is not limited to that method. It is an inherent feature of perturbation theory in general. The two aspects discussed in the preceding paragraph are only a limited facet of the nonuniqueness of the perturbation series.

We demonstrate the freedom in Figure 4.1. The shaded domains denote an n-dimensional sphere around $\mathbf{x}(t)$ with a radius of $O(\varepsilon)$ (defined in a suitable norm). The sphere moves along the trajectory of $\mathbf{x}(t)$ as the latter evolves in time. Any vector function that satisfies all the constraints required by the specific expansion procedure one adopts (e.g., boundedness or having a certain number of derivatives), is a valid zero-order approximation as long (in time) as it remains within a distance of $O(\varepsilon)$ from $\mathbf{x}(t)$. The choice of $\mathbf{x}_0(t)$ determines the functional form of the higher-order corrections $\varepsilon^n \mathbf{x}_n(t)$. Thus, one expects $\mathbf{x}_n$ to depend on $\mathbf{x}_0$. The approach to $\mathbf{x}(t)$ when higher-order corrections are added successively to $\mathbf{x}_0(t)$, defines a "path" (denoted by Σ in the figure). If the perturbation series converges, the "path" will converge onto $\mathbf{x}(t)$. If, on the other hand, the series is an asymptotic one, the "path" will approach $\mathbf{x}(t)$ and then diverge away from it. Namely, the sum of a finite number of terms in the expansion will provide a good approximation for the full solution, whereas the inclusion of additional terms in the expansion causes the sum to diverge away from the full solution. In the figure we show two possible zero-order approximations: $\mathbf{x}_0(t)$ and $\bar{\mathbf{x}}_0(t)$. Of the two, $\bar{\mathbf{x}}_0(t)$ remains a valid approximation for longer times.

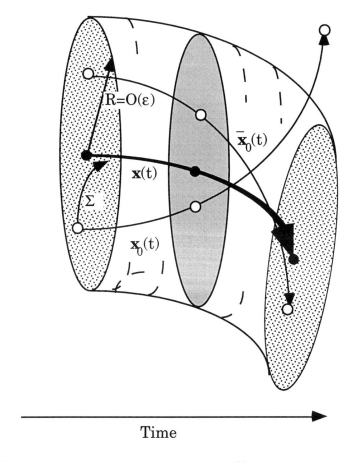

Figure 4.1 Freedom in expansion. $\bar{\mathbf{x}}_0(t)$ is a valid zero-order approximation for longer times than $\mathbf{x}_0(t)$.

FREEDOM IN PERTURBATION EXPANSION

Consider a system of equations for an n-dimensional vector, **x**:

$$\dot{\mathbf{x}} = \mathbf{X}_0 + \varepsilon \mathbf{X}_1 + \varepsilon^2 \mathbf{X}_2 + \cdots, \qquad |\varepsilon| \ll 1, \quad \mathbf{x}(0) = \mathbf{a} \qquad (4.1)$$

where $\mathbf{X}_i$ are *n*-dimensional vector fields that depend on **x** and/or *t*.

Unless otherwise stated, we will assume that $\mathbf{X}_0$ is a linear operator:

$$\mathbf{X}_0 = \mathbf{A}\mathbf{x} \qquad (4.2)$$

For the sake of simplicity, assume that the $n \times n$ matrix **A** is diagonal:

$$\mathbf{A} = \begin{pmatrix} \lambda_1 & 0 & \cdots & 0 \\ 0 & \ddots & \ddots & \vdots \\ \vdots & \ddots & \ddots & 0 \\ 0 & \cdots & 0 & \lambda_n \end{pmatrix} \qquad (4.3)$$

To find an approximation to the solution $\mathbf{x}(t)$, one assumes that a formal expansion in powers of ε (*near-identity transformation*) exists:

$$\mathbf{x} = \mathbf{x}_0 + \varepsilon \mathbf{x}_1 + \varepsilon^2 \mathbf{x}_2 + \cdots \qquad (4.4)$$

When the system is nonautonomous, i.e., when $\mathbf{X}_i$ in Equation (4.1) depend explicitly on time, one intuitively expects $\mathbf{x}_j$ to depend explicitly on time as well. However, even in the case of autonomous systems, $\mathbf{x}_j$ may depend explicitly on time as the examples discussed in Chapter 3 show. Therefore, in general, one should write

$$\mathbf{x}_j = \mathbf{x}_j(\mathbf{x}_0, t) \qquad (4.5)$$

With Equation (4.5) in mind, we substitute Equation (4.4) for **x** in Equation (4.1). Through first order, each component of **x** obeys

$$\dot{x}_0^i + \varepsilon \sum_{j=1}^n \frac{\partial x_1^i}{\partial x_0^j} \dot{x}_0^j + \varepsilon \frac{\partial x_1^i}{\partial t} + \cdots =$$

$$X_0^i(\mathbf{x}_0) + \varepsilon \sum_{j=1}^n \frac{\partial X_0^i(\mathbf{x}_0)}{\partial x_0^j} x_1^j + \varepsilon X_1^i(\mathbf{x}_0) + \cdots \qquad (4.6)$$

In Equation (4.6) all Xs and their derivatives are evaluated at $\mathbf{x}=\mathbf{x}_0(t)$.

In the naive expansion, one equates terms order by order in ε. Through $O(\varepsilon)$ one obtains

$$\dot{x}_0^i = X_0^i(x_0) \tag{4.7a}$$

$$\sum_{j=1}^{n} \frac{\partial x_1^i}{\partial x_0^j}\dot{x}_0^j + \frac{\partial x_1^i}{\partial t} = \sum_{j=1}^{n} \frac{\partial X_0^i(x_0)}{\partial x_0^j} x_1^j + X_1^i(x_0) \tag{4.7b}$$

Note that *the zero-order term is not affected by the perturbation.* As the analysis of the Duffing oscillator has indicated, this is a major shortcoming of the naive expansion. It results in the occurrence of secular terms in $\mathbf{x}_1$ and, similarly, in all $\mathbf{x}_n$, $n \geq 1$, that limit the validity of the approximation to short times.

One wishes to exploit the freedom in the expansion, to find an $x_0(t)$ that better reflects the effect of the perturbation on the time dependence of the solution. For example, for periodic systems, instead of $\mathbf{x}_0(t)$ being a mere $\cos t$ term, one searches for a function with the correctly updated timescale T (see Section 3.2.2), namely, a $\cos[\omega(\varepsilon,a)t]$. However, the dynamical equation for $\mathbf{x}_0$ cannot then be as simple as Equation (4.7a) in the general case, or Equation (3.21a) in the Duffing problem. To avoid spurious secular terms, it must include ε-dependent terms, as postulated by Poincaré:

$$\dot{\mathbf{x}}_0 = \mathbf{U}_0 + \varepsilon \mathbf{U}_1 + \varepsilon^2 \mathbf{U}_2 + \cdots \tag{4.8}$$

Substituting Equation (4.8) in Equation (4.6), and collecting terms order by order, we obtain

$$U_0^i = X_0^i(\mathbf{x}_0) \tag{4.9a}$$

$$U_1^i + \frac{\partial x_1^i}{\partial t} = [\mathbf{X}_0(\mathbf{x}_0), \mathbf{x}_1]^i + X_1^i(\mathbf{x}_0, t) \tag{4.9b}$$

$$U_2^i + \frac{\partial x_2^i}{\partial t} = [\mathbf{X}_0(\mathbf{x}_0), \mathbf{x}_2]^i + X_2^i(\mathbf{x}_0, t)$$

$$+ \sum_j \frac{\partial X_1^i(\mathbf{x}_0,t)}{\partial x_0^j} x_1^j + \tfrac{1}{2} \sum_{j,k} \frac{\partial^2 X_0^i(\mathbf{x}_0)}{\partial x_0^j \partial x_0^k} x_1^j x_1^k \tag{4.9c}$$

$$- \sum_j \frac{\partial x_1^i(\mathbf{x}_0,t)}{\partial x_0^j} U_1^j(\mathbf{x}_0,t)$$

FREEDOM IN PERTURBATION EXPANSION

The structure of the dynamical equations in higher orders is similar to that of Equation (4.9c):

$$U_n^i + \frac{\partial x_n^i}{\partial t} = \begin{cases} \left[\mathbf{X}_0(\mathbf{x}_0), \mathbf{x}_n\right]^i \\ \\ + \text{ coefficient of } \varepsilon^n \text{ in} \\ \text{Taylor expansion of perturbation} \\ \\ - \sum_{\substack{j \\ p+q=n \\ 1 \le p, q \le n-1}} \frac{\partial x_p^i(\mathbf{x}_0, t)}{\partial x_0^j} U_q^j(\mathbf{x}_0, t) \end{cases} \quad (4.9d)$$

The quantity in square brackets is the *Lie bracket*, defined for a pair of vector functions $\mathbf{F}$ and $\mathbf{G}$ as

$$[\mathbf{F}, \mathbf{G}]^i = \sum_j \frac{\partial F^i}{\partial x_0^j} G^j - \sum_j \frac{\partial G^i}{\partial x_0^j} F^j \quad (4.10)$$

Equation (4.9d) is the nth-order version of what is called in the literature the *homological equation*.

To obtain the "coefficient of ε^n in the Taylor expansion of the perturbation" appearing on the r.h.s. of Equation (4.9d), one first expresses $\mathbf{x}$ in terms of its expansion, Equation (4.4), and then computes the desired coefficient. The cases $n=1,2$ are examples of this algorithm. Also, the fact that all quantities are computed at the point $\mathbf{x} = \mathbf{x}_0$ (i.e., $\varepsilon = 0$) is displayed explicitly in Equations (4.9a–d).

Equations (4.9b–d) involve a great deal of freedom. First, the equations merely constitute relationships among U_n^i and x_n^i and leave the latter undetermined. To determine them, one has to impose additional requirements, that depend on what one wishes to achieve. Second, U_n^i and $\partial x_n^i/\partial t$ appear in the combination $\{U_n^i + \partial x_n^i/\partial t\}$. This is a very important feature. It means that, as far as the dynamical equations are concerned, within the framework of a perturbation expansion, in any order n, *we cannot distinguish between the effect of the perturbation* in that order *on the time dependence of* $\mathbf{x}_0$ (U_n^i) *and the explicit time dependence of* $\mathbf{x}_n$.

Thus, in a given order, n, one may split the effect of the r.h.s. of the dynamical equations [e.g., Equations (4.9b,c)] between U_n^i and

($\partial x_n{}^i/\partial t$) arbitrarily. For example, in the naive expansion, all $U_n{}^i$ are chosen to vanish identically. Consequently, in every order, the whole effect of the r.h.s. of the dynamical equations is directed at generating the time dependence of $\mathbf{x}_n$. Choosing the fraction of the r.h.s. of the equations that affects $U_n{}^i$ determines the degree to which the time dependence of $\mathbf{x}_0$ is updated under the influence of the perturbation.

In higher orders, $m>n$, of the expansion, $U_n{}^i$ and $\partial x_n{}^i/\partial t$ do not appear in the same combination of $\{U_n{}^i + \partial x_n{}^i/\partial t\}$. Therefore, the choice for the splitting in nth order will affect the r.h.s. of the dynamical equations in orders higher than n. Hence, it will affect the time dependence of $U_m{}^i$ and of $\partial x_m{}^i/\partial t$ for $m>n$.

Comment. We have expanded $d\mathbf{x}_0/dt$ in a power series in ε and have not done so for $\partial \mathbf{x}_n/\partial t$. Including an expansion for $\partial \mathbf{x}_n/\partial t$ only increases the freedom. The inability to distinguish between the n'th order contributions to $d\mathbf{x}_0/dt$ and to $\partial \mathbf{x}_n/\partial t$ remains.

The perturbation analysis relates two aspects: (*i*) solving Equation (4.8) for the time dependence of the zero-order term, $\mathbf{x}_0$, through some preselected order, n, and (*ii*) computing $\mathbf{x}_k$ for $1 \leq k \leq n$. Our main goal is to simplify the analysis as much as possible. In particular, it is useful to simplify Equation (4.8). For instance, it may be very helpful if $\mathbf{x}_n$ can be chosen so that the Lie brackets on the r.h.s. of Equations (4.9b,c) or, in general, (4.9d), cancel all the remaining terms on the r.h.s. of the equations for *all* $1 \leq k \leq n$:

$$[\mathbf{X}_0, \mathbf{x}_k]^i + \left\{\begin{matrix}\text{coefficient of } \varepsilon^k \text{ in}\\ \text{Taylor expansion of perturbation}\end{matrix}\right\} - \sum_{\substack{j \\ p+q=k \\ 1 \leq p, q \leq k-1}} \frac{\partial x_p{}^i}{\partial x_0{}^j} U_q{}^j = 0 \quad (4.11)$$

where all known terms are computed at the point $\mathbf{x}_0$. If this can be achieved, then both $\mathbf{U}_k$ and $\partial \mathbf{x}_k/\partial t$ can be chosen to be zero for $1 \leq k \leq n$, simplifying significantly the time dependence of $\mathbf{x}_0$.

Superficially, it seems that Equation (4.11) can always be solved for $\mathbf{x}_k$. However, this is not the case. The structure of the Lie bracket allows the elimination of only certain types of terms in the homological equation. Terms that cannot be eliminated by the Lie bracket are called *resonant* terms. Exploiting their effect so as to avoid the generation of secular terms, is the central goal of the method of normal forms.

4.1.1 Contrasting the naive perturbation expansion and Poincaré's normal form expansion

Example 4.1: The Duffing oscillator. To answer the question of how the effect of the perturbation on the time dependence should be split between x_0 and x_n, we resort again to the Duffing equation. We transform Equation (4.7) as follows

$$\dot{x} = y \qquad \dot{y} = -x - \varepsilon x^3 \qquad (4.12a)$$

The linear part of the system of Equation (4.12a) is diagonalized (its eigenvalues $\pm i$ appear explicitly) by using the complex variable

$$z = x + i y$$

The equation for z is (with the complex conjugated equation for z^*)

$$\dot{z} = -i z - \tfrac{1}{8} i \, \varepsilon (z + z^*)^3 \qquad (4.12b)$$

Now expand z in a power series in ε:

$$z = u + \varepsilon T_1 + \varepsilon^2 T_2 + \cdots \qquad (4.13)$$

where T_n are functions of u and u^* and, possibly, of time as well.

As the analysis following Equations (4.9a–d) indicates, there is freedom in the construction of the series of Equation (4.13). This freedom can be eliminated, rendering our expansion scheme fully defined, if we make a particular choice of the T and U functions in all orders of the expansion. The remainder of this chapter is dedicated mainly to the exploitation of the freedom in the search for expansion schemes that have better characteristics than the naive expansion.

i. Naive perturbation theory. In this method, one assumes

$$\dot{u} = -i u \qquad (4.14a)$$

to be correct to all orders in ε. The solution for Equation (4.14a) is

$$u = A \exp(-i \, t)$$

where A is a complex constant. Substituting the expansion, Equation (4.13), for z in Equation (4.12b), we obtain the first-order relation:

$$\frac{\partial T_1}{\partial u}(-i\,u) + \frac{\partial T_1}{\partial u^*}(i\,u^*) + \frac{\partial T_1}{\partial t} = -i\,T_1 - \tfrac{1}{8}i\,(u+u^*)^3 =$$
$$-i\,T_1 - \tfrac{1}{8}i\,(u^3 + 3u^2 u^* + 3uu^{*2} + u^{*3})$$
(4.14b)

In the traditional naive perturbation expansion one assumes that the T functions do not depend explicitly on u and u^*. With this noted, Equation (4.14b) becomes

$$\dot{T}_1 + i\,T_1 = -\tfrac{1}{8}i\,(u+u^*)^3 = -\tfrac{1}{8}i\,(u^3 + 3u^2 u^* + 3uu^{*2} + u^{*3}) \quad (4.14c)$$

As we have seen in Chapter 3 [Equations (3.23) and (3.24)], the solution for T_1 contains a *secular* term. The latter is generated owing to the occurrence on the r.h.s. of Equation (4.14b) of the monomial

$$u^2 u^* = A^2 A^* \exp(-i\,t)$$

which *resonates* with the unperturbed l.h.s. of Equation (4.14c).

ii. A variation on NPT. Secular behavior is not exclusive to the traditional naive expansion. For instance, returning to Equation (4.14b), we modify the naive expansion so that T_1 becomes a function of u and u^* only and does not depend explicitly on time. One then has

$$U_1 = 0 \quad (4.15a)$$

$$\frac{\partial T_1}{\partial t} = 0 \quad (4.15b)$$

(identical constraints are obeyed by T_n and U_n for all $n \geq 1$), yielding

$$\dot{u} = -i\,u \quad (4.15c)$$

Writing $u = \rho \cdot \exp(-i\varphi)$, we find that ρ and φ obey the equations

$$\frac{d\rho}{dt} = 0; \quad \frac{d\varphi}{dt} = 1 \quad (4.15d)$$

Equations (4.15) yield a homological equation of the form

FREEDOM IN PERTURBATION EXPANSION

$$[Z_0, T_1] = -i\, T_1 + i\, u \frac{\partial T_1}{\partial u} - i\, u^* \frac{\partial T_1}{\partial u^*} = \tfrac{1}{8} i\, (u + u^*)^3 \quad (4.16)$$

Using ρ and φ as independent variables, Equation (4.16) becomes

$$-i\, T_1 - \frac{\partial T_1}{\partial \varphi} = \tfrac{1}{8} i\, \rho^3 \left(e^{-3i\varphi} + 3 e^{-i\varphi} + 3 e^{i\varphi} + e^{3i\varphi} \right) \quad (4.17)$$

The solution of this equation is

$$T_1 = \rho^3 \left(\tfrac{1}{16} e^{-3i\varphi} - \tfrac{3}{16} e^{i\varphi} - \tfrac{1}{32} e^{3i\varphi} - \tfrac{3}{8} i\, \varphi e^{-i\varphi} \right) + B e^{-i\varphi}$$

$$= \tfrac{1}{16} u^3 - \tfrac{3}{16} u u^{*2} - \tfrac{1}{32} u^{*3} - \tfrac{3}{8} i\, \varphi u^2 u^* + B e^{-i\varphi} \quad (4.18)$$

Each monomial on the r.h.s. of Equation (4.17) appears in Equation (4.18) with a modified numerical coefficient, except for the *resonant term* $u^2 u^*$. The latter is multiplied by the phase, φ. This is a secular term, since the motion is periodic and φ grows indefinitely, essentially linearly, in time [see Equation (4.15d)]. However, contrary to the case of the usual naive expansion, now T_1 can still be written as a function of u and u^*, since one has

$$\varphi = -\frac{1}{2i} \log \frac{u}{u^*} \quad (4.19)$$

This term cannot be written as a monomial in u and u^* (in general, an entire function of u and u^*).

We will see later on that the problem arises in the naive expansion because the eigenvalues $\lambda_\pm = \pm i$ of the linear unperturbed part of the equations appear in the monomial $u^2 u^*$ in a *resonant* combination:

$$u = \rho e^{-it} = \rho e^{\lambda_- t} \qquad u^* = \rho e^{+it} = \rho e^{\lambda_+ t}$$

$$u^2 u^* = \rho^3 e^{(2\lambda_- + \lambda_+)t} = \rho^3 e^{\lambda_- t} \quad (4.20)$$

This specific monomial resonates with the homogeneous part of the equation for T_1, and generates an unbounded time dependence.

In conclusion, the requirement that the resonant driving term in the first-order part of the perturbation is to affect T_1 (the first-order term in the expansion of the solution) directly, generates a secular term in T_1 in both versions of the naive expansion. This is also the case if *any fraction* of the resonant term affects T_1 directly.

Comment. The freedom in the expansion appears already in this order: Equation (4.18) has a free term, proportional to the constant B. Such freedom will also occur in higher orders.

iii. Traditional normal form expansion. The difficulties just outlined are avoided in this method; the *full* effect of the resonant terms is attributed to U_1, rather than T_1. This choice has the advantage of *fully updating the time dependence of the zero-order term by the $O(\varepsilon)$ effect of the perturbation*. In the same manner, one assigns the full effect of the resonant terms (originally in the perturbation, or generated by the formalism) in higher orders to U_n.

This ansatz, made by Poincaré in the development of the formalism of perturbation theory [1], is well exposed in the books by Arnold [2] and Wiggins [3]. Essentially, this *zero-order updating ansatz* is the starting point of all perturbation methods that avoid secular terms (e.g., the *Poincaré–Lindstedt method* [1,4], the *method of averaging* [5–7], the *method of multiple timescales* [7], and the *method of normal forms* [1–3,8])). Otherwise stated, the ansatz will be adopted throughout this book. Succinctly stated, the ansatz stipulates that, to avoid the emergence of secular terms in $O(\varepsilon^n)$, one requires

$$U_n = \text{All resonant terms on r.h.s. of Eq. (4.9d)} \tag{4.21}$$

Following Poincaré, as in Equation (4.8), we write for u

$$\dot{u} = U_0 + \varepsilon U_1 + \varepsilon^2 U_2 + \cdots \tag{4.22}$$

Inserting Equations (4.13 and 4.22) in Equation (4.12b) and collecting terms in each order separately, we find through second order

$$U_0 = Z_0 = -i\,u \tag{4.23a}$$

$$U_1 + \frac{\partial T_1}{\partial t} = [Z_0, T_1] - \tfrac{1}{8} i \left(u + u^*\right)^3 \tag{4.23b}$$

$$U_2 + \frac{\partial T_2}{\partial t} = [Z_0, T_2] - \left(\tfrac{3}{8}\right) i \left(T_1 + T_1^*\right)(u + u^*)^2$$

$$- \left(\tfrac{3}{8}\right) i \left(T_1 + T_1^*\right)^2 (u + u^*) - U_1 \frac{\partial T_1}{\partial u} - U_1^* \frac{\partial T_1}{\partial u^*} \tag{4.23c}$$

FREEDOM IN PERTURBATION EXPANSION

In the homological equations, the Lie bracket has the form

$$[Z_0, T_n] \equiv T_n \frac{\partial Z_o}{\partial u} + T_n^* \frac{\partial Z_o}{\partial u^*} - Z_0 \frac{\partial T_n}{\partial u} - Z_0^* \frac{\partial T_n}{\partial u^*} \quad (4.24)$$

The problem encountered in the naive expansion can be still "regained." For instance, if in first order one chooses

$$U_1 = \frac{\partial T_1}{\partial u} = \frac{\partial T_1}{\partial u^*} \equiv 0$$

Equation (4.23b) reduces to Equation (4.14c), producing a secular term in T_1.

Comment. In autonomous systems (i.e., systems with equations that do not contain explicitly time-dependent terms), such as the Duffing equation, the zero-order updating ansatz is naturally accompanied by the additional simplification, generalizing Equation (4.15b):

$$\frac{\partial T_n}{\partial t} = 0 \quad (4.25)$$

For instance, consider again Equation (4.23b). There is no explicit time dependence on the r.h.s. of the equation. Hence, there is no need to assume for T_1 such a dependence. However, even in the case of an autonomous system, Equation (4.25) is not a necessary condition.

In general, Equation (4.25) will not hold for *nonautonomous systems*. Still, one is free to adopt the zero-order updating ansatz. Again, to avoid secular terms, all the resonant terms in the dynamical equations are attributed to U_n. The remaining terms (including non-resonant, explicitly time-dependent ones) are eliminated by the (possibly time-dependent) T_n.

In the first-order approximation for the Duffing equation, Equation (4.21) implies that U_1 equals the resonant term on the r.h.s. of Equation (4.16):

$$U_1 = -\left(\tfrac{3}{8}\right) i\, u^2 u^* = -\left(\tfrac{3}{8}\right) i\, A^2 A^* e^{-it} \quad (4.26)$$

Next, we may either assume or abandon Equation (4.18b). If we adopt Equation (4.18b) as well as Equation (4.26), Equation (4.16b) becomes

$$-i\,T_1 + i\,u\frac{\partial T_1}{\partial u} - i\,u^*\frac{\partial T_1}{\partial u^*} = \tfrac{1}{8}i\left(u^3 + 3uu^{*2} + u^{*3}\right) \quad (4.27)$$

Note the absence of the problematic resonant term of Equations (4.19) and (4.20) that led to the rise of a secular term in Equation (4.21).

If, on the other hand, we choose to assume that T_1 depends explicitly on time and *not* on u and u^*, then together with Equation (4.26), Equation (4.23b) now becomes

$$\dot{T}_1 + i\,T_1 = -\tfrac{1}{8}i\left(u^3 + 3uu^{*2} + u^{*3}\right) = \\ -\tfrac{1}{8}i\,\rho^3\left(e^{-3it} + 3e^{it} + e^{3it}\right) \quad (4.28)$$

Note the absence of the problematic resonant term of Equation (4.14c) that led to the appearance of a secular term in NPT (see Example 3.2).

Equations (4.27) and (4.28) are both solved by the same function:

$$T_1 = \tfrac{1}{16}u^3 - \tfrac{3}{16}uu^{*2} - \tfrac{1}{32}u^{*3} + Bu \\ = \rho^3\left(\tfrac{1}{16}e^{-3it} - \tfrac{3}{16}e^{it} - \tfrac{1}{32}e^{+3it}\right) + B\rho e^{-it} \quad (4.29)$$

The Duffing equation is a typical example. The formalism can be applied to any dynamical system with a linear unperturbed part, if resonant terms appear in the perturbation, or are generated by the expansion scheme.

4.2 RESONANT COMBINATIONS OF EIGENVALUES

Before embarking on the development of the formalism of the normal form expansion, we introduce a few concepts that will be useful later in this chapter and throughout the book.

Given a set of N complex numbers $\boldsymbol{\lambda} \equiv (\lambda_1, \lambda_2, ..., \lambda_N)$ and a set of nonnegative integers $\mathbf{m} \equiv (m_1, m_2, ..., m_N)$, we say that the eigenvalues obey a *resonance condition* if one of them (say, λ_i) can be expressed as a linear combination of the form

$$\lambda_i = \sum_{j=1}^{N} \lambda_j m_j \equiv (\boldsymbol{\lambda}, \mathbf{m}) \qquad (m_j \geq 0) \quad (4.30a)$$

Trivial resonance occurs when the perturbation includes a term

RESONANT COMBINATIONS OF EIGENVALUES 85

which is linear in components of **x**, say, x_i. Then one has all $m_j = 0$ for $j \neq i$ except for $m_i = 1$. Nontrivial resonance occurs when Equation (4.30a) involces *at least* two eigenvalues. This happens when

$$|\mathbf{m}| = \sum_{j=1}^{N} m_j \geq 2 \qquad (4.30b)$$

The number of resonance relations obeyed by the set of eigenvalues depends on the components of $\boldsymbol{\lambda}$.

4.2.1 Poincaré domain

In the analysis of nonlinear systems, the eigenvalues of the linear part of the dynamical equations are characterized by their geometrical relation to the origin in the complex plane.

The eigenvalues are said to be in a *Poincaré domain* if their convex hull in the complex plane does not include the origin. An equivalent definition is that a straight line exists that separates all the eigenvalues from the origin. Figure 4.2 provides examples of eigenvalues that are in a Poincaré domain.

Only a finite number of resonant combinations is possible in a Poincaré domain. To see this, denote the distance of the eigenvalue λ_j from the straight line **K** by d_j. Equation (4.30a) for the eigenvalues implies that for some i, these distances obey

$$d_i = (\mathbf{d}, \mathbf{m}) = \sum_{j=1}^{n} d_j m_j \qquad |\mathbf{m}| \geq 2 \qquad (4.31)$$

Since all the d_j and m_j are nonnegative, the following inequality holds:

$$D \equiv \max_{1 \leq k \leq n} d_k \geq d_i \geq \left(\sum_{j=1}^{n} m_j\right) d, \quad \left(d \equiv \min_{1 \leq k \leq n} d_k\right) \qquad (4.32)$$

Equation (4.32) implies that, for all $1 \leq j \leq n$, one has $m_j \leq D/d$. Hence, all integers m_j that participate in the resonance relation, Equation (4.30a), have an upper bound. Thus, the number of resonant combinations is finite. The geometrical meaning is that there is an eigenvalue, λ_p, with the largest distance, $d_p(\equiv D)$, from the line. Resonance combinations should not produce a distance greater than d_p.

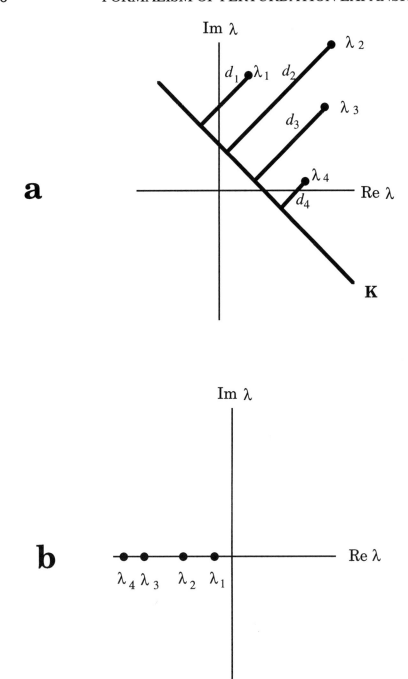

Figure 4.2 Examples of eigenvalues in a Poincaré domain: (a) general case; (b) all eigenvalues on negative real axis.

RESONANT COMBINATIONS OF EIGENVALUES

Example 4.2. Only one resonance relation:

$$\lambda_1 = 1, \quad \lambda_2 = 3, \quad \lambda_3 = 2 + \sqrt{3}\,i, \quad \lambda_4 = 2 - \sqrt{3}\,i$$

$$\lambda_2 = 3\lambda_1, \quad (\mathbf{m} = (3,0,0))$$

Example 4.3. Three resonance relations:

$$\lambda_1 = -1, \quad \lambda_2 = -2, \quad \lambda_3 = -3$$

$$\lambda_2 = 2\lambda_1, \quad (\mathbf{m} = (2,0,0))$$

$$\lambda_3 = 1\cdot\lambda_1 + 1\cdot\lambda_2, \quad (\mathbf{m} = (1,1,0)), \qquad \lambda_3 = 3\lambda_1, \quad (\mathbf{m} = (3,0,0))$$

4.2.2 Siegel domain

The eigenvalues are said to be in a *Siegel domain* if their convex hull includes the origin. Examples are shown in Figure 4.3. In a Siegel domain, if the eigenvalues resonate, the number of resonance relations is infinite. Common examples are: (1) two pure imaginary complex conjugate eigenvalues, $\pm i\omega_0$ (a harmonic oscillator, with unperturbed frequency ω_0); (2) the onset of an instability in a dynamical system. First, all the eigenvalues lie on the left-hand side of the imaginary axis, i.e., with negative real parts. For sufficiently small initial amplitudes, the related degrees of freedom decay exponentially in time. Now some parameter(s) in the problem are changed, causing one or more of the eigenvalues to develop positive real parts and cross the imaginary axis into the right hand side of the plane as in Figures 4.3c,d. (The corresponding coordinates may grow exponentially in time.)

Example 4.4. An infinite number of resonance relations (Figure 4.3b):

$$\lambda_1 = +i\,\omega, \quad \lambda_2 = -i\,\omega$$

$$\lambda_1 = (p+1)\lambda_1 + p\lambda_2, \quad (\mathbf{m} = ((p+1), p))$$

$$\lambda_2 = p\lambda_1 + (p+1)\lambda_2, \quad (\mathbf{m} = (p,(p+1))), \quad p \geq 1$$

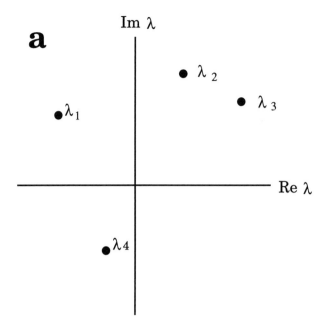

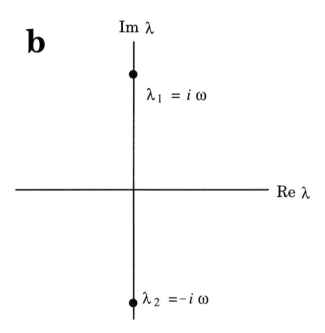

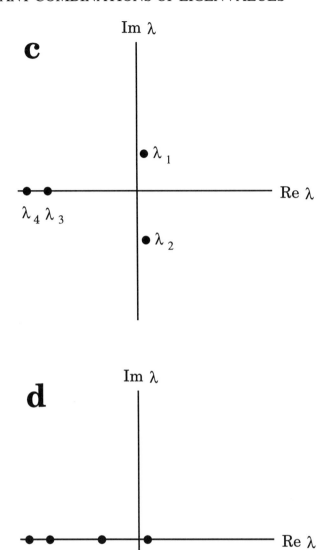

Figure 4.3 Examples of eigenvalues in Siegel domain.: (a) general case; (b) two pure imaginary complex conjugate eigenvalues; (c) two imaginary eigenvalues developing positive real part (Hopf bifurcation); (d) single real eigenvalue becoming positive.

Example 4.5. In Example 4.2, shift the eigenvalues by $+\frac{3}{2}$:

$$\lambda_1 = +\tfrac{1}{2}, \quad \lambda_2 = -\tfrac{1}{2}, \quad \lambda_3 = -\tfrac{3}{2}$$

$$\lambda_1 = (p+1)\lambda_1 + p\lambda_2, \quad p \geq 1, \quad \bigl(\mathbf{m} = \bigl((p+1), p, 0\bigr)\bigr)$$

$$\lambda_2 = p\lambda_1 + (p+1)\lambda_2, \quad p \geq 1, \quad \bigl(\mathbf{m} = \bigl(p, (p+1), 0\bigr)\bigr)$$

$$\lambda_3 = p\lambda_1 + q\lambda_2 + r\lambda_3$$

$$\{p + 3 = q + 3r, \ p,q,r \geq 0, \ p + q + r \geq 2\}, \quad \bigl(\mathbf{m} = (p,q,r)\bigr)$$

Example 4.6. In Example 4.3, shift the eigenvalues by -2. For any integers $p,q,r,s \geq 1$, one then has

$$\lambda_1 = -1, \quad \lambda_2 = 1, \quad \lambda_3 = +\sqrt{3}\,i, \quad \lambda_4 = -\sqrt{3}\,i$$

$$\lambda_1 = (p+1)\lambda_1 + p\lambda_2, \quad \bigl(\mathbf{m} = \bigl((p+1), p, 0\bigr)\bigr)$$

$$\lambda_2 = q\lambda_1 + (q+1)\lambda_2, \quad \bigl(\mathbf{m} = \bigl(q, (q+1), 0\bigr)\bigr)$$

$$\lambda_3 = (r+1)\lambda_3 + r\lambda_4, \quad \bigl(\mathbf{m} = \bigl((r+1), r, 0\bigr)\bigr)$$

$$\lambda_4 = s\lambda_1 + (s+1)\lambda_2, \quad \bigl(\mathbf{m} = \bigl(s, (s+1), 0\bigr)\bigr)$$

The crucial question is whether all possible resonant combinations do occur in a particular normal form expansion. The answer is that they need not. We will see in Chapters 6 and 7 that in a one-degree-of-freedom oscillatory system the occurrence of resonant terms or their absence depends on the functional form of the perturbation as well as on the choice of the free functions in the normal form expansion. In a wide class of harmonic oscillatory problems with nonlinear perturbations, we show that a potentially infinite number of resonant terms may be avoided, and the normal form consists of a very small number of resonant terms (sometimes just one).

Comment. The motivation for the development of the method of normal forms, particularly the expansion ansatz and the zero-order updating ansatz of Poincaré, was the wish to avoid secular terms in cases where we know that the solution does not exhibit such behavior. In the Duffing equation, which we analyze in detail as a generic example for periodic systems, the eigenvalues ($\pm i$) of the unperturbed part are in a Siegel domain. The resulting motion is periodic with an updated frequency, and the occurrence of secular terms (which are not periodic in time) is, evidently, an artifact of the naive expansion procedure. Eigenvalue update is a possibility (although it need not occur in all cases) when the unperturbed eigenvalues lie in a Siegel domain. The source of this capacity for eigenvalue update is traceable to the fact that there are an infinite number of possible resonant combinations, the sum of which may add up to an exponential that updates the unperturbed exponential behavior.

When the eigenvalues lie in a Poincaré domain, there are a finite number of possible resonant combinations. Hence the resonant terms cannot add up to an exponential and eigenvalue update is impossible. Resonant combinations yield terms that are proportional to time (or to other powers of time) in the solutions. These terms cannot be eliminated as they are not artifacts of the expansion procedure. This point will be discussed in detail in Chapter 5.

4.2.3 Small-denominator problem

In the method of normal forms, the quantities

$$\left(\lambda_i - (\lambda, \mathbf{m})\right)$$

occur as denominators of coefficients in the near-identity transformation. If the eigenvalues satisfy a resonance relation with the particular vector $\mathbf{m}$ of integers, then the denominator vanishes. The method of normal forms tells us how to overcome difficulties associated with a vanishing denominator. If, on the other hand, all the eigenvalues (or a subgroup of them) are mutually irrational, then a problem arises, which the method of normal forms cannot address.

Two numbers are defined as *mutually irrational* if their ratio is an irrational number. More specifically, consider the two eigenvalues

$$\lambda_1 = -1, \qquad \lambda_2 = +\sqrt{2}$$

Obviously, these two eigenvalues cannot satisfy a resonance relation: there are no positive integers p and q for which, say,

$$\lambda_1 = p\lambda_1 + q\lambda_2 \tag{4.33}$$

However, the irrational number $\sqrt{2}$ (in fact, any irrational number) can be approximated, to any degree of accuracy, by an infinite sequence of rational numbers p_n/q_n (where p_n and q_n are integers). One can show that, due to the existence of this infinite sequence of approximations to $\sqrt{2}$, there is an infinite number of pairs of integers $\mathbf{m}=(k_i,l_i)$, for which $|\lambda_1 - (\lambda,\mathbf{m})|$ are arbitrarily small. When this happens, the near-identity transformation may not converge.

4.3 THE METHOD OF NORMAL FORMS

Up to this point, we have only exploited the fact that $\mathbf{X}_0$, the zero-order contribution to the "force" in Equation (4.1), is linear in the components of the vector $\mathbf{x}(t)$. The functional form of the remaining $\mathbf{X}_n$ was unspecified. In many applications, the zero-order term in the "force" is linear in the unknown functions, while the perturbation terms $\mathbf{X}_n$ of Equation (4.1) are polynomials or have Taylor expansions in the components of the vector $\mathbf{x}(t)$. For example, in the Duffing equation, the perturbation was a monomial in the unknown function (x^3). In the (diagonalized) complex notation (Equation (4.12b)) it was a polynomial in the variables z and z^*.

The method of normal forms [1–3, 8] is simply the application to this case of the ideas proposed in Section 4.1 for avoiding secular terms. In particular, one incorporates the two ideas of Poincaré.

1. The expansion ansatz, Equation (4.8), which accounts for the effect of the perturbation on the time dependence of the zero-order approximation, $\mathbf{x}_0(t)$. In Equation (4.1), one employs Equation (4.4) for $\mathbf{x}(t)$, with $\mathbf{X}_0$ given by Equations (4.2) and (4.3). The resulting equations, Equations (4.9) and (4.10), are relations between $\mathbf{U}_n$ and $\mathbf{x}_n(t)$.
2. The zero-order updating ansatz: the *full* effect of resonant terms in the dynamical equations is to be ascribed to $\mathbf{U}_n$.

To see how the analysis proceeds, it suffices to study the first-order homological relation, Equation (4.9b). The perturbation term $X_1{}^i(\mathbf{x}_0,t)$ is expanded in powers of components of $\mathbf{x}_0(t)$:

$$X_1^i(\mathbf{x}_0,t) = \sum_{\mathbf{m}} a^{i(\mathbf{m})}(t)\left(x_0^1\right)^{m_1}\left(x_0^2\right)^{m_2}\cdots\left(x_0^n\right)^{m_n} \quad (4.34)$$

where $\mathbf{m}=(m_1,m_2,...,m_n)$ is the vector of nonnegative integer powers appearing in a given monomial. (The number $|\mathbf{m}| = \sum m_k$ is called the *order* of the monomial.) At this stage, we limit our analysis to *autonomous systems*, so that $a^{i(\mathbf{m})}$ are independent of time.

THE METHOD OF NORMAL FORMS

Furthermore, we adopt similar expansions for U_1^i and for x_1^i

$$U_1^i(\mathbf{x}_0, t) = \sum_{\mathbf{m}} b^{i(\mathbf{m})}(t) (x_0^1)^{m_1} (x_0^2)^{m_2} \cdots (x_0^n)^{m_n} \tag{4.35}$$

$$x_1^i(\mathbf{x}_0, t) = \sum_{\mathbf{m}} c^{i(\mathbf{m})}(t) (x_0^1)^{m_1} (x_0^2)^{m_2} \cdots (x_0^n)^{m_n} \tag{4.36}$$

one inserts expansions (4.34)–(4.36) in Equation (4.9b). With Equations (4.2) and (4.3), the Lie bracket in Equation (4.9b) becomes

$$[\mathbf{X}_0, \mathbf{x}_1]^i = \sum_j \frac{\partial X_0^i}{\partial x_0^j} x_1^j - \sum_j \frac{\partial x_1^i}{\partial x_0^j} X_0^j = \lambda_i x_1^i - \sum_j \frac{\partial x_1^i}{\partial x_0^j} \lambda_j x_0^j \tag{4.37}$$

Employing Equations (4.35)–(4.37), Equation (4.9b) becomes an expansion in the monomials. Collecting all the contributions to each monomial in Equation (4.9b), we obtain a relation among the coefficients $a^{i(\mathbf{m})}$, $b^{i(\mathbf{m})}$, and $c^{i(\mathbf{m})}$

$$b^{i(\mathbf{m})} + \frac{dc^{i(\mathbf{m})}}{dt} = (\lambda_i - (\lambda, \mathbf{m})) c^{i(\mathbf{m})} + a^{i(\mathbf{m})} \tag{4.38}$$

with $(\lambda, \mathbf{m})$ defined in Equation (4.30).

In the naive expansion one assumes that the time dependence of $\mathbf{x}_0$ is not affected by the perturbation. In the current notation, this means

$$b^{i(\mathbf{m})} = 0 \tag{4.39}$$

Thus, in naive perturbation theory, Equation (4.38) becomes

$$\frac{dc^{i(\mathbf{m})}}{dt} = (\lambda_i - (\lambda, \mathbf{m})) c^{i(\mathbf{m})} + a^{i(\mathbf{m})} \tag{4.40}$$

If there is no *resonance*, so that Equation (4.30) does not hold, then solution for $c^{i(\mathbf{m})}$ is

$$c^{i(\mathbf{m})} = c_0^{i(\mathbf{m})} \exp((\lambda_i - (\lambda, \mathbf{m}))t) - \frac{a^{i(\mathbf{m})}}{(\lambda_i - (\lambda, \mathbf{m}))} \tag{4.41}$$

Here $c_0^{i(\mathbf{m})}$ is an arbitrary constant. If Equation (4.30) does hold, then from Equation (4.40) we obtain the following:

$$\frac{d}{dt}c^{i(\mathbf{m})} = a^{i(\mathbf{m})} \tag{4.42}$$

Equation (4.42) leads, as expected, to secular behavior:

$$c^{i(\mathbf{m})} = a^{i(\mathbf{m})}t + c_0^{i(\mathbf{m})} \tag{4.43}$$

Alternatively, making the plausible (but not mandatory!) choice that in the autonomous case $c^{i(\mathbf{m})}$ do not depend on time, then, in the naive expansion, *one cannot solve for* $c^{i(\mathbf{m})}$ when **m** is a *resonant vector*.

The zero-order updating ansatz resolves the difficulty. Ascribing the *full* effect of resonant terms to $\mathbf{U}_1$ (which accounts for the effect of the perturbation on the time dependence of $\mathbf{x}_0$ in first order) implies here that Equation (4.38) should be split into two:

$$\lambda_i \neq (\lambda, \mathbf{m}): \quad \begin{cases} \dfrac{\partial c^{i(\mathbf{m})}}{\partial t} = (\lambda_i - (\lambda, \mathbf{m}))c^{i(\mathbf{m})} + a^{i(\mathbf{m})} \\ b^{i(\mathbf{m})} = 0 \end{cases} \tag{4.44}$$

$$\lambda_i = (\lambda, \mathbf{m}): \quad \begin{cases} b^{i(\mathbf{m})} = a^{i(\mathbf{m})} \\ c^{i(\mathbf{m})} = c_0^{i(\mathbf{m})} \end{cases} \tag{4.45}$$

Equation (4.44) is solved by Equation (4.41). If, in addition, one makes the plausible (but not mandatory!) choice that $c^{i(\mathbf{m})}$ do not depend on time, then Equation (4.44) becomes an algebraic equation, yielding

$$c^{i(\mathbf{m})} = -\frac{a^{i(\mathbf{m})}}{(\lambda_i - (\lambda, \mathbf{m}))} \qquad \lambda_i \neq (\lambda, \mathbf{m}) \tag{4.46}$$

In higher orders, the procedure is similar. As in Equation (4.9b), the r.h.s. of Equations (4.9c) and (4.9d) include a Lie bracket and the contribution of $\mathbf{X}_2$ ($\mathbf{X}_n$, respectively). However, they also include contributions from lower orders (that have been solved for in previous stages of the calculation). An expansion such as Equation (4.34) is now applied to the full r.h.s. of Equations (4.9c) and (4.9d). All resonant contributions are ascribed to $\mathbf{U}_2$ ($\mathbf{U}_n$) and the remainder, to $\mathbf{x}_2$ ($\mathbf{x}_n$, respectively). In the autonomous case, again, one adopts the

THE METHOD OF NORMAL FORMS

plausible (but not mandatory!) choice that $c^{i(\mathbf{m})}$ do not depend on time.

Example 4.7: The Duffing oscillator. The first-order perturbation term in Equation (4.23b) (here we denote X_1 by Z_1, x_1 by $T_1(u,u^*)$ where u, u^* are the zero-order terms) is

$$Z_1 = -\tfrac{1}{8} i \left(u + u^* \right)^3 \tag{4.47a}$$

U_1 and T_1 are given by

$$U_1 = -\left(\tfrac{3}{8}\right) i\, u^2 u^*$$

$$T_1 = \tfrac{1}{16} u^3 - \tfrac{3}{16} u u^{*2} - \tfrac{1}{32} u^{*3} + f_1\!\left(uu^*\right)\cdot u \tag{4.47b}$$

where $f_1(uu^*)\cdot u$ is a free function not determined by Equation (4.38).

The coefficients $a(\mathbf{m})$, $b(\mathbf{m})$ and, $c(\mathbf{m})$, obtained through Equations (4.44)–(4.46), are shown in Table 4.1. For the case of u^*, one uses their complex conjugates.

Table 4.1 Coefficients in first-order contribution to normal form expansion in Duffing equation

m	(3, 0)	(2, 1)	(1, 2)	(0, 3)	Any other non resonant (m)	Any other resonant (m)
$a(\mathbf{m})$	$-\tfrac{1}{8}i$	$-\tfrac{3}{8}i$	$-\tfrac{3}{8}i$	$-\tfrac{1}{8}i$	0	0
$b(\mathbf{m})$	0	$-\tfrac{3}{8}i$	0	0	0	0
$c(\mathbf{m})$	$\tfrac{1}{16}$	free	$-\tfrac{3}{16}$	$-\tfrac{1}{32}$	0	free

Through first order, the normal form yields

$$\dot u = -i\,u + \varepsilon U_1 = -i\,u - \varepsilon\left(\tfrac{3}{8}\right) i\, u^2 u^* \tag{4.48}$$

Writing
$$u = \rho \exp(-i\varphi)$$
we find the solution of the normal form equations through first order
$$\rho = \text{const}, \qquad \varphi = \left(1 + \tfrac{3}{8}\varepsilon \rho^2\right)t + \varphi_0$$

The free function in T_1 still has to be chosen.

Let us now check what the approximation
$$z_{\text{approx}} = u + \varepsilon T_1$$
yields, when inserted in the Duffing equation, Equation (4.12b).

Employing Equation (4.47b), we find that all terms in $O(\varepsilon^0)$ and $O(\varepsilon^1)$ cancel exactly. The remainder is $O(\varepsilon^2)$ *independently of the choice of the free function* $f_1 \cdot u$. This should not surprise us. We have constructed U_1 and T_1 so that the dynamical equation is obeyed through first order. This process generates terms that feed the equations for the next-order entities, U_2 and T_2. Thus, the free function does not play any role in how the solution obeys the dynamical equation at this order. In the following orders, f_1 can be exploited in order to affect the manner in which the dynamical equation is obeyed, by modifying the structure of the normal form.

Comment. In many problems of interest, there are several or even infinitely many (e.g., perturbed harmonic oscillators) possible resonant combinations. The question that naturally arises is whether all possible resonant monomials do appear in the normal form. The answer is that they need not. In the following chapters we will show that in many problems, choosing the free functions appropriately, one can eliminate several, sometimes even an infinite number of, resonant terms that might otherwise appear in the normal form.

Example 4.8: Appearance of resonant terms in normal form. When the eigenvalues obey resonance relations, occurrence of the corresponding resonant terms in the normal form depends on the structure of the nonlinear perturbation. Consider the system

$$\frac{dx}{dt} = -x + \varepsilon x y$$

$$\frac{dy}{dt} = -2y + \varepsilon(\alpha x^2 + \beta x y + \gamma y^2)$$

(4.49a)

THE METHOD OF NORMAL FORMS

[The perturbation is $O(\varepsilon^1)$.] Here we have

$$\lambda_1 = -1, \quad \lambda_2 = -2$$

$$X_1 = xy, \quad Y_1 = \alpha x^2 + \beta xy + \gamma y^2$$

(4.49b)

In Equations (4.49a,b), X_1 and Y_1 denote the $O(\varepsilon)$ terms in the x and y equations, respectively. Since λ_1 cannot obey a resonant relation, the normal form for x_0 is just the unperturbed equation (*independent of the form of the perturbation*):

$$\dot{x}_0 = -x_0, \qquad U_1 = U_2 = \cdots = 0$$

where U_n are the terms in the normal form for x_0. In a similar manner, we denote by V_n the terms in the normal form for y_0.

The only possible resonant combination is $\lambda_2 = 2 \cdot \lambda_1$. Therefore, the only resonant combination that can appear in the normal form for y_0 is $(x_0)^2$. In Equation (4.49a), the perturbation includes only quadratic terms. Let us first choose not to add free functions in the near-identity transformation. Then, V_1, x_1, and y_1 will all be quadratic in x_0 and y_0. Equations (4.45) and (4.46) yield

$$V_1 = \alpha x_0^2$$

$$x_1 = -\tfrac{1}{2} x_0 y_0, \quad y_1 = -\beta x_0 y_0 - \tfrac{1}{2}(x_0)^2$$

(4.50)

From Equation (4.50) we see that V_1 does not vanish if $\alpha \neq 0$. In addition, the second-order corrections U_2, V_2, x_2 and y_2 will be cubic in x_0 and y_0. Hence, no resonant terms can appear in U_2 and V_2 and, similarly, in all U_n, V_n, $n \geq 2$. Thus, *the appearance of the resonant term depends on the functional form of the perturbation.*.

Suppose we now add free functions to x_1 and y_1:

$$x_1 = -\tfrac{1}{2} x_0 y_0 + A x_0$$

$$x_y = -\beta x_0 x_y - \tfrac{1}{2}(x_0)^2 + B y_0 + C(x_0)^2$$

(4.50a)

Inserting Equations (4.50a) in Equation (4.9b) and collecting resonant terms in the same manner as in Equations (4.45) and (4.46), we find

$$U_2 = 0$$

$$V_2 = (2A - B)\alpha(x_0)^2$$

Thus, the occurrence of a resonant term in $O(\varepsilon^2)$ depends on both the functional form of the perturbation (i.e., whether α is zero) and the choice of the free functions.

Comment. Adopting the first Poincaré ansatz, one obtains Equation (4.8) for the time dependence of $x_0(t)$. In that equation, each $\mathbf{U}_n$ is computed essentially from the coefficient of ε^n in the expansion of the perturbation in powers of ε. As a result, Equation (4.8) is the only dynamical equation generated that needs to be solved. The equations for $\mathbf{x}_n$ ($n \geq 1$) are direct quadratures. This is where we see the greatest conceptual advantage of the approach delineated here. In other methods, such as the *method of averaging* [5–7] or the *method of multiple timescales* [7], a hierarchy of dynamical equations is generated and has to be solved in each order.

If one also adopts the zero order updating ansatz as well as power expansions for $\mathbf{U}_n$ and $\mathbf{x}_n$ in powers of the components of $\mathbf{x}_0(t)$, one obtains the *method of normal forms*. In autonomous systems, addition of the plausible (but not mandatory!) assumption that $\mathbf{x}_n$ do not depend explicitly on time, converts the equations for $\mathbf{x}_n$ into algebraic ones [see Equations (4.45) and (4.46)]. This makes the method of normal forms particularly suitable for handling by computer programs such as MACSYMA, MATHEMATICA, or MAPLE that are designed for algebraic manipulations.

The analysis of *nonautonomous systems* can be also carried out, starting from Equation (4.40). Now one has to take into account the explicit time dependence of $a^i(\mathbf{m})$. Several examples of the analysis of oscillatory nonautonomous systems are given in Chapter 8.

4.4 THE LAST FREEDOM: FREE RESONANT TERMS IN $\mathbf{x}_n$

The last type of freedom that we have identified in the expansion is stated by Equation (4.45). In the resonant case, the coefficients $c^i(\mathbf{m})$ need not vanish and are arbitrary. Consequently, the first-order term $\mathbf{x}_1$ is not unique. It is determined up to a free function that is composed only of resonant combinations of the components of $\mathbf{x}_0$. The same conclusion is also arrived at in higher orders. In every order, n, $\mathbf{x}_n$ is determined by the dynamical equations up to free functions that depend only on resonant combinations of the components of $\mathbf{x}_0$. These free functions do not affect the dynamical equation in the same order, Equation (4.9d) (they affect only higher-order equations), due to the following lemma:

Lemma 4.1. Let $\mathbf{X}_0(\mathbf{x})$ be given by Equations (4.2) and (4.3). Consider a vector function $\mathbf{F}(\mathbf{x})$. Assume that for each component, $F^i(\mathbf{x})$, $1 \leq i \leq n$, is given by an expansion in monomials

$$\left(x^1\right)^{m_1} \left(x^2\right)^{m_2} \cdots \left(x^n\right)^{m_n}$$

such that *all* the vectors $\mathbf{m}$ of integer powers appearing in $F^i(\mathbf{x})$ obey Equation (4.42) (resonance), for the *same i*. Then $[\mathbf{X}_0(\mathbf{x}), \mathbf{F}(\mathbf{x})]$ vanishes.

Proof: Write

$$F^i(\mathbf{x}) = \sum_\mathbf{m} f^{i(\mathbf{m})} \left(x^1\right)^{m_1} \left(x^2\right)^{m_2} \cdots \left(x^n\right)^{m_n}$$

The Lie bracket becomes

$$[\mathbf{X}_0, \mathbf{F}]^i = \sum_j \frac{\partial X_0^i}{\partial x_0^j} F^j - \sum_j \frac{\partial F^i}{\partial x_0^j} X_0^j = \lambda_i F^i - \sum_j \frac{\partial F^i}{\partial x_0^j} \lambda_j x_0^j =$$

$$\sum_\mathbf{m} f^{i(\mathbf{m})} \left(x^1\right)^{m_1} \left(x^2\right)^{m_2} \cdots \left(x^n\right)^{m_n} \left(\lambda_i - (\lambda, \mathbf{m})\right) = 0$$

The case of no free functions has been extensively employed in the literature. Bruno [8], who calls it the *distinguished* choice, studies the convergence properties of normal form expansion with that choice. There is little information about other choices. Various aspects of the freedom in the expansion have been studied in the literature [9], in particular, for Hamiltonian systems [10–13].

4.5 INTUITIVE MEANING OF LIE BRACKET

The Lie bracket appears in the formalism extensively. It is constructed from the unperturbed part of the equation, $\mathbf{X}_0(\mathbf{x})$, and some vector function, $\mathbf{F}(\mathbf{x})$. We would like to give the Lie bracket

$$[\mathbf{X}_0(\mathbf{x}), \mathbf{F}(\mathbf{x})]^i = \sum_j \frac{\partial X_0^i}{\partial x_0^j} F^j - \sum_j \frac{\partial F^i}{\partial x_0^j} X_0^j \tag{4.51}$$

an intuitive interpretation. We start with the unperturbed linear system,

Equation (4.2), assuming that the matrix **A** is diagonal. The ith component of the solution of Equation (4.2) is given by

$$x_0^i(t) = a_i \exp(\lambda_i t) \qquad (4.52)$$

Equation (4.51) now reads

$$[\mathbf{X}_0(\mathbf{x}), \mathbf{F}(\mathbf{x})]^i = \lambda_i F^i(\mathbf{x}) - \sum_j \frac{\partial F^i}{\partial x_0^j} \lambda_j x_0^j$$

Consider the quantity $\exp(-\mathbf{A}t) \cdot \mathbf{F}(\mathbf{x}_0(t))$. Its time dependence is determined by

$$\frac{d}{dt}\left[\exp(-\mathbf{A}t)\mathbf{F}(\mathbf{x}_0(t))\right]^i =$$

$$\exp(-\lambda_i t)\left\{-\lambda_i F^i(\mathbf{x}_0(t)) + \sum_{j=1}^n \frac{\partial F^i(\mathbf{x}_0(t))}{\partial x_0^j}\dot{x}_0^j\right\} = \qquad (4.53)$$

$$= -\exp(-\lambda_i t)[\mathbf{X}_0(\mathbf{x}_0(t)), \mathbf{F}(\mathbf{x}_0(t))]^i$$

Thus, except for the trivial multiplicative exponential factor, the Lie bracket measures the time dependence of $\exp(-\mathbf{A}t) \cdot \mathbf{F}(\mathbf{x}_0(t))$.

Vanishing of the Lie bracket means that $\exp(-\mathbf{A}t) \cdot \mathbf{F}(\mathbf{x}_0(t))$ is constant *along trajectories of solutions* of the unperturbed system. This means that $\mathbf{F}(\mathbf{x}_0(t))$ can be written as

$$\mathbf{F}(\mathbf{x}_0(t)) = \exp(\mathbf{A}t)\mathbf{F}_0 \qquad (4.54a)$$

and each component of $\mathbf{F}(\mathbf{x}_0(t))$ is of the form

$$F^i(\mathbf{x}_0(t)) = \exp(\lambda_i t) F_0^i \qquad (4.54b)$$

namely, it is proportional to $x_0^i(t)$. $\mathbf{F}_0$ is a constant vector. These results are easy to understand for the case of a *resonant monomial* in components of $\mathbf{x}_0(t)$ (its Lie bracket obviously vanishes):

$$P^{(\mathbf{m})}(t) \equiv \left(x_0^1(t)\right)^{m_1}\left(x_0^2(t)\right)^{m_2}\cdots\left(x_0^n(t)\right)^{m_n} \qquad (4.55)$$

INTUITIVE MEANING OF LIE BRACKET

Given Equation (4.51), Equation (4.55) can be written as

$$P^{(\mathbf{m})}(t) = (a_1)^{m_1} (a_2)^{m_2} \cdots (a_n)^{m_n} \exp((\lambda, \mathbf{m})t) \quad (4.56)$$

For a *resonant* $P^{(\mathbf{m})}(t)$ [that is, the vector $\mathbf{m}$ satisfies Equation (4.30) for some i] we, therefore, have

$$P^{(\mathbf{m})}(t) = P_0 \exp(\lambda_i t) \quad (4.57)$$

where P_0 is a constant. This is the result of Equation (4.54b).

We now add a perturbation, so that $\mathbf{x}(t)$ obeys the *full* Equation (4.1) and $\mathbf{x}_0(t)$ obeys of the normal form, Equation (4.8), rather than the unperturbed equation. As a result, Equation (4.53) is replaced by

$$\frac{d}{dt}\left[\exp(-\mathbf{A}t)\mathbf{F}(\mathbf{x}_0(t))\right]^i =$$

$$\exp(-\lambda_i t)\left\{-\lambda_i F^i(\mathbf{x}_0(t)) + \sum_{j=1}^n \frac{\partial F^i(\mathbf{x}_0(t))}{\partial x_0^j} \dot{x}_0^j\right\} = \quad (4.58)$$

$$= -\exp(-\lambda_i t)\left\{\underline{[\mathbf{X}_0(\mathbf{x}_0(t)),\mathbf{F}(\mathbf{x}_0(t))]^i} + \varepsilon\sum_{j=1}^n \frac{\partial F^i(\mathbf{x}_0(t))}{\partial x_0^j} U_1^j + \cdots\right\}$$

Thus, along the trajectory of the zero-order approximation, the Lie bracket [underlined in Equation (4.58)] provides the zero-order contribution to the time dependence of $\exp(-\mathbf{A}t)\cdot\mathbf{F}(\mathbf{x}_0(t))$. An $\mathbf{F}$ that is constructed from resonant terms only and satisfies the conditions of Lemma 4.1, has a vanishing Lie bracket. Equation (4.58) becomes

$$\frac{d}{dt}\left[\exp(-\mathbf{A}t)\mathbf{F}(\mathbf{x}_0(t))\right]^i = -\exp(-\lambda_i t)\left\{\varepsilon\sum_{j=1}^n \frac{\partial F^i(\mathbf{x}_0(t))}{\partial x_0^j} U_1^j + \cdots\right\}$$

Namely, $\exp(-\mathbf{A}t)\cdot\mathbf{F}(\mathbf{x}_0(t))$ deviates from a constant (which it was in the unperturbed case) by a slow time dependence, the effect of which may become appreciable only over timesscales of $O(1/\varepsilon)$. This, indeed, is characteristic of the free functions that may be added in the near-identity transformation. They do not affect the dynamics in a given order, but only the following orders.

4.6 CONVERGENCE OF NORMAL FORM TRANSFORMATION

In the preceding analysis we assumed that the r.h.s. of Equation (4.1) could be expanded in a power series in ε, as well as in powers of the components of **x**, as in Equation (4.34). These were *formal expansions*. We manipulated powers of ε and monomials in the components of **x**, without worrying about the convergence of the sums. In applications, the question of convergence is very important. For example, in many cases, the "force" on the r.h.s. of Equation (4.1) can be written as a convergent series in powers of ε and powers of the components of **x** in some domain:

$$\|\mathbf{x}\| \le R \qquad |\varepsilon| \le \varepsilon_0 \qquad (4.59)$$

This is certainly true when the perturbation has a simple form, consisting, for example, of only a first-order term [$O(\varepsilon^1)$], given by a single monomial in powers of the components of **x** (e.g., the Duffing equation). In the method of normal forms we assume similar expansions for $\mathbf{U}_n$ and $\mathbf{x}_n$ [e.g., the case $n=1$, in Equations (4.35) and (4.36), respectively]. Unfortunately, in general, these expansions may not converge. A rigorous treatment of this issue is presented in Arnold [2] and Bruno [8]. Here we only mention briefly several points.

4.6.1 General case

If the eigenvalues are in a Poincaré domain, there are a finite number of resonant combinations of integer vectors **m**. Consequently, the normal form contains at most a finite number of resonant monomials. For a wide class of problems, the near-identity transformation converges. (The proofs go back to Poincaré and Dulac.)

When the eigenvalues are in a Siegel domain, convergence is not guaranteed even if they *do not* resonate. Formally, if there are no resonant combinations, the analysis in Section 4.3 provides an algorithm for finding an expansion for $\mathbf{x}(t)$ with $\mathbf{U}_n=0$ for all $n \ge 1$. This means that, *formally*, the normal form degenerates to the unperturbed equation, Equation (4.7a). However, in general, the series expansions for $\mathbf{x}(t)$ may diverge. Technically, this is related to the structure of the $\mathbf{x}_n$. Recall that, in the computation of individual monomial terms in the latter, denominators of the form $\{\lambda_i - (\boldsymbol{\lambda},\mathbf{m})\}$ appear. Although these terms cannot vanish identically when there is no resonance, depending on the particular set of eigenvalues, there may be an infinite number of sets of integers for which the denominators are arbitrarily small. For example, consider a two-dimensional system with eigenvalues $\lambda_1=-1$ and $\lambda_2=\sqrt{2}$ of the linear part. No integer vector $\mathbf{m}=(m_1,m_2)$ satisfies

$$(\lambda,\mathbf{m}) = \lambda_1 m_1 + \lambda_2 m_2 = -m_1 + \sqrt{2} m_2 = \lambda_2 = \sqrt{2}$$

However, since $\sqrt{2}$ can be approximated arbitrarily close by an infinite number of rational numbers (p_n/q_n), there will be an infinite number of vectors $\mathbf{m}=(m_1,m_2)$ for which $|\lambda_2 - (\lambda,\mathbf{m})|$ is arbitrarily small.

Thus, there will be an infinite number of infinitely small denominators in the near-identity transformation [see, e.g., Equations (4.41) and (4.46)]. These make the coefficients, that multiply monomials in an infinite number of the $\mathbf{x}_n$, arbitrarily large. The near-identity expansion of the solution $\mathbf{x}(t)$ [Equation (4.4)] may not converge. In many cases it will be an asymptotic series. Namely, given a required degree of approximation for $\mathbf{x}(t)$, a finite sum of terms in Equation (4.4), will provide the approximation, whereas the infinite sum diverges.

A theorem by Siegel [14] states that the near-identity transformation [Equation (4.4)] is convergent (for sufficiently small |ε| and for some range of initial conditions), if the eigenvalues obey the condition

Two positive real numbers C and ν exist such that for all the eigenvalues λ_i (1≤i≤n) and for all vectors $\mathbf{m}$ of nonnegative integer components one has

$$|\lambda_i - (\lambda,\mathbf{m})| \geq \frac{C}{|\mathbf{m}|^\nu} \tag{4.60}$$

That is, if all possible $(\lambda,\mathbf{m})$ products are separated from all the λ_i by a strip, then the near-identity transformation converges. Note that since the eigenvalues do not obey any resonance relations, the normal form degenerates to the unperturbed equation

As $|\mathbf{m}|$ increases, the strip becomes narrower. Still, it guarantees that the resonance relation is always missed by a nonvanishing separation. The measure of the set of all eigenvalue vectors λ that violate Equation (4.60) vanishes when either: (1) the eigenvalues are complex and $\nu>(n-2)/2$ where n is the dimension of the dynamical system or (2) the eigenvalues are all real and $\nu>(n-1)$. When either of these conditions is satisfied, the near-identity transformation converges for almost all choices of the eigenvalues (except for a set of zero measure).

4.6.2 Two-dimensional case

In the two dimensional case, Bruno [8] provides a more detailed picture. In general, the matrix **A** can be brought to a Jordan form:

$$\mathbf{A} = \begin{pmatrix} \lambda_1 & 0 \\ \sigma & \lambda_2 \end{pmatrix} \tag{4.61}$$

Thus, with an appropriate choice of variables, a general two-dimensional problem can be recast as

$$\dot{x}_1 = \lambda_1 x_1 + \varepsilon \varphi_1(x_1, x_2; \varepsilon)$$
$$\dot{x}_2 = \lambda_2 x_2 + \sigma x_1 + \varepsilon \varphi_2(x_1, x_2; \varepsilon) \tag{4.62}$$

where φ_1 and φ_2 are nonlinear functions.

Comment. When $\lambda_1 \neq \lambda_2$, the matrix **A** can be diagonalized. Then there is no need for a σ term. However, Bruno's analysis addresses a general **A**, where λ_1 and λ_2 may or may not be equal.

Denote by y_1 and y_2 the variables into which the normalizing transformation transforms x_1 and x_2, respectively, and write the normal form as

$$\dot{y}_1 = \lambda_1 y_1 + \varepsilon h_1(y_1, y_2; \varepsilon) \qquad \dot{y}_2 = \lambda_2 y_2 + \varepsilon h_2(y_1, y_2; \varepsilon) \tag{4.63}$$

The series for h_i ($i=1,2$) contain nonlinear resonant monomials only.

For near-identity transformations, where the free functions are always chosen to be zero (the *distinguished* choice), Bruno proves the following results which depend on the values of the ratio

$$\lambda \equiv \frac{\lambda_1}{\lambda_2} \tag{4.64}$$

(with $\lambda_2 \neq 0$ otherwise, one can reverse the roles of λ_1 and λ_2). In the following, "convergence" means convergence for sufficiently small $|\varepsilon|$ and for some range of the initial conditions.

    a. <u>λ complex or real, $\lambda > 0$</u>. The eigenvalues are in a Poincaré domain. There are at most a finite number of resonant combinations. The normalizing transformation converges.

CONVERGENCE OF NORMAL FORM TRANSFORMATION   105

b. $\lambda=0$. If $h_1=0$, then the normalizing transformation converges. If $h_1 \neq 0$, it may diverge.

c. $\underline{\lambda<0,\ a\ rational\ number}$. There are an infinite number of resonant terms. In general, the normalizing transformation may diverge. The exception is problems for which h_i have the form

$$h_i(y_1, y_2; \varepsilon) = \lambda_i\, y_i\, h(y_1, y_2; \varepsilon) \qquad i = 1, 2 \qquad (4.65)$$

where $y_i \cdot h(y_1, y_2; \varepsilon)$ contain only resonant terms. Namely, both functions are proportional to one function, $h(y_1, y_2; \varepsilon)$, with the proportionality factors $\lambda_i y_i$. The normalizing transformation and the normal form are both convergent.

d. $\underline{\lambda<0,\ an\ irrational\ number}$. This falls under the category of the general case (Section 4.6.1) and the normalizing transformation converges if Equation (4.60) is obeyed.

Comment. Convergence of the near-identity transformation may not be related only to the size of ε. Consider, again, the Duffing equation as discussed in Example 4.1. Through first order, *without* a free function, the approximation for the solution, Equations (4.47b) reads

$$z \cong u + \varepsilon T_1 =$$
$$\rho e^{-i\varphi}\left\{1 + \tfrac{1}{32}\varepsilon\rho^2\left(2e^{-2i\varphi} - 6e^{+2i\varphi} - e^{+4i\varphi}\right)\right\} \qquad (4.66)$$

According to Bruno's analysis, the near-identity transformation converges. However, the ratio of the first-order term to the zero-order term involves the product $\varepsilon\rho^2$. This is also true in all higher orders. The ratio of the $O(\varepsilon^{n+1})$ term to the $O(\varepsilon^n)$ term is always of the form

$$\left|\frac{\varepsilon^{n+1} T_{n+1}}{\varepsilon^n T_n}\right| = \varepsilon\rho^2\ \text{function of}\ (\varphi) \qquad (4.67)$$

Bruno's proof shows that, if these ratios are sufficiently small for a range of values of ε *and* ρ, the transformation converges. For example, for $n=0$, one finds that Equation (4.67) obtains its maximal value at $\cos 2\varphi = \sqrt{\tfrac{3}{2}} - 1$, yielding a bound

$$\left|\frac{\varepsilon T_1}{u}\right| \leq |\varepsilon|\rho^2\left(\tfrac{9}{32} + \tfrac{3}{2}\sqrt{\tfrac{3}{2}}\right) \cong 2.1|\varepsilon|\rho^2$$

Thus, roughly speaking, convergence requires that

$$|\varepsilon|\rho^2 \ll \frac{1}{2.1}$$

Example 4.9. Convergent near-identity transformation.
Consider

$$\dot{z} = -i\,z - \tfrac{3}{2}\varepsilon\left(z^2 - z^{*2}\right) - \tfrac{1}{2}i\,\varepsilon^2\left(z + z^*\right)^3 \qquad (4.68)$$

Equation (4.68) belongs to case c in Bruno's analysis (eigenvalues: $\pm i$). It is solved by

$$z = \frac{A\sin t}{1 - 2\varepsilon A\cos t} + i\,\frac{A\cos t - 2\varepsilon A^2}{(1 - 2\varepsilon A\cos t)^2} \qquad (4.69)$$

with A an arbitrary constant. (Changing the initial conditions by replacing by $t+t_0$ does not affect the following discussion.)

Although the eigenvalues are in a Siegel domain, the solution has no frequency update. Moreover, writing

$$u = i\,A\exp(-i\,t) \qquad (4.70)$$

We can cast Equation (4.69) in the form

$$z = \frac{u + i\,\varepsilon\left(\tfrac{1}{2}u^2 - \tfrac{1}{2}u^{*2} - 2uu^*\right)}{\left[1 + i\,\varepsilon(u - u^*)\right]^2} \qquad (4.71)$$

Expanding Equation (4.71), we obtain through second order:

$$z = u + \varepsilon\left\{-\tfrac{3}{2}i\,u^2 - \tfrac{1}{2}i\,u^{*2}\right\} + \varepsilon^2\left\{-2u^3 + u^{*3} + u^2 u^*\right\} + \cdots \qquad (4.72)$$

Clearly, the expansion converges for $2|\varepsilon A| < 1$.

Through second order, the normal form expansion is found to be

$$U_1 = 0, \quad T_1 = -\tfrac{3}{2}i\,u^2 - \tfrac{1}{2}i\,u^{*2} + f_1(uu^*)\cdot u$$
$$U_2 = 0, \quad T_2 = -2u^3 + u^{*3} - 3i\,f_1 u^2 - i\,f_1^* u^{*2} + f_2(uu^*)\cdot u \qquad (4.73)$$

CONVERGENCE OF NORMAL FORM TRANSFORMATION   107

One proves by induction that $U_n=0$ for all $n\geq 1$. (In all odd orders, the contributions to U_n contain only polynomials that are even in u and u^*, which cannot produce resonant terms. Thus, one has to show only that in odd n no resonant terms are generated.) Therefore, the normal form is the unperturbed equation:

$$\dot{u} = -i\, u$$

which is solved by Equation (4.70) and confirms our observation that there is no frequency update.

Note that the normal form satisfies Equation (4.65), which in case c of Bruno's analysis, guarantees convergence of the *distinguished* choice (i.e., no free functions). However, from Equation (4.73) we see that this choice does not reproduce the direct expansion of the full solution, as given by Equation (4.72). To obtain that expansion requires a particular choice of the free functions. For instance, through second order, with the choice

$$f_1 = 0, \qquad f_2 = uu^*$$

the near-identity expansion coincides with Equation (4.72) through $O(\varepsilon^2)$. With an appropriate choice of the free functions, one can make the near-identity expansion coincide with that of Equation (4.72) to any desired order. Therefore, in this case, the expansion converges for a range of values of $|\varepsilon|A$ both in the *distinguished* choice (no free functions) as well as with this particular choice of the free functions.

Example 4.10: **Divergent near-identity transformation**–case b. This is an example due to Euler. Consider the coupled equations

$$\dot{x}_1 = \varepsilon x_1^2, \qquad\qquad \dot{x}_2 = x_2 - x_1 \qquad (4.74)$$

This set is equivalent to the single equation

$$\varepsilon x_1^2 \frac{dx_2}{dx_1} = x_2 - x_1 \qquad (4.75)$$

If we solve for x_2 in a power series in x_1, we find that the series is

$$x_2 = \sum_{k=1}^{\infty} (k-1)!\, \varepsilon^{k-1} x_1^k \qquad (4.76)$$

This series diverges for all $x_1 \neq 0$. It can be also obtained as a formal power expansion of the solution of Equation (4.75) with the initial condition $x_2(x_1=0)=0$:

$$x_2 = -\frac{1}{\varepsilon}\exp\left(-\frac{1}{\varepsilon x_1}\right)\int_0^{\varepsilon x_1}\frac{\exp(1/u)}{u}du \qquad (4.77)$$

This function is nonanalytic, but infinitely differentiable at $x_1=0$.

Denoting the zero-order approximations of x_1 and x_2 by y_1 and y_2, respectively, the normalizing transformation for Equations (4.74) is

$$x_1 = y_1, \qquad x_2 = y_2 + \sum_{k=1}^{\infty}(k-1)!\varepsilon^{k-1}y_1^k \qquad (4.78)$$

The resulting normal form is

$$\dot{y}_1 = \varepsilon y_1^2, \qquad \dot{y}_2 = y_2 \qquad (4.79)$$

Thus, a formal series exists for the normalizing transformation, which yields a simple normal form. However, that series is divergent. In terms of Bruno's analysis, Equation (4.74) belongs to case b. In the notation of Equation (4.63), $h_1(y_1,y_2;\varepsilon)=y_1^2\neq 0$. Thus, the condition for convergence of the transformation is not met.

Developing the formalism in this chapter, we have assumed that $\mathbf{X}_0$, the zero-order term in Equation (4.1), is linear. We did so because, usually, one expands problems of interest around a linear zero-order term. The formalism can be applied also when $\mathbf{X}_0$ is not linear in the components of the vector $\mathbf{x}$. A resonant function is then defined by the requirement that its Lie bracket with $\mathbf{X}_0$ vanishes.

4.7 EIGENVALUE UPDATE: CASE c IN BRUNO'S ANALYSIS

In case c of Bruno's analysis, updating the timescale that characterizes the full motion manifests itself in a simple manner—the unperturbed eigenvalues are updated.

4.7.1 Energy-conserving oscillations: Behavior near a center

Case c covers all problems of a single harmonic oscillator that is perturbed by an *energy-conserving* perturbation:

EIGENVALUE UPDATE: CASE c IN BRUNO'S ANALYSIS

$$\ddot{x} + x + \varepsilon f(x;\varepsilon) = 0 \qquad (4.80)$$

For the sake of simplicity, we have chosen the unperturbed frequency to be $\omega_0=1$. Due to energy conservation, the solution of Equation (4.80) is periodic, its frequency ω differing from ω_0 by $O(\varepsilon)$. In the diagonalized (complex) notation, Equation (4.80) becomes

$$\dot{z} = -i\,z - i\,\varepsilon f\left(\tfrac{1}{2}(z+z^*);\varepsilon\right), \quad (z \equiv x + i\,\dot{x}) \qquad (4.81)$$

Here f is a real function of $(z+z^*)$. The equation for z^* is

$$\dot{z}^* = i\,z^* + i\,\varepsilon f\left(\tfrac{1}{2}(z+z^*);\varepsilon\right) \qquad (4.82)$$

Denoting the zero-order term by u, one finds that the normal form has the generic form (to be discussed in detail in Chapter 6):

$$\dot{u} = -i\,u - i\,\varepsilon u\, g(uu^*;\varepsilon)$$
$$\dot{u}^* = +i\,u^* + i\,\varepsilon u^*\, g(uu^*;\varepsilon). \qquad (4.83)$$

Excluding free functions in the near-identity transformation, g is a real function of uu^*. By Bruno's analysis, the transformation converges.

If we write $u=\rho\cdot\exp(-i\varphi)$, Equation (4.83) yields, as expected

$$\dot{\rho} = 0, \quad \Rightarrow \quad \rho = \text{const}$$
$$\omega = \dot{\varphi} = 1 + \varepsilon g(\rho^2;\varepsilon) \qquad (4.84)$$

This result is characteristic of conservative oscillations in the plane. It is an example of eigenvalue update. The structure of the normal form, Equation (4.83) (a special example of case c) is responsible for the results of Equation (4.84): (1) the amplitude is constant and (2) the basic time dependence is modified by a frequency update. The frequency is only slightly modified relative to the unperturbed one.

4.7.2 Behavior near a saddle fixed point

The features discussed above are not specific to periodic problems. They are typical of case c problems of a wider class. Consider, for example, the two-dimensional *real* system:

FORMALISM OF PERTURBATION EXPANSION

$$\dot{x} = x + \varepsilon F(x, y; \varepsilon) \qquad \dot{y} = -y + \varepsilon G(x, y; \varepsilon) \quad (4.85)$$

F and G are real nonlinear functions that can be expanded in convergent power series in x, y, and ε. This problem is different from the oscillatory one. The unperturbed eigenvalues are real and the origin is not a center around which periodic motion evolves, but a saddle point. However, from the point of view of Bruno's analysis, the two problems *are* similar. In both cases the unperturbed eigenvalues are in a Siegel domain and obey similar resonance relations. Here we have $\lambda_1=+1$ and $\lambda_2=-1$, satisfying

$$\lambda_1 = (n+1)\lambda_1 + n\lambda_2, \qquad \lambda_2 = n\lambda_1 + (n+1)\lambda_2 \quad (4.86)$$

for any integer $n \geq 1$. The structure of the normal form will be relatively simple. If we write the near-identity transformations as

$$x = u + \varepsilon T_1 + \cdots, \qquad y = v + \varepsilon S_1 + \cdots \quad (4.87)$$

then the normal form equations for u and v will contain monomials of the form $u^{n+1}v^n$ and $u^n v^{n+1}$, respectively. Factoring out a u from the u equation and a v from the v equation, the two can be written as

$$\dot{u} = u\{+1 + \varepsilon f((u \cdot v); \varepsilon)\} \qquad \dot{v} = v\{-1 + \varepsilon g((u \cdot v); \varepsilon)\} \quad (4.88)$$

To make Equation (4.88) comply with Equation (4.65), so that, by Bruno's analysis (remember that he only discusses the case without free functions!), the near-identity transformation converges, we need

$$g((u \cdot v); \varepsilon) = -f((u \cdot v); \varepsilon) \quad (4.89)$$

If this happens, the simple structure of the normal form implies an eigenvalue update *also in this case*! Equations (4.88) now become

$$\dot{u} = -u\{1 + \varepsilon f((u \cdot v); \varepsilon)\} \qquad \dot{v} = -v\{1 + \varepsilon f((u \cdot v); \varepsilon)\} \quad (4.90)$$

It is easy to show that Equations (4.90) imply

$$u \cdot v = u_0 \cdot v_0 = \text{const}$$

Equations (4.90) are, therefore, solved by

$$u = u_0 \exp(\lambda t) \quad v = v_0 \exp(-\lambda t) \quad \lambda = 1 + \varepsilon f((u_0 \cdot v_0); \varepsilon) \quad (4.91)$$

Thus, the eigenvalues are updated by the perturbation according to

$$\pm 1 \quad \Rightarrow \quad \pm\{1 + \varepsilon f((u_0 \cdot v_0); \varepsilon)\}$$

This is similar to the conservative oscillatory case where we had

$$u \cdot u^* = \rho^2 = \text{const}$$

and the eigenvalues were updated [Equation (4.84)]:

$$\pm i \quad \Rightarrow \quad \pm i \{1 + \varepsilon g(\rho^2; \varepsilon)\}$$

4.8 TIME VALIDITY OF THE TRUNCATED EXPANSION

A major aspect of any perturbation method is the validity in time of the approximations it generates for the full solution. Naive perturbation theory has been found to be valid over a limited timespan. The method of normal forms, developed in this chapter, is expected to have an improved time validity. In this section, we develop estimates of the time validity of the approximations generated by the method of normal forms. Gronwall's lemma-type inequalities (discussed in Chapter 2) are used repeatedly. We derive estimates for a sequence of examples of increasing complexity for autonomous single-oscillator systems.

The full dynamical equation is written in complex notation as

$$\dot{z} = -i\, z + \varepsilon Z_1(z, z^*) + \varepsilon^2 Z_2(z, z^*) + \cdots \quad (4.92)$$

We assume the following:

1. Z_n are polynomials in z and z^* or have convergent Taylor expansions in the two variables in some domain $|z| \leq C$.
2. Z_n and their derivatives are bounded in the same domain.
3. The power expansion in ε in equation (4.92) converges for $|\varepsilon| \leq \varepsilon_0$ in this domain.

As the system is autonomous, we make the plausible (but not mandatory!) assumption that in all orders in the near-identity transformation the generators, T_n ($n \geq 1$), are independent of time and depend only on u, u^*.

For a given approximation, z_{approx}, the error, R, is defined as

$$R = z - z_{approx}$$

We will repeatedly use a generalization of a trivial identity

$$f(x+h) - f(x) = f'(x)h + h^2 \int_0^1 (1-s) f''(x+sh) ds \quad (4.93)$$

4.8.1 Naive perturbation theory

We analyze the uninteresting but simple case of NPT, to gain insight into the method of derivation of the bounds. Here we have

$$z_{approx} = u \quad (4.94a)$$

The dynamical equation for the approximation is

$$\dot{z}_{approx} = \dot{u} = -i\,u \quad (4.94b)$$

solved by

$$u = \rho \exp(-i\varphi) \qquad \rho = \text{const} \qquad \varphi = t + \varphi_0 \quad (4.94c)$$

Using a generalization of Equation (4.93), we expand Z_1 around u, u^*:

$$Z_1(z, z^*) = Z_1(u, u^*)$$
$$+ \left\{ \frac{\partial Z_1(u, u^*)}{\partial u} R + \frac{\partial Z_1(u, u^*)}{\partial u^*} R^* \right\} \quad (4.94.d)$$
$$+ K(u, u^*, R, R^*)$$

where

$$K(u,u^*,R,R^*) \equiv \int_0^1 (1-s) \left\{ \begin{array}{l} \left(\dfrac{\partial^2 Z_1(u+sR,u^*+sR^*)}{\partial u^2} \right) R^2 \\ + 2 \left(\dfrac{\partial^2 Z_1(u+sR,u^*+sR^*)}{\partial u \partial u^*} \right) RR^* \\ + \left(\dfrac{\partial^2 Z_1(u+sR,u^*+sR^*)}{\partial u^{*2}} \right) R^{*2} \end{array} \right\} ds$$

(4.94e)

We use Equation (4.9b) to express $Z_1(u,u^*)$ in terms of U_1 and $[Z_0,T_1]$. We lump the contribution of all the terms with ε^n $n \geq 2$, in an $O(\varepsilon^2)$ term. Their detailed structure is irrelevant at this level of approximation.

The equation for the time dependence of R is

$$\dot{R} = -i R + \varepsilon \{U_1(u,u^*) - [Z_0,T_1(u,u^*)]\} + O(\varepsilon^2)$$

$$+ \varepsilon \left\{ \dfrac{\partial Z_1(u,u^*)}{\partial u} R + \dfrac{\partial Z_1(u,u^*)}{\partial u^*} R^* \right\} + \varepsilon K(u,u^*,R,R^*)$$

We now define

$$R = S \exp(-i t)$$

The equation for S is

$$\dot{S} = \exp(i t) \varepsilon \{U_1(u,u^*) - [Z_0,T_1(u,u^*)]\} + O(\varepsilon^2)$$

$$+ \exp(i t) \varepsilon \left\{ \dfrac{\partial Z_1(u,u^*)}{\partial u} R + \dfrac{\partial Z_1(u,u^*)}{\partial u^*} R^* \right\}$$

$$+ \exp(i t) \varepsilon K(u,u^*,R,R^*)$$

which is formally solved by integration:

$$S = S_0 + \varepsilon \int_0^t \{\exp(i\,\tau)\{U_1(u,u^*) - [Z_0, T_1(u,u^*)]\}\} d\tau$$

$$+ \int_0^t \{\exp(i\,\tau) O(\varepsilon^2)\} d\tau$$

(4.95)

$$+ \varepsilon \int_0^t \left\{\exp(i\,\tau)\left\{\left[\frac{\partial Z_1(u,u^*)}{\partial u}R + \frac{\partial Z_1(u,u^*)}{\partial u^*}R^*\right]\right\}\right\} d\tau$$

$$+ \int_0^t \exp(i\,\tau) d\tau\, K(u,u^*,R,R^*) d\tau$$

For the error estimate, we will study absolute values:

$$W = |S| = |R| \qquad W_0 = |S_0| = O(\varepsilon)$$

The constant W_0 is $O(\varepsilon)$, as it represents the deviation of the initial condition of the approximation from the full solution. However, we must first estimate several terms in Equation (4.95).

First integral in Equation (4.95). The first integral contains two contributions: the resonant U_1 and the nonresonant Lie bracket.

The resonant term is constructed from monomials of the form

$$u^{n+1} u^{*n} = \rho^{2n+1} \exp(-i\,(\tau + \varphi_0))$$

Note that all the monomials have the same time dependent exponential factor, the remainder being time independent. [See Equation (4.94c).] Therefore, the integral over the part involving U_1, will be of the form

$$\exp(-i\,\varphi_0) f(\rho) t$$

where $f(\rho)$ is a function of ρ, arising from all the resonant monomials and bounded in absolute value by some positive constant a.

The Lie bracket is constructed from nonresonant monomials:

TIME VALIDITY OF THE TRUNCATED EXPANSION

$$u^n u^{*m} = \rho^{n+m} \exp\left(i (m-n)(\tau + \varphi_0)\right) \qquad (m - n \neq -1)$$

Each such monomial will contribute in the integral a term of the form

$$\rho^{n+m} \exp\left(i (m-n)\varphi_0\right) \frac{\exp\left(i (m-n)t\right) - 1}{\left(i (m - n + 1)\right)}$$

These terms add up to a quantity bounded by some positive constant b. (This is obvious if all the Z_n are polynomials. It is a bit more difficult to prove if they have convergent Taylor expansions.) As a result, the absolute value of the first integral in equation (4.95) is bounded by

$$at + b$$

Second integral in Equation (4.95). The detailed structure of the second integral is irrelevant. It suffices to note that it will also be a sum of resonant and nonresonant terms and, hence, be bounded by

$$ct + d$$

with c and d, some positive constants.

Third and fourth integrals in Equation (4.95). These integrals are totally unknown, as they include the unknown error R. However, they can be bounded in absolute value as follows.

The third integral is bounded by

$$\left| \int_0^t \left\{ \exp(i \tau) \left\{ \left\{ \frac{\partial Z_1(u,u^*)}{\partial u} R + \frac{\partial Z_1(u,u^*)}{\partial u^*} R^* \right\} \right\} \right\} d\tau \right| \leq M \int_0^t W\, d\tau$$

M represents a bound for the first derivatives of Z_1. W is the absolute value of R. If we denote a bound for the second derivatives of Z_1 appearing in $K(u,u^*,R,R^*)$ by N, then the following inequality can be derived from Equation (4.95):

$$W \leq W_0 + \varepsilon(at + b) + \varepsilon^2(ct + d) + \varepsilon M \int_0^t W\, d\tau + \varepsilon N \int_0^t W^2\, d\tau \quad (4.96)$$

The inequality in Equation (4.96) is solved as follows. We take a

116   FORMALISM OF PERTURBATION EXPANSION

time derivative of both sides of the equation. As all the quantities on the r.h.s. are positive, the inequality also holds for the derivative:

$$\dot{W} \leq \varepsilon a + \varepsilon^2 c + \varepsilon M W + \varepsilon^2 N W^2 \qquad (4.97)$$

Defining

$$F = W + \frac{M}{2N} \qquad F_0 = W_0 + \varepsilon b + \varepsilon^2 d + \frac{M}{2N}$$

the equation for F is

$$\dot{F} \leq \varepsilon N(F^2 - q^2) \qquad q^2 = \left(\frac{M^2}{4N^2} - \frac{a}{N} - \varepsilon \frac{c}{N} \right) \qquad (4.98)$$

For $q^2 > 0$, Equation (4.98) is solved by (the procedure for $q^2 < 0$ is similar)

$$F \leq q \frac{(F_0 + q) + (F_0 - q)\exp(\varepsilon A t)}{(F_0 + q) - (F_0 - q)\exp(\varepsilon A t)} \qquad (A = 2Nq) \qquad (4.99)$$

For $t = O(1/\varepsilon)$, both the numerator and denominator are $O(1)$. Hence, the error W is $O(1)$. For $t = O(1)$, on the other hand, one can expand the exponential in Equation (4.99) as its argument is small:

$$\exp(\varepsilon A t) \cong 1 + \varepsilon A t$$

(At the present level of accuracy, there is no need to go beyond this order in the expansion.) Equation (4.99) becomes

$$F \leq \frac{F_0 + ((F_0 - q)/2)\varepsilon A t}{1 - ((F_0 - q)/2q)\varepsilon A t}$$

yielding for W the bound

$$W \leq \frac{(W_0 + \varepsilon b + \varepsilon^2 d) + ((F_0 - q)/2)\varepsilon A t \left(1 + \left(\frac{M}{2N} \middle/ q \right) \right)}{1 - ((F_0 - q)/2q)\varepsilon A t} = O(\varepsilon)$$

TIME VALIDITY OF THE TRUNCATED EXPANSION

Thus, a Gronwall-type inequality tells us what we anticipated: that, in the naive expansion, the zero-order term is an $O(\varepsilon)$ approximation for the full solution only for $t=O(1)$. For $t=O(1/\varepsilon)$, the error may degenerate to at least $O(1)$, making the approximation useless.

In a similar manner, the extension to higher-order approximations:

$$z_{approx} = u + \sum_{k=1}^{n} \varepsilon^k T_k$$

yields that R is now bounded by an $O(\varepsilon^{n+1})$ error for $t=O(1)$ only. For $t=O(1/\varepsilon)$, the error degenerates to at least $O(\varepsilon^n)$.

4.8.2 Normal form error estimates for planar oscillatory systems

In this section we follow the method of analysis presented in Ref. 15.

4.8.2.1 First-order normal form.
In the first improvement over the naive expansion, we retain the zero-order approximation, Equation (4.94a), but replace Equation (4.94b) of the unperturbed system by a first-order normal form:

$$\dot{u} = -i\,u + \varepsilon U_1(u, u^*) \qquad (4.100)$$

Following the steps of the naive expansion, we arrive at a new equation for S. It differs from Equation (4.96) by three key elements.

First, the inclusion of U_1 in Equation (4.100) leads to its *elimination* from the first integral in Equation (4.95). Only the nonresonant Lie bracket remains. Thus, the effect of the first-order resonant term on the error estimate is eliminated and the term $a \cdot t$ disappears from Equations (4.96) and (4.97).

Second, now that the first-order effect is bounded, we have to examine the effect of the $O(\varepsilon^2)$ terms in the original equation in greater detail.

Third, as the normal form equation includes some of the dynamics, we have to distinguish in our analysis between conservative and dissipative systems. In dissipative systems we study only those cases that produce bounded motion, namely, systems that decay to the origin or that evolve onto a limit cycle.

The equation for S now becomes

FORMALISM OF PERTURBATION EXPANSION

$$S = S_0 - \varepsilon \int_0^t \{\exp(i\ \tau)[Z_0, T_1(u,u^*)]\} d\tau$$

$$+ \varepsilon^2 \int_0^t \{\exp(i\ \tau)\{U_2(u,u^*) - [Z_0, T_2(u,u^*)]\}\} d\tau$$

$$+ \int_0^t O(\varepsilon^3) \exp(i\ \tau) d\tau \qquad (4.101\text{a})$$

$$+ \varepsilon \int_0^t \left\{ \exp(i\ \tau) \left\{ \left\{ \frac{\partial \bar{Z}_1(u,u^*)}{\partial u} R + \frac{\partial \bar{Z}_1(u,u^*)}{\partial u^*} R^* \right\} \right\} \right\} d\tau$$

$$+ \varepsilon \int_0^t \exp(i\ \tau) \bar{K}(u,u^*,R,R^*) d\tau$$

where the effect of Z_2 is included in $\bar{Z}_1$:

$$\bar{Z}_1 = Z_1 + \varepsilon Z_2 \qquad (4.101\text{b})$$

and $\bar{K}$ is defind as K in Equation (4.94e) with $\bar{Z}_1$ replacing Z_1.

First integral in Equation (4.101a). The first integral has only nonresonant contributions.

For *conservative* systems, we have

$$u = \rho \exp(-i\ \varphi) \quad \rho = \text{const} \quad \varphi = (1 + \varepsilon \omega_1)t + \varphi_0$$

where ω_1 is the first-order frequency update. The contribution of a nonresonant monomial is of the form

$$\int_0^t u^n u^{*m} \exp(i\ \tau) d\tau =$$

$$\rho^{n+m} \exp((m-n) i\ \varphi_0) \int_0^t \exp(i\ (m-n+1+(m-n)\varepsilon \omega_1)\tau) d\tau =$$

TIME VALIDITY OF THE TRUNCATED EXPANSION

$$\rho^{n+m} \exp((m-n) i\, \varphi_0) \frac{\exp\bigl(i\,(m-n+1+(m-n)\varepsilon\omega_1)t\bigr) - 1}{i\,(m-n+1+(m-n)\varepsilon\omega_1)} = O(1)$$

Thus, the first integral in Equation (4.101a) is bounded by some constant b.

For *dissipative* systems, ρ is not constant and the phase is given by

$$\varphi = \tau + \varepsilon\varphi_1(\rho)$$

Where φ_1 is the first-order correction to the phase. Its time dependence is generated through its dependence on the variable amplitude, ρ. The integral is now evaluated by integration by parts:

$$\int_0^t u^n u^{*m} \exp(i\,\tau) d\tau =$$

$$\int_0^t \rho^{m+n} \exp\bigl(i\,\varepsilon(m-n)\varphi_1(\rho)\bigr) \exp\bigl(i\,(m-n+1)\tau\bigr) d\tau =$$

$$\left[\rho^{m+n} \exp\bigl(i\,\varepsilon(m-n)\varphi_1(\rho)\bigr) \frac{\exp\bigl(i\,(m-n+1)\tau\bigr) - 1}{i\,(m-n+1)}\right]_0^t$$

$$- \int_0^t \frac{d}{d\tau}\Bigl[\rho^{m+n} \exp\bigl(i\,\varepsilon(m-n)\varphi_1(\rho)\bigr)\Bigr] \frac{\exp\bigl(i\,(m-n+1)\tau\bigr)}{i\,(m-n+1)} d\tau$$

The first term is bounded by $O(1)$. In the second term, $(d\rho/dt)$ is generated only through U_1 in Equation (4.100). Hence, the second term will be proportional to ε. With no additional information about the functions in the integral, it can be bounded only by some constant, c. Thus, the second integral in Equation (4.101a) is bounded by

$$\underbrace{b}_{\text{conservative and dissipative}} + \underbrace{\varepsilon c t}_{\text{dissipative}}$$

Second integral in Equation (4.101a). The analysis is more involved then that of Equation (4.95), as, again, we must distinguish between conservative and dissipative systems.

For *conservative* systems, the contribution of a monomial in the U_2 (resonant) term is of the form

$$\exp(-i\,\varphi_0)\rho^{2n+1}\int_0^t \exp(-i\,\varepsilon\omega_1\tau)d\tau =$$

$$\exp(-i\,\varphi_0)\rho^{2n+1}\frac{\exp(-i\,\varepsilon\omega_1 t)-1}{-i\,\varepsilon\omega_1}$$

For $t \approx O(1)$, this is bounded by a constant of $O(1)$ times t (simply expand the exponential), while for $t \approx O(1/\varepsilon)$, it is bounded by $O(1/\varepsilon)$.

For *dissipative* systems, the integral over a monomial becomes

$$\exp(-i\,\varphi_0)\int_0^t \rho^{2n+1}\exp(-i\,\varepsilon\varphi_1(\rho))d\tau$$

In the absence of any further information, the integrand can be bounded simply by a positive constant, f, yielding a bound of the form

$$f t$$

Therefore, the overall effect of the U_2 integral in Equation (4.101a) can be represented by a bound of the form

$$\underbrace{dt}_{\substack{\text{conservative}\\t=O(1)}} + \underbrace{e/\varepsilon}_{\substack{\text{conservative}\\t=O(1/\varepsilon)}} + \underbrace{ft}_{\text{dissipative}}$$

We now turn to the Lie bracket contribution to the second integral in Equation (4.101a). Its contribution is bounded by the same type of bound as of the first integral in Equation (4.101a):

$$\underbrace{g}_{\text{conservative and dissipative}} + \underbrace{\varepsilon h t}_{\text{dissipative}}$$

Third integral in Equation (4.101a). This is the $O(\varepsilon^3)$ contribution. It may contain resonant and nonresonant terms. Thus, based on the preceding analysis, it may contain a $1/\varepsilon$ term.

The last two integrals are, again, bounded in absolute value by some constants, denoted by M and N for the first and second derivatives of $\bar{Z}_1$, respectively.

Equation (4.101a) then yields an inequality for W, the absolute value of R, similar to Equation (4.96):

TIME VALIDITY OF THE TRUNCATED EXPANSION

$$W \leq W_0 + \varepsilon(b + \varepsilon c t) + \varepsilon^2\left((d + f + \varepsilon h)t + g + e/\varepsilon\right)$$
$$+ \varepsilon^3 O(1/\varepsilon) + \varepsilon M \int_0^t W \, d\tau + \varepsilon N \int_0^t W^2 \, d\tau \qquad (4.102)$$

Equation (4.102) is similar in structure to Equation (4.97). Therefore, there is no need to repeat all the steps of the analysis. Again, we define a function F and a constant q. The bound for W has *exactly the same form* as in Equation (4.99), except that now

$$F_0 = W_0 + \varepsilon(b + e) + \varepsilon^2 g + \frac{M}{2N} \qquad W_0 = O(\varepsilon)$$

where the $\varepsilon^3 \cdot O(1/\varepsilon)$ term is absorbed in g, and

$$q^2 = \left(\frac{M^2}{4N^2} - \varepsilon \frac{c + d + f + \varepsilon h}{N}\right)$$

A *crucial* difference between the present case and the naive one is that, in the negative contribution to q^2, there is no $O(1)$ term [such as a in Equation (4.98)). Only an $O(\varepsilon)$ term has remained. This is due to the elimination of U_1 in the first integral of Equation (4.101). This difference will make it possible to extend the bound to times of $O(1/\varepsilon)$. (We were unable to do so in the naive case.) Now, we have

$$q \cong \frac{M}{2N} - \varepsilon \frac{c + d + f + \varepsilon h}{M}$$

Rather then repeating the steps that follow Equation (4.99), we point out that now, to leading order, q and F_0 are equal to $(M/2N)$ so that

$$F_0 + q = \frac{M}{N} + O(\varepsilon) \qquad F_0 - q = O(\varepsilon)$$

Subtracting $(M/2N)$ from Equation (4.99), to obtain a bound for W, the leading $(M/2N)$ term in the numerator of Equation (4.99) is canceled, and we obtain

$$W \leq \frac{O(\varepsilon) + \exp(\varepsilon A t) O(\varepsilon)}{1 - \exp(\varepsilon A t) O(\varepsilon)} \qquad (4.103)$$

Thus, as long as t does not exceed $O(1/\varepsilon)$, the error is bounded by $O(\varepsilon)$. Namely, the validity of the approximation provided by u, the zero-order term, is extended beyond the $t=O(1)$ limitation of the naive theory.

4.8.2.2 First-order approximation and first-order normal form. We now combine the first-order normal form, Equation (4.100), with a first-order approximation for the solution

$$z_{approx} = u + \varepsilon T_1 \qquad (4.104)$$

With knowledge of T_1 taken into account, one change occurs in Equation (4.101)–the first-order Lie bracket $([Z_0,T_1])$ is eliminated, yielding the following equation for S

$$S = S_0 + \varepsilon^2 \int_0^t \left\{ \exp(i\,\tau) \left\{ U_2(u,u^*) - [Z_0, T_2(u,u^*)] \right\} \right\} d\tau$$

$$+ \int_0^t O(\varepsilon^3) \exp(i\,\tau) d\tau$$

$$+ \varepsilon \int_0^t \left\{ \exp(i\,\tau) \left\{ \left[\frac{\partial \bar{Z}_1(u,u^*)}{\partial u} R + \frac{\partial \bar{Z}_1(u,u^*)}{\partial u^*} R^* \right] \right\} \right\} d\tau \quad (4.105)$$

$$+ \varepsilon \int_0^t \exp(i\,\tau) \bar{K}(u,u^*,R,R^*) d\tau$$

with $\bar{Z}_1$ and $\bar{K}$ defined previously [see Equation (4.101b)]. With the elimination of the first-order Lie bracket term, there is one term less to generate an $O(\varepsilon)$ error. However, the remaining $O(\varepsilon^2)$ term *is* a source for such errors because it includes the resonant contribution, U_2, which for $t=O(1/\varepsilon)$, may yield a term of $O(1/\varepsilon)$. (See discussion of second integral in the previous case.) The ε^2 multiplicative factor turns this into an error source of $O(\varepsilon)$ for $t=O(1/\varepsilon)$. Thus, higher precision in the approximate solution *without* an accompanying refinement in the normal form (which must include one additional order over the one included in z_{approx}) does not improve the error bound beyond the one found in Section 4.8.2.1.

TIME VALIDITY OF THE TRUNCATED EXPANSION

One can gain better insight regarding this statement in the case of conservative systems. With Equation (4.100), u has the form

$$u = \rho \exp\left(-i\left(1+\varepsilon\omega_1\right)t - i\,\varphi_0\right) \quad \rho = const. \quad (4.106)$$

Here ω_1 is the first-order correction to the frequency, induced by U_1. Using Equation (4.104) for z_{approx}, has the potential to leave an error of $O(\varepsilon^2)$. However, this really fixes the initial condition (and hence the amplitude ρ) with an error of $O(\varepsilon^2)$. As the frequency is updated only through $O(\varepsilon)$, we should keep in mind that the full expression for u differs from Equation (4.106) in the following manner:

$$u(\text{full}) = \rho \exp\left(-i\left(1+\varepsilon\omega_1 + O(\varepsilon^2)\right)t - i\,\varphi_0\right) \quad \rho = const$$

The unknown $O(\varepsilon^2)$ term will generate an $O(\varepsilon)$ error for $t=O(1/\varepsilon)$.

There is an exception to the last conclusion. It involves exploitation of the freedom of choice in the expansion. In the case of an oscillatory system, the equation for U_2 [Equation (4.9c)] becomes

$$U_2 = \left[Z_0(u,u^*),T_2\right] + \\ + Z_2(u,u^*) + T_1\frac{\partial Z_1}{\partial u} + T^*_1\frac{\partial Z_1}{\partial u^*} - U_1\frac{\partial T_1}{\partial u} - U^*_1\frac{\partial T_1}{\partial u^*} \quad (4.107)$$

where T_1 has been determined in the previous order up to a free function. In many cases, the latter can be chosen to make U_2 vanish. (This idea is discussed extensively in Chapters 6, 7 and 10, in the context of *minimal normal forms*.) For some systems, the free function does not destroy the boundedness of T_1, so that the first-order term in z_{approx} remains $O(\varepsilon)$. As U_2 is eliminated in Equation (4.105), the error remains $O(\varepsilon^2)$ for $t=O(1/\varepsilon)$. However, as argued in the following section, this is no surprise. Making U_2 vanish is equivalent to the inclusion of U_2 in the normal form for other choices of the free function, which guarantees that the error is $O(\varepsilon^2)$ as far as $t=O(1/\varepsilon)$.

4.8.2.3 Generalization to higher orders. We now combine the first-order approximation for the solution, Equation (4.104), with a second-order normal form:

$$\dot{u} = -i\,u + \varepsilon U_1(u,u^*) + \varepsilon^2 U_2(u,u^*)$$

Inclusion of U_2 in the normal form leads to the elimination of the U_2 term in Equation (4.105). Now, we write the $O(\varepsilon^3)$ term explicitly. Also, the definition of $\bar{Z}_1$ has to be modified into

$$\bar{Z}_1 = Z_1 + \varepsilon Z_2 + \varepsilon^2 Z_3$$

with a corresponding modification in the definition of $\bar{K}(u,u^*,R,R^*)$. The equation for S becomes

$$S = S_0 - \varepsilon^2 \int_0^t \{\exp(i\ \tau)[Z_0, T_2(u,u^*)]\} d\tau$$

$$+ \varepsilon^3 \int_0^t \{\exp(i\ \tau)\{U_3(u,u^*) - [Z_0, T_3(u,u^*)]\}\} d\tau$$

$$+ \int_0^t O(\varepsilon^4) \exp(i\ \tau) d\tau$$

(4.108)

$$+ \varepsilon \int_0^t \left\{\exp(i\ \tau)\left\{\left\{\frac{\partial \bar{Z}_1(u,u^*)}{\partial u} R + \frac{\partial \bar{Z}_1(u,u^*)}{\partial u^*} R^*\right\}\right\}\right\} d\tau$$

$$+ \varepsilon \int_0^t \exp(i\ \tau) \bar{K}(u,u^*,R,R^*) d\tau$$

The pattern becomes obvious, and we need not repeat the remainder of the analysis. The $O(\varepsilon^2)$ term generates an error of $O(\varepsilon^2)$. So does the $O(\varepsilon^3)$ term, owing to the occurrence of a resonant contribution, U_3.

Comment. From the point of view of the present section, the minimal normal form result mentioned in the previous section [see discussion following Equation (4.107)] is no surprise. Making U_2 vanish is tantamount to saying that we include U_2 in the normal form.

The generalization to higher orders is straightforward. For an nth-order approximation for the solution

$$z_{approx} = u + \sum_{k=1}^{n} \varepsilon^k T_n$$

and a truncation of the normal form at $(n+1)$st order

$$\dot{u} = -iu + \sum_{k=1}^{n+1} \varepsilon^k U_k$$

the equation for the error is

$$S = S_0 - \varepsilon^{n+1} \int_0^t \{\exp(i\tau)[Z_0, T_{n+1}(u, u^*)]\} d\tau$$

$$+ \varepsilon^{n+2} \int_0^t \{\exp(i\tau)\{U_{n+2}(u, u^*) - [Z_0, T_{n+2}(u, u^*)]\}\} d\tau$$

$$+ \int_0^t O(\varepsilon^{n+3}) \exp(i\tau) d\tau \qquad (4.109)$$

$$+ \varepsilon \int_0^t \left\{\exp(i\tau)\left\{\left[\frac{\partial \overline{Z}_1(u, u^*)}{\partial u} R + \frac{\partial \overline{Z}_1(u, u^*)}{\partial u^*} R^*\right]\right\}\right\} d\tau$$

$$+ \varepsilon \int_0^t \exp(i\tau) \overline{K}(u, u^*, R, R^*) d\tau$$

with

$$\overline{Z}_1 = Z_1 + \varepsilon Z_2 + \cdots + \varepsilon^{n+2} Z_{n+2}$$

and $\overline{K}$ modified accordingly. Leading to an error of $O(\varepsilon^{n+1})$ for $t=O(1)$ and of $O(\varepsilon^n)$ for $t=O(1/\varepsilon)$.

4.8.2.4 Extension to longer times? Superficially, in the case of conservative systems, we may erroneously expect to be able to extend the validity of the approximations for longer times in the following manner. As an example, we return to u, the zero-order term, as our approximation. However, we include a second-order frequency update by adopting the second-order normal form

$$\dot{u} = -iu + \varepsilon U_1 + \varepsilon^2 U_2$$

yielding

$$u = \rho \exp\left(-i\left(1 + \varepsilon \omega_1 + \varepsilon^2 \omega_2\right)t - i\varphi_0\right) \qquad \rho = \text{const}$$

If ρ is chosen so that the initial conditions are satisfied within $O(\varepsilon)$, we may be misled to expect u to constitute an $O(\varepsilon)$ approximation for the full solution up to $t=O(1/\varepsilon^2)$. There is no additional error in the amplitude, and the phase generates new errors only when $t=O(1/\varepsilon^3)$! This is, of course, a wrong statement, since by fixing ρ with an error of $O(\varepsilon)$, the ω_1 term in the phase may generate an unknown $O(\varepsilon^2)$ error, which spoils the precision of the phase.

If we choose to force u to obey the initial conditions *exactly*, then this requires a particular choice of the free functions (this issue will be discussed at great length in Chapter 6) in each order. Therefore, to obtain the values of ω_1 and ω_2 that are correct in this choice, we must include T_1 in z_{approx}, which we haven't done. Thus, ω_2 will not be known precisely and will generate an error of $O(1)$ for $t=O(1/\varepsilon^2)$.

In general, it is impossible to extend to longer times the validity of error estimates in a method that is based on Gronwall-lemma type inequalities for absolute value of the error. Take, for instance, the present example. The equation for the error, R, is

$$\dot{R} = -i\,R - \varepsilon[Z_0, T_1] + \varepsilon^2\{Z_2(u,u^*) - U_2(u,u^*)\} + O(\varepsilon^3)$$

$$+ \varepsilon\left\{\frac{\partial \bar{Z}_1(u,u^*)}{\partial u}R + \frac{\partial \bar{Z}_1(u,u^*)}{\partial u^*}R^*\right\} \quad (4.110)$$

$$+ \varepsilon \bar{K}(u,u^*,R,R^*)$$

with $\bar{Z}_1$ as in Equation (4.101) and $\bar{K}$ defined accodringly. There are two obstacles that prevent one from extending the test to longer times.

First, the $O(\varepsilon^2)$ term contains resonant contributions, which reduce its accuracy from $O(\varepsilon^2)$ to $O(\varepsilon)$. This observation is a reflection of the statement made above, concerning the error generated by the ω_1 part of the phase.

Second, and crucial, is the fact that the last two terms in Equation (4.110), involving the error R, are multiplied by ε. This factor enters into the exponential term $\exp(\varepsilon A t)$ that appears in *all* the bounds for the absolute value of the error [see Equations (4.99) and (4.103)]. It only enables us to extend the analysis to times of $O(1/\varepsilon)$. For $t=O(1/\varepsilon^2)$, the exponential is no more of $O(1)$, and the error becomes large.

Thus, it is this weakness of the Gronwall-type analysis (the only general one we have) that prevents us from extending the error analysis to longer times. For most cases, it yields a good bound, namely, one which faithfully reflects the effect of the perturbation on the quality of the approximation and does not blow up the error above what it really is. However, the method is limited to $t=O(1/\varepsilon)$.

It is possible to extend the error bound to longer timescales only if the factor multiplying the terms involving $\bar{Z}_1$ is changed from ε to ε^2. This is certainly the case if in Equation (4.92) of the full system one has

$$Z_1 \equiv 0$$

Then the expansion starts with ε^2, and this factor will replace ε in front of the $\bar{Z}_1$ terms. ($\bar{Z}_1$ then starts with Z_2.) A somewhat more intricate possibility occurs if

$$U_1(u,u^*) \equiv 0$$

In this case, one can show that, despite the fact that the $\bar{Z}_1$ terms are multiplied by ε, the vanishing of U_1 enables the extension of the error analysis to $t=O(1/\varepsilon^2)$. This case has been proved within the context of the method of multiple timescales [15,16].

REFERENCES

1. Poincaré, H., *New Methods of Celestial Mechanics* [originally published as *Les Méthodes Nouvelles de la Méchanique Celeste*, Gauthier-Villars, Paris (1892)], American Institute of Physics (1993).
2. Arnold, V. I., *Geometrical Methods in the Theory of Ordinary Differential Equations*, 2nd ed., Springer-Verlag, New York (1988).
3. Wiggins, S., *Introduction to Applied Nonlinear Dynamical Systems and Chaos*, Springer-Verlag, New York (1990).
4. Lindstedt, A., "Über die Integration einer für die störungstheorie wichtigen Differentialgleichung," *Astron. Nach.* **103**, 211 (1882).
5. Bogoliubov, N. M., and Ya. Mitropolsky, *Asymptotic Methods in the Theory of Nonlinear Oscillations*, Gordon & Breach, New York (1961).
6. Sanders, J. A., and F. Verhulst, *Averaging Methods in Nonlinear Dynamical Systems*, Springer-Verlag, New York (1985).
7. Nayfeh, A. H., *Perturbation Methods*, Wiley, New York (1973).
8. Bruno, A. D., *Local Methods in Nonlinear Differential Equations*, Springer-Verlag, New York (1989).
9. Guckenheimer, J., and P. Holmes, *Nonlinear Oscillations, Dynamical Systems and Bifurcations of Vector Fields*, Springer-Verlag, New York 1983).
10. Kummer, M., "How to Avoid Secular Terms in Classical and Quantum Mechanics," *Nuov. Cim.* **1B**, 123 (1971).

11. Van der Meer, J. C., *The Hamiltonian Hopf Bifurcation*, Springer-Verlag, New York (1985).
12. Baider, A., and R. C. Churchill, "Uniqueness and Non-Uniqueness of Normal Forms for Vector Fields." *Proc. Roy. Soc.* (Edinburgh) **A108**, 27 (1988).
13. Baider, A., "Unique Normal Forms for Vector Fields and Hamiltonians," *J. Diff. Eqns.* **78**, 33 (1989).
14. Siegel, C. L., "Über die analytische Normalform analyticher Differentialgleichenung in der Nähe einer Gleichgewichtslösung," *Nachr. Akad. Wiss. Gött. Math.-Phys.* **Kl**, 21 (1952).
15. Smith, D. R., *Singular-Perturbation Theory*, Cambridge University Press, Cambridge, UK (1985).
16. Murdock, J., "Some Fundamental Issues in Multiple Scale Theory," *Appl. Anal.* **53**, 157 (1994).

5

PROBLEMS WITH EIGENVALUES THAT HAVE NEGATIVE REAL PART

The normal form expansion takes a particularly simple form if all the eigenvalues of the unperturbed problem lie in a Poincaré domain. By a rotation, it is possible to locate all of them either in the right or left half-plane, and by a change in the direction of the arrow of time, one can discuss the case of the left-half plane as a representative of the general one. As a direct consequence of the fact that the eigenvalues lie in a Poincaré domain, for problems in this class, the normal form expansion is easy to construct. There are only a finite number of possible resonant terms. The complexity associated with problems in which "eigenvalue update" is possible is absent. (These latter problems will form the basis of the discussion in the following chapters.) The eigenvalues can be ordered and the u equations–solved sequentially. This leads to the result that, for problems with eigenvalues in a Poincaré domain, the naive perturbation expansion yields *precisely* the same terms as the normal form expansion. The T functions can be found to the desired order and the recovery of the solution to the original problem, through the desired order in the perturbation expansion, is straightforward.

Nonlinear equations have the capacity to exhibit *finite-time blowup*. When the solution exhibits this behavior, the phenomenon may or may not be recovered by the formal perturbation expansion. This point may cause some confusion. We introduce the phenomenon in Section 5.4 and discuss it through several examples.

To illustrate a variety of aspects of the normal form expansion, we begin with the problem of a single nonzero eigenvalue. This is the simplest possible case; resonance cannot occur, the generators of the expansion are always solutions to the unperturbed problem, and the normal form expansion is obtained in a straightforward manner.

Often one studies problems that involve a set of coupled equations. However, the number of equations that model the system may be larger than what one may perceive as convenient for analysis. (For example, when the solution of a nonlinear partial-differential equation is expanded in a Fourier series, one obtains an infinite set of coupled Orinary Differential Equations for the time dependence of the Fourier coefficients.) One then tends to truncate the set, thereby lowering the dimension of the space in which motion takes place. In Example 5.8, we illustrate difficulties that may arise owing to truncation. It turns out that, unless due care is exercised, one may obtain solutions that produce an incorrect picture of the motion of the system.

5.1 BASIC RESULT: REVIEW

Consider the equation

$$\frac{d\mathbf{x}}{dt} = \mathbf{A}\mathbf{x} + \varepsilon F(\varepsilon, \mathbf{x}) \tag{5.1}$$

The eigenvalue spectrum of the matrix $\mathbf{A}$ of Equation (5.1) plays a central role in the normal form expansion. Throughout this chapter, we restrict ourselves to problems in which the eigenvalues are in a Poincaré domain and all have either positive or negative real parts. As noted in Chapter 4, in this case, the number of resonant combinations is finite. The monomials in the normal form will be of degree no higher than the largest |**m**|. This situation is discussed by Bruno [1], pages 193–194. Introducing the near-identity transformation

$$\mathbf{x} = \mathbf{y} + \Sigma \varepsilon^n \mathbf{T}_n(\mathbf{y})$$

in Equation (5.1), we obtain a set of normal form equations:

$$\frac{dy_k}{dt} = \lambda_k y_k + \varepsilon G_k(\mathbf{y}; \varepsilon), \qquad k = 1, \cdots, n$$

The G_k exclusively contain monomials $[y_1]^{m_1}[y_2]^{m_2}\cdots[y_n]^{m_n}$, such that $(\mathbf{m}, \boldsymbol{\lambda}) = \lambda_k$.

It is necessary to order the eigenvalues. (Coincident eigenvalues do not cause any difficulties.) We have chosen to order them according to their real parts:

$$|\operatorname{Re}\lambda_1| \le |\operatorname{Re}\lambda_2| \le |\operatorname{Re}\lambda_3| \le \ldots \le |\operatorname{Re}\lambda_n|$$

We solve the equations sequentially. We order the components of $\mathbf{y}$ in the order of the corresponding eigenvalues of the lincar part of Equation (5.1). With this ordering, λ_1 cannot resonate with any combination of λ_k, $k \geq 1$. Then, the normal form equation for y_1 cannot contain any resonant terms; it must be linear. In general, a given λ_s cannot resonate with any combination of λ_k, $k \geq s$. Consequently, the normal form equation for a given component y_s will contain resonant terms involving at most y_k, with $k \leq s-1$. The general structure of the normal form is

$$\frac{dy_1}{dt} = \lambda_1 y_1$$

$$\frac{dy_2}{dt} = \lambda_2 y_2 + \varepsilon G_2(y_1; \varepsilon)$$

$$\vdots$$

$$\frac{dy_n}{dt} = \lambda_n y_n + \varepsilon G_n(y_1, \ldots, y_{n-1}; \varepsilon)$$

The equation for y_1 is solved explicitly by an exponential. Substituting the solution for y_1 into the equation for y_2, the latter becomes a linear equation for y_2 with a known time-dependent forcing perturbation. (It should be noted that if λ_1 and λ_2 do not resonate, the equation for y_2 will also be linear and independent of y_1.) Once y_2 is found, one can substitute the solutions for y_1 and y_2 into the equation for y_3, rendering it a linear equation in y_3 with a known time-dependent forcing perturbation. The procedure is extended in an obvious manner to higher orders. The generators of the normal form transformation are given in terms of the y's. Therefore, one can obtain an expansion to any desired order. However, in many cases, the transformation is not convergent, and care must be exercised in the analysis. (This point is illustrated in Example 5.7, where we discuss an exactly solvable model. The exact solution is given and one sees that the convergence of the normal form expansion depends on the initial conditions.)

Comment. If the eigenvalues of the matrix **A** lie in a Poincaré domain, the normal form equations can be solved sequentially. Then the transformation that relates the original variable $\mathbf{x}(t)$ to the transformed variable $\mathbf{y}(t)$ can be expressed in terms of known functions, $y_1(t), \ldots, y_n(t)$. The eigenvalues are not updated. This means that the linear parts of original equations for $\mathbf{x}(t)$ and of the normal form equations for $\mathbf{y}(t)$ have the same eigenvalues. Hence, the generators of the normal form expansion and the naive expansion are the same; they are the solutions of the linear unperturbed equations.

Example 5.1. Consider the system

$$\dot{x} = -x + \varepsilon x^2 y$$
$$\dot{y} = -2y + \varepsilon\{x^2 + xy\}$$
(5.2a)

Denoting the zero-order terms for x and y by u and v respectively, the normal form equations are

$$\dot{u} = -u$$
$$\dot{v} = -2v + \varepsilon u^2$$
(5.2b)

The solution of these equations is

$$u = u_0 \exp(-t)$$
$$v = v_0 \exp(-2t) + \varepsilon t u_0^2 \exp(-2t)$$
(5.2c)

Observe that the zero-order solution contains a term that is proportional to t. This is typical of problems in which the eigenvalues of the unperturbed system are in a Poincaré domain.

In Chapter 2 we encountered such terms when we employed naive perturbation theory in the analysis of conservative oscillatory systems. They were called secular terms and limited the validity of the perturbation expansion to times of $O(1)$. We then resorted to the method of normal forms in order to avoid these terms, thereby extending the validity of the expansion to longer times [e.g., $t \approx O(1/\varepsilon)$]. Here, because the unperturbed eigenvalues lie in a Poincaré domain, such terms are present in the naive expansion as well as in the method of normal forms.

Comment. Since the small parameter, ε, in our problems serves as the basis of a perturbation expansion, the "appropriate" eigenvalue spectrum to consider is that of the unperturbed operator. In other words, in our considerations, for the system to have all its eigenvalues in a Poincaré domain, they all must be $O(1)$ and cannot include $O(\varepsilon)$ terms that are introduced by the perturbation. For example, in the case of the Van der Pol oscillator

$$\frac{dx^2}{dt^2} + x = \varepsilon(1 - x^2)\frac{dx}{dt}, \qquad 0 < \varepsilon \ll 1$$

the eigenvalues of the unperturbed problem are $\pm i$ and hence *do not* lie in a Poincaré domain. The perturbed system has a linear term. If we insist on including that term in the "unperturbed part," the analysis becomes more cumbersome. It is best to see this in the complex notation: $z = x + i(dx/dt)$. The Van der Pole equation then becomes

$$\dot{z} = \left(-i + \tfrac{1}{2}\varepsilon\right)z - \tfrac{1}{2}\varepsilon z^* - \tfrac{1}{8}i\,\varepsilon\left(z + z^*\right)^2\left(z - z^*\right)$$

with the complex conjugated equation for z^*. First, there is an ambiguity in the definition of what is the "unperturbed part." If, in the equation for z, we choose to include only the term $(\varepsilon/2)z$ (and the corresponding term in the equation for z^*), then the eigenvalues acquire a small positive real part. They are now equal to $\pm i + (\varepsilon/2)$. If we choose to include also the term $-(\varepsilon/2)z^*$, then further diagonalization is required. The diagonalized linear system now is

$$w = z - \frac{\tfrac{1}{2}i\varepsilon}{1 + \sqrt{1 - \tfrac{1}{4}\varepsilon^2}} z^*$$

and w^*. The eigenvalues of the derived unperturbed part are now

$$\tfrac{1}{2}\varepsilon \mp i\sqrt{1 - \tfrac{1}{4}\varepsilon^2}$$

In either choice, the eigenvalues of the resulting linear part lie in a Poincaré domain. Therefore, a formal transformation exists that reduces the u equation to a linear one.

However, one is now confronted with difficulties. In either case, the new "zero-order term" describes a spiral that diverges away from the origin. The solution of the full equation is a closed periodic orbit (which, for small ε, is a slightly distorted circle). Clearly, the structure of these spiraling solutions makes them worse candidates for generating the expansion than the simple periodic solutions of the customary linear term. As a result, in the near-identity transformation of the normal form expansion, the new "unperturbed terms" generate small denominators [that go like $(1/\varepsilon)$] as well as complicated expressions in ε. The latter may lead to errors in the consistent computation in each order, unless one is extremely careful. To summarize, while we do have the freedom of deciding what we call quantities of $O(1)$, we must make a judicious choice that is convenient to implement. This will become clearer following the extensive discussion of the Van der Pol oscillator in Chapter 7.

5.2 ONE COORDINATE WITH A NEGATIVE REAL EIGENVALUE: NO POSSIBLE RESONANCES

We apply the method of normal forms to the analysis of this simple class of equations, to illustrate both how the method works and why it is an effective perturbation technique. We will show three separate aspects of the general method as applied to this problem:

1. Construction of the T functions to kill all of the U_i for $i > 0$
2. The "inversion problem" that is required to be addressed in order to satisfy the initial conditions
3. The role free functions play in satisfaying the initial conditions

Our equation is of the form

$$\frac{dx}{dt} = \lambda x + \varepsilon f(x), \qquad \text{Re}\lambda = O(1) \neq 0; \quad |\varepsilon| \ll 1 \qquad (5.3)$$

where $f(x)$ is a polynomial (or has a Taylor series expansion) in x, the lowest power being at least 2.

For equations of this type, with one eigenvalue of the unperturbed problem with nonzero real part, in the complex (eigenvalue) plane one can draw a circle of $O(1)$ centered at the origin that will *not* contain the eigenvalue. It is this aspect that we use to generate the normal form expansion that forms the basis of our perturbative solution.

5.2.1 Procedure

We have an equation for $x(t)$ and we seek a transformation to a new variable, $u(t)$, that satisfies the unperturbed linear equation associated with $x(t)$ when $\varepsilon = 0$. The near identity expansion is written as

$$x = u + \sum_{n \geq 1} \varepsilon^n T_n \qquad (5.4)$$

with the T functions chosen to have a structure similar to that of $f(x)$. Since we have one nonzero eigenvalue and $k \geq 2$, there cannot be resonant combinations of eigenvalues and, therefore, no resonant monomials in the expansion. From Poincaré's theorem, it follows that a transformation exists that yields a $u(t)$ that satisfies the unperturbed linear equation. Namely, we can achieve $U_i = 0$ for $i > 0$ by an appropriate choice of T_i. We begin by writing

ONE COORDINATE WITH NEGATIVE EIGENVALUE

$$\frac{du}{dt} = U_0 = \lambda u \tag{5.5a}$$

$$U_i = 0, \qquad\qquad i > 0 \tag{5.5b}$$

$$u(t) = u(0)\exp(\lambda t) \tag{5.5c}$$

As the equation for u is known, one concentrates on obtaining a perturbative expansion of $x(t)$ through some order in ε by finding the T functions through that order. A critical point to note is the dynamics associated with $u(t)$ is that of the unperturbed problem. The nonlinear structure enters only through the T functions.

Consider Equation (5.3) and write it as

$$\left[\frac{dx}{dt} - \lambda x\right] = \varepsilon f(x) \tag{5.6}$$

We employ the near-identity transformation [Equation (5.4)] in Equation (5.6). Since the T functions depend only on u, we have

$$\frac{dT_n}{dt} = \left(\frac{dT_n}{du}\right)\left(\frac{du}{dt}\right) = \left(\frac{dT_n}{du}\right)\lambda u$$

Therefore, in order ε^n, the Lie bracket takes the form

$$[Z_0, T_n] = \lambda T_n - \lambda u\left(\frac{dT_n}{du}\right) \tag{5.7}$$

equal to a known function computed from terms that have been determined in the orders $1 \leq k \leq (n-1)$.

Let us assume that $f(x)$ is a polynomial of the form

$$f(x) = \sum_{k \geq 2}^{N} b_k x^k \tag{5.8a}$$

and that T_n is a similar polynomial in u, namely

$$T_n(u) = \sum_{k \geq 2}^{N} a_{n,k} u^k \tag{5.8b}$$

With Equations (5.8a,b), Equation (5.7) becomes

$$[Z_0, T_n] = \lambda \sum_{k \geq 2}^{N} a_{n,k}(1-k)x^k \qquad (5.8c)$$

Note. From Equation (5.8c), it is clear that the $k=1$ term vanishes. However, no other terms in the Lie bracket need vanish. This structure of the Lie bracket is (must be) the same in each order of the expansion, as it is determined entirely by the linear part of the problem.

5.3 ANALYSIS OF A SPECIFIC EXAMPLE

Example 5.2. Consider the equation

$$\frac{dx}{dt} = \lambda x - \varepsilon a x^2, \qquad |\varepsilon| \ll 1 \qquad (5.9)$$

We develop the normal form expansion for this equation for a variety of choices of the free functions.

5.3.1 Case 1: Free functions set = 0

Equation (5.9) can be solved in closed form [see Equation (5.14a)]. Thus, it is straightforward to obtain the T functions for all orders by expanding the solution in a power series in ε. As our goal is to understand aspects of the perturbation expansion itself, we proceed as if the full solution were not available to us. We carry out the expansion given in Equations (5.4)–(5.8c) through $O(\varepsilon^2)$ and show that one can realize $U_i=0$, $i=1,2$, by an appropriate choice of T_i.

Substituting Equation (5.4), for x in terms of u in Equation (5.9), we obtain through second order

$$\frac{du}{dt}\left[1 + \varepsilon \frac{dT_1}{du} + \varepsilon^2 \frac{dT_2}{du}\right] =$$

$$\lambda[u + \varepsilon T_1(u) + \varepsilon^2 T_2(u)] - \varepsilon a[u + \varepsilon T_1(u) + \varepsilon^2 T_2(u)]^2 \qquad (5.10)$$

$$\frac{du}{dt} = U_0 + \varepsilon U_1 + \varepsilon^2 U_2$$

Equating equal powers of ε, we obtain the following:

ANALYSIS OF A SPECIFIC EXAMPLE

$$O(\varepsilon^0): \quad U_0 = \lambda u \tag{5.11a}$$

$$O(\varepsilon^1): \quad U_1 = \left[\lambda T_1 - \lambda u \frac{dT_1}{du}\right] - au^2 \tag{5.11b}$$

As u^2 is not a resonant term, U_1 can be made to vanish by choosing

$$\left[\lambda T_1 - \lambda u \frac{dT_1}{du}\right] = au^2 \tag{5.11c}$$

Writing $T_1 = b \cdot u^2$ in Equation (5.11c), we find

$$\lambda[b - 2b]u^2 = au^2 \tag{5.11d}$$

Thus, with $b = -(a/\lambda)$, we eliminate the $O(\varepsilon)$ term in the normal form. In second order we have

$$O(\varepsilon^2): \quad U_2 = \left[\lambda T_2 - \lambda u \frac{dT_2}{du}\right] - U_1 \frac{dT_1}{du} - 2auT_1 \tag{5.11e}$$

Clearly, U_2 can be eliminated by $T_2 = c \cdot u^3$ with $c = (a/\lambda)^2$, once U_1 and T_1 are known. Note that the Lie brackets on the r.h.s. of Equations (5.11e) and (5.11b) are of the same form:

$$\left[\lambda T_i - \lambda u \frac{dT_i}{du}\right] \qquad i = 1, 2$$

This structure is maintained in the higher order calculations and the remainder of the r.h.s. contains a combination of terms that have been determined in previous orders.

To recapitulate, with the choice

$$T_1 = -\left(\frac{a}{\lambda}\right)u^2 \tag{5.11f}$$

$$T_2 = +\left(\frac{a}{\lambda}\right)^2 u^3 \tag{5.11g}$$

we have obtained $U_1 = U_2 = 0$, so that

$$\frac{du}{dt} = \lambda u \quad \Rightarrow \quad u = u(0)\exp(\lambda t) \qquad (5.11\text{h})$$

The approximation for $x(t)$ is then

$$x(t) = u(t) - \varepsilon\left(\frac{a}{\lambda}\right)u^2 + \varepsilon^2\left(\frac{a}{\lambda}\right)^2 u^3 + O(\varepsilon^3) \qquad (5.12)$$

Inserting Equation (5.11h) for $u(t)$, we have

$$x(t) = u(0)\exp(\lambda t)\left[\begin{array}{l}1 - \varepsilon\left(\dfrac{a}{\lambda}\right)u(0)\exp(\lambda t) \\ + \varepsilon^2\left(\dfrac{a}{\lambda}\right)^2 u(0)^2 \exp(2\lambda t) + O(\varepsilon^3)\end{array}\right] \qquad (5.13)$$

5.3.2 Comparison with an expansion of the exact solution

Solving Equation (5.9) is straightforward yielding

$$x(t) = \frac{x(0)\exp(\lambda t)}{\{1 + \varepsilon(a/\lambda)x(0)[\exp\lambda t - 1]\}} \qquad (5.14\text{a})$$

Expanding through second order, we have

$$x(t) = x(0)\exp(\lambda t)\left\{\begin{array}{l}1 - \varepsilon\left(\dfrac{a}{\lambda}\right)x(0)(\exp(\lambda t) - 1) \\ + \varepsilon^2\left(\dfrac{a}{\lambda}\right)^2 x(0)^2(\exp(\lambda t) - 1)^2 + O(\varepsilon^3)\end{array}\right\} \qquad (5.14\text{b})$$

We see that Equations (5.13) and (5.14b) are similar, but not identical. We learn how to handle this in the next section.

5.3.3 Implementation of the initial conditions

It is important to understand how the initial conditions are treated by the various approximation schemes.

The solution method we have outlined uses a power-series expansion, with $u(0)$ as a "seed." We need to connect the results with those obtained from an expansion of the exact solution, given by

ANALYSIS OF A SPECIFIC EXAMPLE 139

Equations (5.14). Using Equation (5.13), we evaluate the $x(t)$ at $t=0$:

$$x(0) = u(0)\left[1 - \varepsilon\left(\frac{a}{\lambda}\right)u(0) + \varepsilon^2\left(\frac{a}{\lambda}\right)^2 u(0)^2 + O(\varepsilon^3)\right] \quad (5.15a)$$

This equation can be inverted to yield $u(0)$ in terms of $x(0)$:

$$u(0) = x(0)\left[1 + \varepsilon\left(\frac{a}{\lambda}\right)x(0) + \varepsilon^2\left(\frac{a}{\lambda}\right)^2 x(0)^2 + O(\varepsilon^3)\right] \quad (5.15b)$$

Comment. The full solution, Equation (5.14a), obeys the initial condition in a specific manner. The numerator produces the initial condition, while the coefficient of ε in the denominator vanishes at $t = 0$. Therefore, Equation (5.14b), being a direct expansion of Equation (5.14a), yields the initial condition by its zero-order term and all higher-order terms vanish at $t = 0$. However, in the expansion for $x(t)$ as given by Equation (5.13), $u(t)$ does not satisfy the initial condition. In fact, the higher-order terms in Equation (5.13) *do not* vanish at $t = 0$. For the two representations of the solution to coincide, the connection between $u(0)$ and $x(0)$ must be as in Equations (5.15a) and (5.15b). Thus, when carrying out a perturbation expansion, one needs to invert the expansion in order to satisfy the initial conditions.

5.3.4 Case 2: Free functions

It is possible to generate a perturbation expansion where the zero-order term, $u(t)$, obeys the initial condition and all higher-order terms vanish at $t = 0$. However, this requires specific nonzero free functions. We, therefore, redevelop the method of normal forms, exploiting the *freedom of choice* associated with the T functions, and show how the free terms affect the implementation of the initial conditions.

Note that the Lie bracket

$$\lambda\left[T_i - u\frac{dT_i}{du}\right]$$

is unchanged if we add to T_i a linear term, $d \cdot u$, because the Lie bracket of the latter vanishes. Consequently, inclusion of a linear term in T_1 does not help "kill" anything in the *same* order; in the expression for U_1 this term appears in the Lie bracket and contributes zero. However, such a function will affect the results in second and higher orders. We include this free term and denote the generator by S_1:

$$S_1 = -\left(\frac{a}{\lambda}\right)u^2 + du \qquad (5.16)$$

This leads to a modification of the equation for U_2. The generator, now called S_2, has the form

$$S_2 = +\left(\frac{a}{\lambda}\right)^2 u^3 - 2\left(\frac{a}{\lambda}\right)du^2 + e \cdot u \qquad (5.17)$$

where we have included a free term, $e \cdot u$.

The perturbation expansion is now given as

$$x(t) = u + \varepsilon S_1 + \varepsilon^2 S_2 + O(\varepsilon^3) \qquad (5.18)$$

Retaining terms through second order, we obtain

$$\begin{aligned} x(t) = u &+ \varepsilon\left[-\left(\frac{a}{\lambda}\right)u^2 + du\right] \\ &+ \varepsilon^2\left[\left(\frac{a}{\lambda}\right)^2 u^3 - 2\left(\frac{a}{\lambda}\right)du^2 + eu\right] \end{aligned} \qquad (5.19a)$$

$$x(t) = u(0)\exp\lambda t \left\{ \begin{aligned} &1 + \varepsilon\left[-\left(\frac{a}{\lambda}\right)u(0)\exp\lambda t + d\right] \\ &+ \varepsilon^2 \begin{bmatrix} \left(\frac{a}{\lambda}\right)^2 u(0)^2 \exp 2\lambda t \\ -2\left(\frac{a}{\lambda}\right)du(0)\exp\lambda t + e \end{bmatrix} \end{aligned} \right\} \qquad (5.19b)$$

Equations (5.14b) and Equation (5.19b) become identical if we write

$$u(0) = x(0), \quad d = \left(\frac{a}{\lambda}\right)x(0), \quad e = \left(\frac{a}{\lambda}\right)^2 x(0)^2 \qquad (5.20)$$

Thus, the freedom of adding a resonant term (i.e., one with a vanishing Lie bracket) to the generator can be used to facilitate the implementation of the initial conditions. We also need to know how the initial condition, $x(0)$, varies as $\varepsilon \to 0$. The inversion procedure,

ANALYSIS OF A SPECIFIC EXAMPLE

yielding $u(0)$ from $x(0)$ depends on whether $x(0)$ depnds on ε.

Comment. The generators T_i have undetermined terms. As a result, one is free to choose the *origin* around which to generate the near-identity transformation that provides the perturbation solution. This freedom can be exploited in a comparison with solutions obtained by different methods or in the way one satisfies the initial conditions.

Example 5.3. Consider the one-dimensional equation

$$\frac{dx}{dt} = \lambda x + \sum_{k=1}^{3} x\{\varepsilon^k c_k x^k\} \tag{5.21}$$

where λ and the coefficients c_k are real. The normal form is linear. Through third order, the normal form is

$$U_1 = \lambda T_1 - \lambda u \frac{dT_1}{du} + c_1 u^2 \tag{5.22a}$$

$$U_2 = \lambda T_2 - \lambda u \frac{dT_2}{du} - U_1 \frac{dT_1}{du} + c_2 u^3 + 2c_1 u T_1 \tag{5.22b}$$

$$U_3 = \lambda T_3 - \lambda u \frac{dT_3}{du} - U_2 \frac{dT_1}{du} - U_1 \frac{dT_2}{du} \tag{5.22c}$$
$$+ c_3 u^4 + 2c_1 u T_2 + 3c_2 u^2 T_1 + c_1 T_1^2$$

The structure that appears in the low orders is repeated in all higher orders. When the generator functions are expressed as appropriate homogeneous polynomials (no free functions included), each equation is a linear equation for the coefficients of the T functions. We obtain

$$U_1 = U_2 = U_3 = 0$$
$$\frac{du}{dt} = \lambda u, \Rightarrow u(t) = u_0 \exp \lambda t$$

$$x(t) = u + \varepsilon \left(\frac{c_1}{\lambda}\right) u^2 + \varepsilon^2 \left[\frac{1}{2}\left(\frac{c_2}{\lambda}\right) + \left(\frac{c_1}{\lambda}\right)^2\right] u^3$$
$$+ \varepsilon^3 \left[\frac{1}{3}\left(\frac{c_3}{\lambda}\right) + \frac{4}{3}\left(\frac{c_1}{\lambda}\right)\left(\frac{c_2}{\lambda}\right) + \left(\frac{c_1}{\lambda}\right)^3\right] u^4 \tag{5.23}$$

To construct the full solution, we need to determine the initial value, $u(0)$, of the zero-order approximation, in terms of $x(0)$, the initial value of the original coordinate.

Example 5.4. We now consider a three-component problem. The eigenvalues of the unperturbed problem are

$$\lambda_1 = -1, \quad \lambda_2 = -2, \quad \lambda_3 = -3.$$

The nonlinear terms are quadratic in x_i with coefficients proportional to ε. We develop the normal form expansion through $O(\varepsilon)$. Two aspects become immediately clear:

1. Computation of the general expansion is straightforward, but complicated enough that use of a symbolic manipulation program is advisable.
2. The presence or absence of particular terms in the original equation can affect the structure of the normal form in a substantial way.

The equations for this example are written as:

$$\frac{dx_m}{dt} = \lambda_m x_m + \varepsilon \sum_{i+j+k=2} a[m;i,j,k] x_1^{\,i} x_2^{\,j} x_3^{\,k}$$

Notation and conditions. The components of the vector are denoted by the index m ($m=1,2,3$). The powers of the various components are indicated by i,j,k ($i+j+k=2$). The coefficients of the $O(\varepsilon)$ terms are written as $a[m; i,j,k]$.

Through first order, the near-identity transformation is written as

$$x_1 = u + \varepsilon T^1{}_1(u,v,w)$$

$$x_2 = v + \varepsilon T^2{}_1(u,v,w)$$

$$x_3 = w + \varepsilon T^3{}_1(u,v,w)$$

Comments

1. We order the eigenvalues according to their absolute value.
2. The normal form for u (eigenvalue = -1) must be the unperturbed equation, since no monomial in u, v, or w with power ≥ 2 can resonate with u.

ANALYSIS OF A SPECIFIC EXAMPLE

3. In a given order, n, if no free functions are included, the U and T functions are homogeneous polynomials of degree $n+1$.
4. Except for a a linear term, the normal form for v (eigenvalue = -2) can only have a u^2 term. Hence, it must terminate in $O(\varepsilon)$.
5. In addition to a linear term, the normal form for w can have only a $u \cdot v$ term [quadratic, hence $O(\varepsilon)$] and a u^3 term [hence $O(\varepsilon^2)$]. Therefore, it must terminate in $O(\varepsilon^2)$.
6. The generators are simple to compute, but rather cumbersome in form. Therefore, the expressions found for them are omitted.
7. The generators may include free terms that have vanishing Lie brackets. These terms can be chosen to eliminate the $O(\varepsilon^2)$ term in the normal form equation for w. Similarly, the free terms in higher orders can be chosen to guarantee that this compact structure of the normal form equations is maintained, i.e., that they terminate at $O(\varepsilon)$.

With this choice, the *full* normal form equations are

$$\frac{du}{dt} = -u$$

$$\frac{dv}{dt} = -2v + \varepsilon a[2;2,0,0]u^2$$

$$\frac{dw}{dt} = -3w + \varepsilon a[3;1,1,0]uv$$

5.4 FINITE-TIME BLOWUP

One of the more intriguing new phenomena that occurs in nonlinear differential equations, goes by the name of *finite-time blowup*.

It is well known that the solutions of the linear differential equation

$$\frac{d\mathbf{x}}{dt} = \mathbf{A}\mathbf{x},$$

with a constant matrix $\mathbf{A}$, may blow up only when $t \to \infty$. If an eigenvalue has a positive real part, the trajectory associated with that eigenvalue grows exponentially, obtaining arbitrarily large values as $t \to \infty$.

However, consider the differential equation

$$\frac{dx}{dt} = \alpha x^2 \tag{5.24}$$

which is solved by

$$x(t) = \frac{x_0}{(1 - \alpha x_0 t)} \tag{5.25}$$

If $\alpha x_0 > 0$, $x(t)$ becomes infinite when $t = t_c = 1/(\alpha x_0)$. The solution exists only for a finite time. The solution [Equation (5.25)] is said to possess a *spontaneous singularity* at finite time.

Comment. For linear differential equations, the singularity structure of the solutions is associated only with the character of the equations. In nonlinear differential equations, the singularity structure of the solutions may be a affected not only by the character of the equation but also by the initial conditions.

To understand finite-time blowup, two misconceptions regarding the phenomenon need to be dispelled. One is the common assumption that if all the eigenvalues of the linear part of the equation have negative real parts, then the solution is attracted to the origin. The other, less obvious one, is the assumption that the analysis of the solutions in terms of their singular behavior in time yields the finite time at which a blowup occurs. We address each of the two notions separately and provide examples that dispel the misconceptions.

Attraction to the origin when all eigenvalues have negative real parts. Consider an equation of the form

$$\frac{d\mathbf{x}}{dt} = [\text{a linear term} + \text{a polynomial in } x_1, \cdots, x_n],$$

where n is the dimension of the vector. Usually, one identifies the fixed points and studies the behavior of the trajectories in the neighborhood of these points. To be specific, consider the case where all the eigenvalues have negative real parts and the origin is the only fixed point in the range of the variables. *Not always* will the system be attracted to the origin for any initial condition. One needs to be sufficiently close to the fixed point, i.e., within *its basin of attraction*. In particular, attraction is guaranteed if, for *all* the components, the slopes (dx_i/dt) are negative at $t=0$. This is the essence of the Poincaré–Lyapunov theorem. If one or more of the components have positive derivatives at $t=0$, finite-time blowup may or may not occur. This

FINITE-TIME BLOWUP

point is studied in Example 5.5 and in special cases derived from it.

Example 5.5. Consider the equation

$$\frac{dx}{dt} = \lambda x + \alpha x^2 \tag{5.26}$$

[Example 5.2 is a special case of Equation (5.26).] The solution is

$$x(t) = \frac{x_0}{1 + (\alpha/\lambda) x_0 (1 - \exp \lambda t)} \tag{5.27}$$

A finite-time blowup occurs when the denominator vanishes. The eight possible combinations depending on the sign and magnitude relationship among the parameters are presented in Table 5.1. The solution, $x(t)$, becomes infinite at the critical time, t_c. Observe that in some cases, the existence of such a time is connected to both the magnitude and the sign of each of the parameters.

5.4.1 Illustrative examples

Example 5.5a. Refer to Equation (5.26), with a negative real eigenvalue, $\lambda = -\mu$ ($\mu > 0$) and $\alpha = -\varepsilon \cdot b^2$:

$$\frac{dx}{dt} = -\mu x - \varepsilon b^2 x^2 \qquad x(0) = C \qquad 0 < \varepsilon \ll 1$$

In this equation the nonlinear term enhances or modifies the exponential decay. (We have introduced the irrelevant parameter b just to identify how various terms enter.)

Example 5.5b. The logistic equation. The logistic equation

$$\frac{dx}{dt} = \lambda x - \varepsilon b x^2, \qquad x(0) = C, \quad \lambda, b, > 0, \quad 0 < \varepsilon \ll 1$$

plays an important role in the modeling of many biological and physical systems. For small initial amplitudes (i.e., near the fixed point at $x=0$) the system is dominated by the linear term that generates an exponential growth. However, as the system evolves in time, the linear and nonlinear terms exchange roles, in terms of importance. As the amplitude increases, the quadratic "saturation" term becomes as important. It modifies the exponential growth near the origin, and the

EIGENVALUES THAT HAVE NEGATIVE REAL PART

system evolves asymptotically to the fixed point at $x = (\lambda/\varepsilon b)$.

Example 5.5c. Consider

$$\frac{dx}{dt} = x + \varepsilon x^2, \quad x(0) = A, \quad |\varepsilon| \ll 1, \quad \varepsilon A > 0$$

The solution is

$$x(t) = \frac{A \exp t}{1 - \varepsilon A (\exp t - 1)}$$

The nonlinearity of the differential equation affects two aspects of the exact solution. First, contrary to our experience with linear differential equations (where the solutions are valid for an infinite time), here the solution possesses a *spontaneous singularity*. The singularity is associated with the initial condition and not just the structure of the equation. The solution "blows up" and fails to exist beyond a finite time $t=t_c$, where

$$t_c = \ln\left(\frac{1+\varepsilon A}{\varepsilon A}\right) \approx \ln\left(\frac{1}{\varepsilon A}\right)$$

Second, the perturbative analysis of the solution is valid only for times appreciably shorter than t_c. As $t \to t_c$, the perturbation, $\varepsilon \cdot x^2$, ceases to be small compared to the unperturbed linear term, x. The unperturbed solution, $x_0(t)=A \cdot \exp(t)$, grows in time, and eventually the perturbation becomes comparable to the linear term, no matter how small ε is. When $x_0 \approx \varepsilon \cdot x_0^2$ [$t \approx \ln(1/\varepsilon A)$], the problem can no longer be regarded as a slightly perturbed linear one.

Example 5.5d. We have just seen that it is not sufficient to find the fixed points of an equation. The initial condition affects the behavior of the solution, in particular when the equation has several fixed points. The realization of a fixed point depends on the initial amplitude. For example, the equation

$$\frac{du}{dt} = -u + u^2 \qquad u(0) = b$$

which is solved by

FINITE-TIME BLOWUP

$$u(t) = \frac{b}{b - (b-1)\exp t}$$

has two fixed points: $u = 0$ (stable) and $u = 1$ (unstable). Each fixed point has a *basin of attraction*, determined by b:

<u>$b = 0$</u>: The solution remains at 0 forever.
<u>$b < 1$</u>: The solution exists for all times; $u(t)$ approaches 0 as $t \to \infty$.
<u>$b > 1$</u>: The solution becomes infinite at the finite time $t_c = \ln[b/(b-1)]$.
<u>$b = 1$</u>: The solution remains at 1 forever.

Table 5.1 Results of Example 5.5 and its special cases

λ	α	x_0	Finite-time blowup	$x(\infty)$
> 0	> 0	> 0	Yes	
> 0	> 0	< 0	No	$x(\infty) = 0_-$
> 0	< 0	> 0	No	$x(\infty) = 0_+$
> 0	< 0	< 0	Yes	
< 0	> 0	> 0	If $x_0 > \|\lambda\|/\alpha$: yes If $x_0 \leq \|\lambda\|/\alpha$: no	If $x_0 < \|\lambda\|/\alpha$: $0 < x(\infty) < +\infty$ If $x_0 = \|\lambda\|/\alpha$: $x(\infty) = +\infty$
< 0	> 0	< 0	No	$-\infty < x(\infty) < 0$
< 0	< 0	> 0	No	$0 < x(\infty) < +\infty$
< 0	< 0	< 0	If $\|x_0\| > \lambda/\alpha$: yes If $\|x_0\| \leq \lambda/\alpha$: no	If $\|x_0\| < \lambda/\alpha$: $-\infty < x(\infty) < 0$ If $\|x_0\| = \lambda/\alpha$: $x(\infty) = -\infty$

Example 5.5e. Consider an equation of the form

$$\frac{du}{dt} = +u - u^2$$

The system has a fixed point at $u = 1$. We expand about that fixed point, by writing $u = 1+z$, to obtain

$$\frac{dz}{dt} = -z - z^2$$

Deviations decay and the system stays near the stable fixed point $u = 1$.

Leading singularity analysis. Consider Equation (5.26) for the case $\alpha, \lambda > 0$ and $x_0 > 0$. In the neighborhood of the blowup, the nonlinear term *must* be greater than the linear one. There, one can neglect the linear term and conclude that the amplitude $x(t)$ approximately satisfies Equation (5.24), whose solution is given by Equation (5.25). This observation leads us to consider *scaling* as a way to determine the character of the solution as it exhibits finite-time blowup. Such a scaling analysis is often developed in the search for a possible blowup behavior. It only indicates the *potential* for blowup. Depending on parameter values and initial conditions, this behavior may or may not be realized. This is shown in Examples 5.6 and 5.7.

Example 5.6. Consider Equation (5.24) with $x_0 > 0$. Rather than solving it, assume that a finite-time blowup may occur. Therefore, a time, t_c, must exist for which

$$x(t) \sim B(t - t_c)^\beta \qquad t \approx t_c \qquad \beta < 0 \qquad (5.28a)$$

To determine β, we insert Equation (5.28a) in Equation (5.24), finding

$$\beta B s^{\beta-1} = \alpha B^2 s^{2\beta} \qquad s \equiv t - t_c \qquad (5.28b)$$

For Equation (5.28b) to hold, $\beta = -1$ and $B = -1/\alpha$ are required. Thus, the parameter α, which characterizes the growth of the solution near the blowup point, is related to the amplitude in that neighborhood.

Example 5.7: ***Two-dimensional example.*** Consider the system

$$\frac{dx}{dt} = x\left[-\lambda(1+y) + \tfrac{1}{2}Bx^2\right] \qquad x(0) = x_0 \qquad (5.29a)$$

$$\frac{dy}{dt} = (1+y)\left[-2\lambda y + Bx^2\right] \quad y(0) = y_0 \quad (5.29b)$$

with $\lambda > 0$. The fixed points are at $(0,0)$ and $(0,-1)$. The exact solutions are

$$x(t) = x_0 D^{-1/2} \exp(-\lambda t)$$
$$y(t) = \left(y_0 + x_0^2 Bt\right) D^{-1} \exp(-2\lambda t) \quad (5.30)$$
$$D = \left[1 + y_0 - (y_0 + x_0^2 Bt)\exp(-2\lambda t)\right]$$

This model is exactly solvable. The denominator, D, is constructed from contributions of resonant terms. Since both eigenvalues are negative, resonance results in a multiplicative factor of t. Also, note that, if the initial amplitudes are "too large" and/or, correspondingly, the eigenvalues are "too small," the denominator, D, may vanish at some finite time. For given initial conditions, there is a critical value, B_0 such that for $B > B_0$, the solution exhibits finite-time blowup and the system does not reach the origin.

To perform a perturbation expansion, we need to identify the small parameters in the problem. From Equations (5.29a,b) we see that B should be small and so should the amplitude of y. Assuming this to be the case, the normal form expansion, although cumbersome, can be carried out to any desired order.

Denoting the zero-order approximations for x and y by u and v, respectively, we first note that the normal form for u must terminate in zeroth order, since no higher-degree monomial can resonate with u. Also, the only resonant term in the normal form equation for v is u^2. Such a term arises from the first-order interaction in Equation (5.29b) (the x^2 term). It may also appear in higher orders, depending on the choice of the free functions in the generators of the near-identity transformation. One can guarantee that this does not happen. (This is the case, for example, if all the free functions are chosen to vanish.) As a result, the *full* normal form obtains a compact structure:

$$\frac{du}{dt} = -\lambda u \quad \frac{dv}{dt} = -2\lambda v + Bu^2$$

yielding

$$u = u_0 \exp(-\lambda t) \quad v = \left(v_0 + u_0^2 Bt\right)\exp(-2\lambda t)$$

With these results, the *full* near-identity transformation becomes

$$x = u(1-v)^{-1/2} \qquad y = v(1-v)^{-1}$$

$$u_0 = \frac{x_0}{(1+y_0)^{1/2}} \qquad v_0 = \frac{y_0}{1+y_0}$$

Numerical results. We have computed the solution, using $x_0 = 1$, $y_0 = 0.1$, and $\lambda = 0.1$. In Fig. 5.1 we present graphs of $D(t)$ for several values of B. For $B \leq (2\lambda y_0/x_0^2)$ (=0.02), D is a monotonic function, rising from the value of 1.0 at $t=0$, to its asymptotic value of $1+y_0$ (=1.1 in the present example). For $(2\lambda y_0/x_0^2) < B < B_0$, $D(t)$ has a minimum at $t = 1/(2\lambda) - y_0/(x_0^2 B)$. (For the parameters chosen here, $B_0 = 0.578$.) For $B > B_0$, $D(t)$ vanishes at some point t_0 and finite-time blowup occurs.

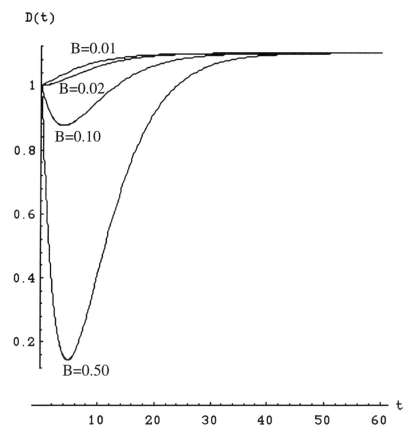

Figure 5.1 Time dependence of the denominator $D(t)$.

From Fig. 5.1 we see that, as B is increased, $D(t)$ develops a minimum at intermediate values of time. For a relatively large B (here, $B=0.50$), $D(t)$ is significantly smaller than unity near the minimum. Both $x(t)$ and $y(t)$ become large and the perturbation scheme fails for such times. In particular, the zero-order terms constitute very poor approximations for the solutions. Still, although the solutions increase initially, they eventually decay toward the origin. The initial condition in this case is still within the basin of attraction of the stable fixed point (0,0). Near $B=B_0$ (≈ 0.578 in the present calculation), the system is very sensitive to small changes in the parameters. For example, changing B from 0.5 to 0.6, the solutions blow up at $t = t_0 \approx 3.599$. Clearly, the perturbation scheme breaks down at a finite time also in this case.

In Fig. 5.2 we compare plots of $x(t)$ and $y(t)$ with their zero-, first- and second-order approximations in the normal form analysis, the details of which were given above. This is done for several values of B (except for $B=0.6$). As long as B is sufficiently small, the zero order terms, $u(t)$ and $v(t)$, are close to $x(t)$ and $y(t)$, respectively. The higher-order corrections improve the approximations even further. The perturbation expansion is uniformly valid in time, since as time progresses both $x(t)$ and $y(t)$ decrease. Therefore, in the normal form expansion, the higher-order terms become progressively smaller.

To determine the exponents that characterize the blowup *if* it occurs, we assume that a time, t_0, exists where the solution blows up and write the leading time dependence of the solution for t near t_0 as

$$x \sim \xi_0(t-t_0)^p \qquad y \sim \eta_0(t-t_0)^q$$

with p or $q < 0$. We introduce $s = t - t_0$ and observe that as $s \to 0$,

$$\frac{dx}{dt} \sim p\xi_0 s^{p-1} \qquad \frac{dy}{dt} \sim q\eta_0 s^{q-1}$$

We seek to identify the negative exponents that yield the most singular behavior. Inserting the singular expression for x and y in Equations (5.29a,b), the equations for the leading terms become

$$p\xi_0 s^{p-1} \sim -\lambda \xi_0 s^p - \lambda \xi_0 \eta_0 s^{p+q} + \tfrac{1}{2} B \xi_0^3 s^{3p} \qquad (5.31\text{a})$$

$$\begin{aligned} q\eta_0 s^{q-1} \sim\ & -2\lambda \eta_0 s^q - 2\lambda \eta_0^2 s^{2q} \\ & + B\xi_0^2 s^{2p} + B\xi_0^2 \eta_0 s^{2p+q} \end{aligned} \qquad (5.31\text{b})$$

In Equations (5.31a,b), equality must hold among terms with the

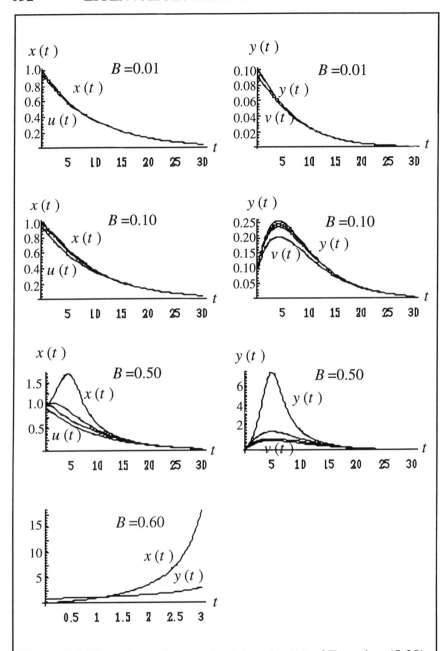

Figure 5.2 Time dependence of $x(t)$ and $y(t)$ of Equation (5.29) for several values of B, along with zero-, first- and second-order approximations. The zero-order terms are denoted by $u(t)$ and $v(t)$. The first- and second-order approximations successively approach $x(t)$ and $y(t)$.

FINITE-TIME BLOWUP

leading exponents. If this requirement is not satisfied, then the coefficient of the highest term is equal to zero and hence one does not have a possible blowup.

We neglect the terms $\xi_0 \cdot s^p$ and $\eta_0 \cdot s^q$ in Equations (5.31a) and (5.31b), respectively, since they are surely not the leading singular terms (except when $p=0$ or $q=0$ – cases that we are not interested in). Now we have to determine the leading singularities in each equation. For each choice, we obtain constraints on the integers p and q. For example, if the s^{3p} term on the r.h.s. of Equation (5.31a) is to be the leading term, then its exponent, $3p$, must be equal to $(p-1)$, the exponent on the l.h.s. of the equation, and more negative than the remaining exponent $(p+q)$: $p-1 = 3p \leq p+q$.

In Table 5.2, we enumerate the conditions that emerge in the search for a consistent set. The top row presents the constraints derived from Equation (5.31b) and the left column, those derived from Equation (5.31a). The inner entries show the results p and q when pairs of constraints are imposed simultaneously. An empty entry signifies that the constraints are inconsistent. Discarding the case with $q=0$, we see that the only set of fully determined values for the exponents is

$$p = -\tfrac{1}{2}, \qquad q = -1.$$

Table 5.2 leading singularity analysis of Equations (5.31a,b)

Equation (5.31b)	$q-1=$ $2q\leq$ $2p\leq$ $2p+q$	$q-1=$ $2q\leq$ $2p+q$ $\leq 2p$	$q-1=$ $2p\leq$ $2q\leq$ $2p+q$	$q-1=$ $2p\leq$ $2p+q$ $\leq 2q$	$q-1=$ $2p+q$ $\leq 2p$ $\leq 2q$	$q-1=$ $2p+q$ $\leq$ $2q\leq 2p$
Equation (5.31a)						
$p-1=$ $3p\leq$ $p+q$	----	$p=-\tfrac{1}{2}$ $q=-1$	----	$p=-\tfrac{1}{2}$ $q=0$	$p=-\tfrac{1}{2}$ $-1/2\leq q$ ≤ 0	$p=-\tfrac{1}{2}$ $-1\leq q \leq -\tfrac{1}{2}$
$p-1=$ $p+q$ $\leq 3p$	----	$p\geq -\tfrac{1}{2}$ $q=-1$	----	----	----	$p=-\tfrac{1}{2}$ $q=-1$

The preceding analysis only tells us how $x(t)$ and $y(t)$ behave *if* a finite-time blowup occurs. To determine if it does occur, additional aspects of the problem must be studied. To this end, we substitute in Equations (5.31a,b) the values obtained for p and q and collect all terms with the leading singular behavior as $s \to 0$. We find

$$-\tfrac{1}{2}\xi_0 s^{-3/2} \sim \lambda \xi_0 \eta_0 s^{-3/2} + \tfrac{1}{2} B \xi_0^{\,3} s^{-3/2} + \text{h.o.t.} \quad (5.32a)$$

$$-\eta_0 s^{-2} \sim 2\lambda \eta_0^{\,2} s^{-2} + B \xi_0^{\,2} \eta_0 s^{-2} + \text{h.o.t.} \quad (5.32b)$$

where h.o.t. stands for higher-order terms. (Equations (5.32a,b) follow from Equations (5.31a,b), respectively.) We now equate the coefficients of the singular terms on the l.h.s. and r.h.s. of Equations (5.32a,b). In the present problem this yields only one independent constraint, a relation between ξ_0 and η_0:

$$B\xi_0^{\,2} = 2\lambda \eta_0 - 1 \quad (5.33a)$$

In particular, for a blowup to occur one must have

$$\frac{2\lambda \eta_0 - 1}{B} > 0 \quad (5.33b)$$

The system discussed here constitutes an example for the following rule: In general, the initial conditions, and possibly other parameters of the problem, must satisfy constraints that are obtained in the same manner that led to Equations (5.33). Otherwise, although a set of consistent powers may be found for the leading singular behavior, the latter does not materialize and a finite-time blowup does not occur.

Example 5.8. Truncation can lead to an incorrect analysis. Consider a system of the type encountered in the construction of amplitude equations for parabolic partial-differential equations:

$$\frac{dx}{dt} = x\{\lambda - \varepsilon\lambda x - 2\mu(\varepsilon y)^2 - 3\sigma(\varepsilon z)^3\}$$

$$\frac{dy}{dt} = y\{\mu - \tfrac{1}{2}\varepsilon\lambda x - \mu(\varepsilon y)^2 - \tfrac{3}{2}\sigma(\varepsilon z)^3\} \quad (5.34)$$

$$\frac{dz}{dt} = z\{\sigma - \tfrac{1}{3}\varepsilon\lambda x - \tfrac{2}{3}\mu(\varepsilon y)^2 - \sigma(\varepsilon z)^3\}$$

FINITE-TIME BLOWUP

This is an important example to chew on, as it will illustrate how truncation can lead to incorrect results. The system is solved by

$$x(t) = x(0)\exp(\lambda t) D^{-1} \quad y(t) = y(0)\exp(\mu t) D^{-1/2}$$
$$z(t) = z(0)\exp(\sigma t) D^{-1/3} \quad (5.35a)$$

$$D = \begin{cases} 1 - \varepsilon x(0)[1 - \exp(\lambda t)] \\ -(\varepsilon y(0))^2 [1 - \exp(2\mu t)] - (\varepsilon z(0))^3 [1 - \exp(3\sigma t)] \end{cases} \quad (5.35b)$$

We choose the following numerical values of the parameters:

<u>Fixed points:</u> The only nonnegative fixed point is at (0,0,0).
<u>Parameter values:</u> $\varepsilon = 0.1$; $\lambda = -1$; $\mu = -2$; $\sigma = -10$.
<u>Initial conditions:</u> $x(0) = 8$; $y(0) = 4.3$; $z(0) = 2.8$.

Here, there is finite-time blowup at $t \cong 4.7600713$. This occurrence is explained as follows. With the given parameter values, we compute the values of the initial derivatives:

$$\frac{dx(0)}{dt} = 9.59 \quad \frac{dy(0)}{dt} = -3.87 \quad \frac{dz(0)}{dt} = -25.95$$

Only the x derivative is positive at $t = 0$. Precisely because it is positive, finite-time blowup is possible. If all the derivatives were negative, the system would decay to the origin.

We now truncate Equations (5.34) by setting z equal to zero. This is often done in the literature, using the argument that the linear part of the z equation has such a large negative eigenvalue [more precisely, we have a small value of $z(0)$ and a large negative derivative at $t=0$], that z must decay at an early time, and never be of any significance. We find that *the truncated system does not blow up!* Hence, it is not a good approximation to the full system for all times. Figure 5.3 presents the result for the solution of Equation (5.34) and its truncated version (with z omitted), for the parameter values and initial conditions detailed above. We see that, for a limited duration, the truncated system provides a good approximation to $x(t)$ and $y(t)$, but that it fails miserably as one approaches the blowup time.

The finite-time blowup occurs because the denominator, D, vanishes at a finite time. Naturally, one would like to know why the system is so sensitive to the omission of $z(t)$, that has such small values for most of the time. To answer this, we need to study the solution for $z(t)$ in greater detail. As long as time is not *extremely close* to the blowup

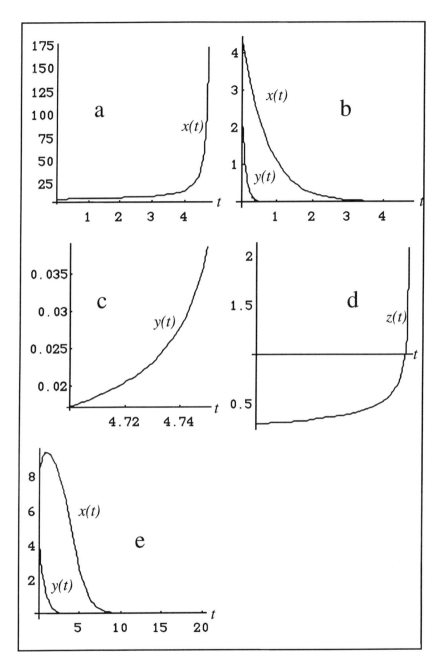

Figure 5.3 Solutions of Equations (5.34): (a) $x(t)$; (b) $y(t)$ and $z(t)$ far from blowup point; (c) $y(t)$ in the vicinity of blowup point; (d) $z(t)$ near blowup point ($t > 4.76007$); (e) $x(t)$ and $y(t)$ in truncated system – no finite-time blowup.

FINITE-TIME BLOWUP

point, the numerator in the expression for $z(t)$ decays much faster than the denominator. The latter does not become as small as the former until extremely close to the blowup time.

To see this, we compute $z(4.75)$. The denominator is small ($D=6.936\times10^{-5}$, $D^{1/3}=0.04109$), but the numerator is *much* smaller: $2.8\cdot\exp(-47.5)$ ($=6.6\times10^{-21}$). For $z(t)$ to achieve a reasonable magnitude, say, $O(1)$, the denominator D has to become comparable to $\exp(-3\times47.5)=9.6\times10^{-63}$, which is extremely small. Thus, z continues to decay until immediately before blowup, "justifying" the idea of truncation. (It seems to be lying in wait.)

When blowup occurs, it is $z(t)$ that pulls the others with it, and brings the entire system to infinity. It is a cooperative effect. If we omit any of the three variables from the system (in particular, if we omit z), *without modifying any parameter or initial condition related to the remaining variables*, the denominator D does not vanish ever. This is the reason why truncation destroys the finite-time blowup.

Moral. When truncating a system of equations, one must be extremely careful. Even if truncation is a correct procedure, one should not change any parameter values in the truncated system. One must return to an analysis of the full set of equations.

For example, in the present case, in the truncated system (i.e., if z is kept at zero), one has

$$\frac{dx(0)}{dt} = 4.3 \qquad \frac{dy(0)}{dt} = -5.29$$

Although one of the derivatives is positive at $t=0$, no blowup occurs. Thus, a positive derivative at $t=0$ is only an indication that the potential for a finite-time blowup exists. A more detailed study must be carried out to determine whether the blowup occurs.

5.5 CONCLUDING REMARKS

In Section 5.1 we derived the normal form equations for systems in which the eigenvalues lie in a Poincaré domain. We saw in Example 5.1 that terms proportional to t or powers of t appear naturally in the normal form expansion. In Chapter 4 we mentioned that the method of normal forms originated with the study of periodic motion, in which naive perturbation theory yielded spurious secular terms. The ability to eliminate these terms in the periodic case is intimately related to the occurrence of an infinite number of resonant eigenvalue combinations. In this chapter we have discussed problems in which there are at most a finite number of resonant combinations. Hence, factors of t or powers of t are inherent in the solutions and cannot be eliminated.

We emphasize once again the difference between problems in which the eigenvalues lie in a Poincaré domain as contrasted with a Siegel domain. In the former case, there is no *updating* of the eigenvalues due to the effect of the nonlinear perturbation, so that one can use the unperturbed solutions as the full generators of the normal form expansion. In the latter case, the potential for eigenvalue updating exists. Using the unperturbed solutions as the basis functions of the perturbation expansion, one encounters spurious secular terms.

REFERENCES

1. Bruno, A. D., *Local Methods in Nonlinear Differential Equations*, Springer-Verlag, New York (1989).

6

NORMAL FORM EXPANSION FOR CONSERVATIVE PLANAR SYSTEMS

6.1 THE PERTURBED SIMPLE HARMONIC OSCILLATOR

Perturbed simple harmonic oscillators occur in a wide variety of systems and display a sufficiently broad spectrum of phenomena to warrant extensive consideration. For example, equations of this type occur in models of mechanical and electric oscillations, vibrations, biological and chemical reactions, etc. The basic equation has the form

$$\frac{d^2x}{d\tau^2} + \omega_0^2 x = \varepsilon' g\left(x, \left(\frac{dx}{d\tau}\right); \varepsilon'\right) \qquad |\varepsilon'| \ll 1$$

It is useful to introduce a new time variable, $t = \omega_0 \tau$. Our equation becomes

$$\frac{d^2x}{dt^2} + x = \varepsilon f\left(x, \left(\frac{dx}{dt}\right); \varepsilon\right)$$

$$\left(\varepsilon \equiv \frac{\varepsilon'}{\omega_0^2} \quad f\left(x, \left(\frac{dx}{dt}\right); \varepsilon\right) = g\left(x, \left(\frac{dx}{d\tau}\right); \varepsilon'\right)\right)$$

The next step is to introduce $y = dx/dt$ so that our second-order equation becomes a first-order equation for a two-component vector:

$$\frac{dx}{dt} = y \qquad \frac{dy}{dt} = -x + \varepsilon f(x, y; \varepsilon)$$

The unperturbed system is called *simple harmonic* or a *simple harmonic oscillator* system. [The potential energy is $V(x) = \tfrac{1}{2}x^2$.] The period of the motion is independent of the amplitude, and the motion is

159

described by a single harmonic term. If the added perturbation is a function of x only, the perturbed system is *conservative*, and energy is conserved. The motion has a well-defined period. In the phase plane it follows a closed curve, that, for a small perturbation, is "close to a circle." The solution is often given as an expansion in harmonics. The strength of the nonlinearity affects the size of the harmonic components.

There is a marked difference between linear and nonlinear systems regarding the determination of the coefficients of the Fourier expansion. In *linear perturbed systems,* the Fourier expansion yields a final determination of the coefficients. That is, the numerical value of any component of the harmonic expansion is determined once and for all at a particular order in the expansion. For example, say that one has

$$x(t) = A_1 \cos(t) + A_2 \cos(2t) + A_3 \cos(3t) + \cdots$$

where the coefficients A_1, A_2, and A_3 have been found. Finding the next coefficient, A_4, is accomplished without altering the previously determined coefficients. However, in the case of *nonlinear perturbed systems,* a harmonic expansion coupled with a perturbation expansion, reveals that the coefficients are generally *updated* or *redetermined* in each order of the expansion. This aspect of perturbation schemes, associated with nonlinear systems, will become clear in the analysis.

We outline the application of the method of normal forms to the study of systems in which the unperturbed component is simple harmonic, that are perturbed by weak polynomial nonlinearities (or functions that have a Taylor series expansion in their variables). We first analyze the Duffing oscillator, a particularly simple conservative system, describing motion in a quartic potential (i.e., symmetric about $x=0$). The solution to this problem can be expressed in terms of elliptic functions. In our discussion, however, we ignore this important point and proceed as if the problem were amenable only to a perturbation analysis, using a power series in a small parameter. We then apply the method to more general perturbations.

6.2 THE DUFFING OSCILLATOR

To provide a complete treatment of the problem, we review the analysis of the Duffing oscillator:

$$\frac{d^2x}{dt^2} + x = -\varepsilon x^3 \qquad x(0) = A \qquad \frac{dx(0)}{dt} = 0 \qquad (6.1a)$$

THE DUFFING OSCILLATOR

which has been discussed extensively in previous chapters.

The normal form analysis is best carried out in the diagonalized (complex) notation:

$$\frac{dz}{dt} = -iz - i\,\varepsilon\left(\tfrac{1}{8}\right)\left(z + z^*\right)^3 \qquad \left(z \equiv x + i\,\frac{dx}{dt}\right) \qquad (6.1b)$$

with the complex conjugate equation satisfied by $z^* = [x - i(dx/dt)]$. At times, it is convenient to write Equation (6.1b) in the form

$$\dot{z} = \sum_{n \geq 0} \varepsilon^n Z_n(z, z^*)$$

which, for the Duffing oscillator, corresponds to

$$Z_0 = -i\,z \qquad Z_1 = -i\left(\tfrac{1}{8}\right)\left(z + z^*\right)^3 \qquad (6.1c)$$

Comments. As the system is conservative, the motion is periodic (the phase-plane orbit is closed). Moreover, E (the total energy), A (the amplitude at the turning point), and T (the period) are all related. Also, given the total energy, the position and velocity at each moment are related to one another and to E. Without loss of generality, we may study the motion starting at the turning point, $x(0)=A$. As this is an autonomous system, the displacement, $x(t)$, starting with any initial condition, is related to the displacement, obtained from any other initial condition, by a simple change in the initial phase.

We wish to calculate the displacement as a function of time and the frequency or period. Both quantities are expressed as power series in ε. We agree in advance on the number of terms to be included in the expansion. The frequency can be calculated to any desired accuracy by solving a definite integral. However, often one obtains an expression for the frequency as a power series that is computed in parallel with the computation of $x(t)$. The benefits of performing a separate calculation for the frequency will be discussed in the general discussion that will also include a comparison with the exact solution. (A general perturbative method for finding the frequency to any desired accuracy is presented in the Appendix.)

6.2.1 Normal form expansion

The eigenvalues of the unperturbed system, $\pm i$, lie in a Siegel domain. There is potentially an eigenvalue update (hence a frequency update) in each order of the expansion. We will show that, for the Duffing

162   NORMAL FORMS: CONSERVATIVE PLANAR SYSTEMS

oscillator, it is possible to choose the free functions so that the frequency in only updated in first order and the resulting expression for the displacement is, especially in lower orders, more accurate than that obtained from other choices of the free functions.

We compute the first few orders in the normal form expansion as developed in Chapter 4. Our first objective is to obtain a purely resonant U_1. To accomplish this, one constructs a T_1 that kills all the nonresonant terms. One adjoins to T_1 a free function composed of resonant terms only (the Lie bracket of the latter vanishes). In particular, and without loss of generality, we can choose the free term to be of the form $\alpha u^2 u^*$. In second order, one constructs a T_2 that yields a purely resonant U_2. There is a contribution in U_2 from the free term included in T_1. Also, T_2 has a free term, that will affect, for the first time, U_3. The pattern persists in higher orders. We obtain

$$U_1 = -\tfrac{3}{8} i\, u^2 u^* \tag{6.2a}$$

$$\begin{aligned} T_1 &= \tfrac{1}{16} u^3 - \tfrac{3}{16} u u^{*2} - \tfrac{1}{32} u^{*3} + \alpha u^2 u^* \\[4pt] U_2 &= \left(\tfrac{51}{256}\right) i\, u^3 u^{*2} - \left(\tfrac{3}{8}\right) i\, (\alpha + \alpha^*) u^3 u^{*2} \\[4pt] T_2 &= \tfrac{3}{1024} u^5 + \left(-\tfrac{15}{256} + \tfrac{3}{16}\alpha\right) u^4 u^* \\[4pt] &\quad + \left\{\tfrac{69}{512} - \tfrac{3}{8}\left(\tfrac{1}{2}\alpha + \alpha^*\right)\right\} u^2 u^{*3} \\[4pt] &\quad + \left(\tfrac{21}{1024} - \tfrac{3}{32}\alpha^*\right) u u^{*4} - \tfrac{1}{512} u^{*5} + \beta u^3 u^{*2} \end{aligned} \tag{6.2b}$$

Continuing the calculation we find in third order

$$\begin{aligned} T_3 &= \tfrac{1}{8192} u^7 + \left(-\tfrac{189}{32{,}768} + \tfrac{15}{1024}\alpha\right) u^6 u^* \\[4pt] &\quad + \left\{\tfrac{1065}{16{,}384} - \tfrac{15}{64}(\alpha + \tfrac{1}{4}\alpha^*) + \tfrac{3}{16}\alpha^2 + \tfrac{3}{16}\beta\right\} u^5 u^{*2} \\[4pt] &\quad + \left\{ \begin{aligned} &-\tfrac{2139}{16{,}384} + \tfrac{69}{256}\alpha + \tfrac{207}{512}\alpha^* \\ &-\tfrac{3}{16}(2\alpha\alpha^* + \alpha^{*2}) - \tfrac{3}{16}(\beta + 2\beta^*) \end{aligned} \right\} u^3 u^{*4} \\[4pt] &\quad + \left\{-\tfrac{561}{32{,}768} + \tfrac{21}{256}\left(\tfrac{1}{4}\alpha + \alpha^*\right) - \tfrac{3}{32}\alpha^{*2} - \tfrac{3}{32}\beta^*\right\} u^2 u^{*5} \\[4pt] &\quad + \left(\tfrac{29}{8192} - \tfrac{5}{512}\alpha^*\right) u u^{*6} - \tfrac{3}{32{,}768} u^{*7} + \gamma u^4 u^{*3} \end{aligned} \tag{6.2c}$$

$$U_3 = \left\{-\tfrac{1419}{8192} + \tfrac{51}{128}(\alpha + \alpha^*) - \tfrac{3}{8}\alpha\alpha^* - \tfrac{3}{8}(\beta + \beta^*)\right\} i\, u^4 u^{*3} \tag{6.2d}$$

THE DUFFING OSCILLATOR

One also needs the generators for u^*, T^*_j, $j = 1,...,n$. As the equation for z^* is just the complex conjugate of the equation for z:, one has

$$[T^*]_j = [T_j]^* \tag{6.2e}$$

6.2.2 Further discussion

From Equations (6.2a–e), we see that the coefficients in U_i, $i=0,1,2,3$ are pure imaginary, a consequence of the conservative nature of the system. From this it follows that u executes circular motion in the phase plane. Additional aspects of the z and, consequently, the x motion, particularly the harmonic structure, are determined by the T functions.

The basic elements of the method are clear already at this order. The implementation of higher-order corrections is straightforward. The general expression will include sines and cosines of θ, 3θ, 5θ, and 7θ with combinations of the free coefficients α, β, γ, α^*, β^*, γ^*, allowing for the most general initial condition. For the sake of simplicity, we show here the solution for real α, β, γ, corresponding to a zero initial velocity $dx/dt(0)=0$. Using Equation (6.2a) and writing

$$x = \tfrac{1}{2}(z + z^*), \qquad u = \rho \exp(-i\theta)$$

we have

$$\begin{aligned}
x(t) &= \left[\rho + \varepsilon\left(-\tfrac{3}{16} + \alpha\right)\rho^3 + \varepsilon^2\left(\tfrac{69}{512} - \tfrac{9}{16}\alpha + \beta\right)\rho^5 \right.\\
&\qquad \left. + \varepsilon^3\left(-\tfrac{2139}{16{,}384} + \tfrac{345}{512}\alpha - \tfrac{9}{16}\alpha^2 - \tfrac{9}{16}\beta + \gamma\right)\rho^7\right]\cos\theta \\
&\quad + \left[\varepsilon\tfrac{1}{32}\rho^3 + \varepsilon^2\left(-\tfrac{39}{1024} + \tfrac{3}{32}\alpha\right)\rho^5 \right.\\
&\qquad \left. + \varepsilon^3\left(\tfrac{2569}{32{,}768} - \tfrac{195}{1024}\alpha + \tfrac{3}{32}\alpha^2 + \tfrac{3}{32}\beta\right)\rho^7\right]\cos 3\theta \\
&\quad + \left[\varepsilon^2\tfrac{1}{1024}\rho^5 + \varepsilon^3\left(-\tfrac{73}{32{,}768} + \tfrac{5}{1024}\alpha\right)\rho^7\right]\cos 5\theta \\
&\quad + \varepsilon^3\tfrac{1}{32{,}768}\rho^7\cos 7\theta
\end{aligned} \tag{6.3}$$

We note that the numerical coefficients in U_n are pure imaginary for all n. (We have shown this explicitly through third order, but one can prove it for all orders by induction.) Therefore, the normal form has the following simple structure (here g is a real function)

$$\dot{u} = -i\, u g(uu^*;\varepsilon)$$

With

$$u = \rho \exp(-i\theta)$$

it is easy to show that

$$\frac{d\rho}{dt} = 0 \tag{6.4a}$$

Namely, the radius, ρ, is constant. Hence, the U functions affect only the frequency, ω, which is updated in each order of the expansion. Using the expressions for U_i, $i \le 3$, one writes

$$\frac{d\theta}{dt} = \omega = 1 + \varepsilon \tfrac{3}{8}\rho^2 - \varepsilon^2 \left\{ \tfrac{51}{256} - \tfrac{3}{8}(\alpha + \alpha^*) \right\} \rho^4$$
$$- \varepsilon^3 \left\{ -\tfrac{1419}{8192} + \tfrac{51}{128}(\alpha + \alpha^*) - \tfrac{3}{8}\alpha\alpha^* - \tfrac{3}{8}(\beta + \beta^*) \right\} \rho^6 + O(\varepsilon^4) \tag{6.4b}$$

Choosing the coefficients α and β real, one has

$$\omega = 1 + \tfrac{3}{8}\varepsilon\rho^2 - \tfrac{3}{4}\varepsilon^2 \left(\tfrac{17}{64} - \alpha \right) \rho^4$$
$$- \tfrac{3}{4}\varepsilon^3 \left(-\tfrac{473}{2048} + \tfrac{68}{64}\alpha - \tfrac{1}{2}\alpha^2 - \beta \right) \rho^6 + O(\varepsilon^4) \tag{6.4c}$$

With the choice

$$\alpha = \tfrac{17}{64} \qquad \beta = \tfrac{131}{8192}$$

one obtains $U_2 = U_3 = 0$; hence the frequency is updated only in first order. Furthermore, one can show that the free functions, introduced in each order, can be used to kill all higher-order U s [1,2]. Thus, the equation for the zero-order approximation is determined in first order, i.e., functionally the frequency is only updated at $O(\varepsilon)$. We call the resulting u equation the *minimal normal form* (MNF).

Comment: Calculation of the period. The period can be computed by evaluating a definite integral over one cycle in phase space:

THE DUFFING OSCILLATOR

$$T = \oint \frac{dx}{\sqrt{2(E - V(x))}} \quad (6.5a)$$

where E is the total energy and $V(x)$ is the potential. This, obviously, determines the frequency. Inserting the potential and the total energy

$$V(x) = \tfrac{1}{2}x^2 + \tfrac{1}{4}x^4 \qquad E = \tfrac{1}{2}A^2 + \tfrac{1}{4}A^4$$

where A is the turning point, the integral becomes

$$T = 4 \int_0^A \frac{dx}{\sqrt{(A^2 - x^2) + \tfrac{1}{2}\varepsilon(A^4 - x^4)}} \quad (6.5b)$$

In the case of the Duffing oscillator, this integral can be converted to a complete elliptic integral of the first kind. Evaluating the integral thorough $O(\varepsilon^4)$, the frequency $\omega = (2\pi/T)$ is written as

$$\omega = 1 + \varepsilon \tfrac{3}{8} A^2 - \varepsilon^2 \tfrac{21}{256} A^4 + \varepsilon^3 \tfrac{81}{2048} A^6 - \varepsilon^4 \tfrac{6549}{262,144} A^8 + O(\varepsilon^5) \quad (6.6)$$

In Equation (6.6), ω is expressed in terms of A, the amplitude at the turning point, while in Equation (6.4c), it is expressed in terms of ρ, the radius of the zero-order approximation. To properly compare the two expressions, we need to find the relationships among ρ, the initial condition $x(0)$ and A. One obtains this relation through the implementation of the initial conditions. We will see, in Section 6.2.3, how various choices of the free functions affect this relationship as well as how the updating of the phase is accomplished. [For example, Equations (6.4b) and (6.6) become identical for $\alpha = \tfrac{5}{32}$.]

Comment. Let us say that we are interested in finding the frequency within a certain level of precision, for example, through second order. There is no problem in achieving this goal through Equation (6.6), as A is a given number. However, if we choose to use Equation (6.4c), where ω is expressed in terms of ρ, we must first evaluate ρ to an appropriate level of precision. In equation (6.4c), ρ first appears in the combination $\varepsilon\rho^3$, it is sufficient to know ρ through an $O(\varepsilon)$ precision [i.e., allowing for an $O(\varepsilon^2)$ error in ρ]. Therefore, it is sufficient to compute $z(0)$ [or, equivalently, $x(0)$ and $dx/dt(0)$] through $O(\varepsilon)$. Thus, although T_2 is required in order to eliminate all potential nonresonant contributions in the expression for U_2, at the specified level of precision of $\omega\ O((\varepsilon^2))$, there is no need to compute T_2 explicitly.

Comment. Today, with powerful computers, one can compute the period of oscillations of a planar conservative system to any desired accuracy from Equation (6.5a). Why, then, bother at all to develop the formalism for the expansion of the frequency in a perturbation theory? First, the perturbation expansion is an alternative to the integration procedure (which has subtle problems of its own, such as integration near the turning points). Second, and probably more important, the normal form expansion provides us with a systematic expansion procedure that reflects the essence of the effect of the nonlinearity on the solution. While the T_n are, essentially, higher-harmonics corrections to the time dependence of the solution, the normal form provides the frequency update. Third, one should view the analysis of conservative two-dimensional systems as preparation for the study of more complicated problems, e.g., of higher-dimensional systems and systems with dissipation. In particular, the simplicity of the analysis of conservative two-dimensional systems enables us to point out an important element in the theory: the freedom inherent in the expansion, that will play an important role in the analysis of more complex systems. In particular, the choice of *minimal normal forms* yields approximations that seem to be valid for timespans that are longer than those attained in the usual perturbation expansion.

6.2.3 Choice of the free functions

We discuss five common choices for the free functions and give numerical comparisons for specific values of the parameter and specific initial conditions. For the sake of simplicity, we assume that the coefficients $\alpha, \beta, \gamma, \ldots$ in the free terms are real. We begin by determining the values of the various free functions. Refer to Equations (6.2a–d).

Case a: Usual choice. No free functions. In Equations (6.2a–d), the coefficients of all the free terms are set to zero, so that no terms with a vanishing Lie bracket are included in T_n. It is the choice most often made in the literature. In the case of the Duffing equation, this leads to *all U_i being $\neq 0$.* Therefore, the frequency and hence the phase are updated in each order of the expansion. (In the general case, this choice will typically lead to the same result.)

Case b: Canonical transformations. In the study of Hamiltonian systems, one tends to restrict oneself to canonical near-identity transformations. This is the common practice in the literature. From

THE DUFFING OSCILLATOR

our point of view, requiring that the transformation is canonical amounts to just one specific choice of the free functions in each order.

For a canonical transformation, the determinant of the Jacobian matrix of the transformation is unity. The condition is concise in the complex variable notation. Thus if we write

$$z = u + \sum_{n \geq 1} \varepsilon^n T_n \qquad z^* = u^* + \sum_{n \geq 1} \varepsilon^n T_n^*$$

the requirement that the transformation be canonical takes the form

$$\det \begin{bmatrix} \dfrac{\partial z}{\partial u} & \dfrac{\partial z}{\partial u^*} \\ \dfrac{\partial z^*}{\partial u} & \dfrac{\partial z^*}{\partial u^*} \end{bmatrix} = \det \begin{bmatrix} 1 + \sum_{n \geq 1} \dfrac{\varepsilon^n \partial T_n}{\partial u} & \sum_{n \geq 1} \dfrac{\varepsilon^n \partial T_n}{\partial u^*} \\ \sum_n \dfrac{\varepsilon^n \partial T_n^*}{\partial u} & 1 + \sum_n \dfrac{\varepsilon^n \partial T_n^*}{\partial u^*} \end{bmatrix} = 1$$

In first order the requirement takes the form

$$\frac{\partial T_1}{\partial u} = -\frac{\partial T_1^*}{\partial u^*} \tag{6.7a}$$

Referring to Equation (6.2a), we find that the free function should obey

$$\alpha + \alpha^* = 0$$

and for real α, this implies $\alpha = 0$.

In nth order, the requirement that the transformation is canonical yields

$$\frac{\partial T_n}{\partial u} + \frac{\partial T_n^*}{\partial u^*} + \sum_{p=1}^{n-1} \left[\frac{\partial T_p}{\partial u} \frac{\partial T_{n-p}^*}{\partial u^*} - \frac{\partial T_p}{\partial u^*} \frac{\partial T_{n-p}^*}{\partial u} \right] = 0 \tag{6.7b}$$

Employing Equations (6.2a–d), we choose the free terms to have real coefficients and denote them as f_k, $k = 1,2,3,4,5$, where the index k denotes the order of the expansion. We find

$$f_1 = 0 \qquad f_2 = \tfrac{27}{2048} \qquad f_3 = -\tfrac{531}{32,768} \qquad f_4 = \tfrac{162,755}{8,388,608} \qquad f_5 = -\tfrac{1,604,347}{67,108,864}$$

etc. Note that in the case of a canonical near-identity transformation the

168 NORMAL FORMS: CONSERVATIVE PLANAR SYSTEMS

values of the coefficients do not depend on the initial conditions.

Case c: ***"Zero–zero" initial conditions.*** In the literature, it is also common to require that the initial conditions be satisfied by the zero-order approximation. This means that all T_i ($i \geq 1$) must vanish at $t=0$. That is, both their real and imaginary parts (their contributions to $x(t)$ and dx/dt, respectively) must vanish at $t=0$. This is why we call this the case of *zero–zero* initial conditions. At at $t=0$, we thus have

$$x(0) = \tfrac{1}{2}\left(z(0) + z^*(0)\right) = \tfrac{1}{2}\left(u(0) + u^*(0)\right)$$

$$\dot{x}(0) = \dot{x}(0) = \tfrac{1}{2i}\left(z(0) - z^*(0)\right) = \tfrac{1}{2i}\left(u(0) - u^*(0)\right)$$

Clearly, in this case, coefficients in the free functions depend on the initial conditions. For the sake of simplicity, we assume that the initial condition is at the turning point, $(A,0)$, i.e., with a vanishing velocity. Referring to Equations (6.2a–d) we have, through fifth order

$$T_1(t=0) = 0 \Rightarrow \alpha = \alpha^* \equiv f_1 = \tfrac{5}{32}$$

$$T_2(t=0) = 0 \Rightarrow \beta = \beta^* \equiv f_2 = -\tfrac{25}{1024}$$

$$T_3(t=0) = 0 \Rightarrow \gamma = \gamma^* \equiv f_3 = \tfrac{281}{32,768}$$

$$f_4 = -\tfrac{571}{131,072}, \qquad f_5 = \tfrac{23,563}{8,388,608}$$

In summary, when the full solution obeys the $(A,0)$ initial condition, this choice of the free functions equates the radius of the zero-order approximation, $u(0)$, with the maximum amplitude, $x(0)=A$. The expression for the period becomes identical with Equation (6.6).

Case d: ***Minimal normal forms.*** We now choose the free functions so that the u equation is terminated after U_1. Referring to Equations (6.2a–d), we find that to achieve $U_i = 0$, $i = 2\text{–}6$, we need

$$\alpha = \alpha^* = f_1 = \tfrac{17}{64} \qquad \beta = \beta^* = f_2 = \tfrac{131}{8192}$$

$$\gamma = \gamma^* = f_3 = -\tfrac{667}{524,288} \qquad f_4 = \tfrac{22,691}{134,217,728}$$

$$f_5 = \tfrac{274,383}{8,589,934,592}$$

THE DUFFING OSCILLATOR

Comment. Referring to Equation (6.4b), we observe that in the MNF, and only in this choice of the free functions, the frequency is determined entirely by its first-order expression:

$$\omega = 1 + \tfrac{3}{8}\varepsilon\rho^2 \qquad (6.8)$$

As we have argued, the frequency can be obtained by the evaluation of the integral in Equation (6.5b) that relates the period to the potential to any desired accuracy. Since the potential is symmetric, it can be shown that this relationship is unique [3]. Thus, in the MNF case, ρ is determined by the known value of ω and the MNF expression for ω.

Comment. It is most convenient to choose the coefficients of the free terms to be real since a choice of complex coefficients would lead to "spurious" dissipation terms.

One now proves by induction that, for any $n > 1$, the free terms in the T functions can be chosen so that $U_n = 0$. First, we note that the relative simplicity of the nonlinear interaction enables us to write expression for U_{n+1} explicitly:

$$U_{n+1} = [Z_0, T_{n+1}]$$

$$+ (T_n + T_n^*)\frac{\partial Z_1}{\partial u} + \left[\begin{array}{l}(T_{n-1} + T_{n-1}^*)(T_1 + T_1^*) \\ + (T_{n-2} + T_{n-2}^*)(T_2 + T_2^*)\end{array}\right]\frac{\partial^2 Z_1}{\partial^2 u}$$

$$+ \tfrac{1}{2}(T_{n-2} + T_{n-2}^*)(T_1 + T_1^*)^2\frac{\partial^3 Z_1}{\partial^3 u} \qquad (6.9a)$$

$$- \sum_{k=1}^{n}\left\{U_k \frac{\partial T_{n+1-k}}{\partial u} + U_k^* \frac{\partial T_{n+1-k}}{\partial u^*}\right\}$$

At this stage, all the U_k and T_k, $k \leq n$ have been computed in previous orders, including the free functions in T_k, $k \leq n-1$, but not in T_n. The latter has been computed in the nth-order calculation, except for its free function, which is still undetermined.

We note that both Z_0 and Z_1 are constructed from monomials in u and u^* with imaginary coefficients. Limiting ourselves to free resonant functions in T_n that are monomials with real coefficients, we now make the induction assumption.

1. All T_k, $1 \le k \le n$, are polynomials of degree $2k+1$ in u and u^* with real coefficients.
2. For all $1 \le k \le n$, $U_k = 0$.

As a result, all computable terms in Equation (6.9a) (second and third lines) are polynomials in u and u^* of degree $2n+1$ with purely imaginary coefficients. Therefore, T_{n+1} can be computed (except for its free function, which does not contribute to the equation and will be determined in the next order, $n+2$). From the definition of the Lie bracket, we conclude that it will be a polynomial in u and u^* of degree $2n+3$ with purely real coefficients.

Once T_{n+1} has been computed, the only terms left on the r.h.s. of Equation (6.9a) are as follows: a resonant monomial of degree $2n+3$ and with a purely imaginary coefficient that we denote by $i\alpha_n$, and the resonant contribution the contribution of the term involving the yet unknown free function in T_n. Denote that free function by $F_n(uu^*) \cdot u$. Equation (6.9a) yields

$$U_{n+1} = -\tfrac{3}{8} i \left(F_n + F_n^* \right) u^2 u^* + i \, \alpha_n u^{n+2} u^{*n+1} \qquad (6.9b)$$

Hence, with the choice (we have limited ourselves to free functions with real coefficients)

$$F_n = \tfrac{4}{3} \alpha_n u^n u^{*n}$$

we can make U_{n+1} vanish as well. Note that this choice of $F_n(uu^*)$ preserves the structure of T_n as a polynomial of degree $2n+1$ with purely real coefficients. This completes the proof by induction.

Comment. Without free functions, for each n, T_n are polynomials of degree $2n+1$ in u and u^* with purely real coefficients and all U_n are monomials of the same degree with purely imaginary coefficients.

Case e: Killing the fundamental. The free functions offer a broader range of options than we have presented here. For example, it may be useful to develop an expansion in which the fundamental $\cos(\theta)$ component, which first appears in the zero-order term, is not updated in higher orders (namely, it does not appear in higher orders). This choice makes the higher-order corrections *orthogonal* to the fundamental component, a feature that some investigators find advantageous. Thus, referring to Equation (6.3), this is accomplished by the real coefficients

THE DUFFING OSCILLATOR

$$\alpha = \tfrac{3}{16} \quad \beta = -\tfrac{15}{512} \quad \gamma = -\tfrac{123}{16{,}384} \quad \text{etc.}$$

6.2.4 Implementation of initial conditions: Equivalence of different choices of free functions

Keep in mind that just because you have solved some equation and obtained a "result" doesn't mean that you can immediately compare it with an expression found by a different perturbation procedure or an expansion that uses a different zero-order approximation.

Referring to Equation (6.3), we note that at $\theta=0$ (turning point) the displacement, $x(0)$ obtains its maximal value, A, and is related to the radius of the zero-order term, ρ, through $O(\varepsilon^3)$ by

$$x(0) = \rho + \left(-\tfrac{5}{32} + \alpha\right)\varepsilon\rho^3 + \left(\tfrac{25}{256} - \tfrac{15}{32}\alpha + \beta\right)\varepsilon^2\rho^5 \\ + \left(-\tfrac{1781}{32{,}768} + \tfrac{125}{256}\alpha - \tfrac{15}{32}\alpha^2 - \tfrac{15}{32}\beta + \gamma\right)\varepsilon^3\rho^7 \quad (6.10)$$

Thus, different choices of the free coefficients will result in different values for ρ, given the initial value $x(0)$ for the full solution.

However, physical quantities are not affected by the choice of the free coefficients. We will show this through second order. Inverting Equation (6.10), we obtain for ρ the expansion through $O(\varepsilon^2)$:

$$\rho = x(0) + \varepsilon\left(\tfrac{5}{32} - \alpha\right)x(0)^3 + \varepsilon^2\left(-\tfrac{25}{1024} - \tfrac{15}{32}\alpha + 3\alpha^2 - \beta\right)x(0)^5 \\ + O(\varepsilon^3) \quad (6.11)$$

Inserting Equation (6.11) in the general expansion of the frequency, Equation (6.4c), the expression obtained for ω is

$$\omega = 1 + \varepsilon\tfrac{3}{8}x(0)^2 - \varepsilon^2\tfrac{21}{256}x(0)^4 + O(\varepsilon^3)$$

which, through second order, is identical to Equation (6.6).

Thus, different choices of the free functions yield different expansions for the solution through the order for which the calculation is carried out. However, the computation of measurable quantities is not affected by this choice through the same order.

6.2.5 Numerical comparisons

To illustrate various aspects associated with the implementation of the initial conditions, we consider a specific example discussed by

172 NORMAL FORMS: CONSERVATIVE PLANAR SYSTEMS

Davis [4], pages 291-297, and by Helleman and Montroll [5]. To enable us to make comparisons with the "exact" results of Refs. 4 and 5, we choose their initial conditions and parameter values:

$$x(0) = \frac{\pi}{3} \qquad \frac{dx}{dt}(0) = 0 \qquad \varepsilon = -\frac{1}{6}$$

With this information, we can find the frequency precisely by evaluating the integral given by Equation (6.5b). The angular frequency, correct to six decimal places, is $\omega = 0.928447$. We use the method of normal forms to obtain an approximation for the displacement, $x(t)$ through *second order* in ε.

Comment. In our analysis we assume that the frequency is known to any desired accuracy and do not evaluate it *approximately* by the perturbation expansion. This method is convenient for this example, but may not be so in a more general case.

Comment. This is a good point to mention again that secular terms are artifacts of the expansion procedure. Using the unperturbed frequency in a perturbation expansion yields secular terms, starting in first order. If we update the frequency through $O(\varepsilon^n)$ in the expansion procedure, then secular terms will begin to appear in $O(\varepsilon^{n+1})$. The secular terms, being proportional to t, grow indefinitely in time.

If the other hand, we use an approximate value for the corrected frequency in a numerical calculation of the solution, then the computed solution will deviate from the exact one due to the imprecision in the frequency. The deviation will be linear in time for short times (just as with secular terms) but may change in character for longer times. For example, in periodic problems, if one computes the solution by a Fourier expansion with an approximate value for the frequency, the deviation will be bounded for long times because all terms in the expansion are periodic functions that are bounded.

We now assess the quality of the approximation for the free functions associated with cases a–e, of Section 6.2.3. In cases a–c, we obtain the relationship between the radius of the zero-order approximation, ρ, and the maximum displacement, $x(0)=A$, by implementing the initial conditions. Using Equation (6.10) only through second order, we obtain

$$x(0) = \rho - \tfrac{1}{6}\left(-\tfrac{5}{32} + \alpha\right)\rho^3 + \tfrac{1}{36}\left(\tfrac{25}{256} - \tfrac{15}{32}\alpha + \beta\right)\rho^5 \qquad (6.12)$$

On inversion this yields

THE DUFFING OSCILLATOR

$$\rho = x(0) + \tfrac{1}{6}\left(\alpha - \tfrac{5}{32}\right)x(0)^3 + \tfrac{1}{36}\left(3\alpha^2 - \beta - \tfrac{15}{32}\alpha - \tfrac{25}{1024}\right)x(0)^5 \quad (6.13)$$

This relationship is "updated" in each order as we continue the perturbation expansion. Now we can write explicit expressions for T_1 and T_2, using the now known value of ρ.

In MNF, it is better to obtain the relationship between the radius of the zero-order approximation and the frequency by Equation (6.8):

$$\omega = 1 + \tfrac{3}{8}\varepsilon\rho^2 = 1 - \tfrac{1}{16}\rho^2 \quad (6.14)$$

Since we know $\omega = 0.928447189...$, we have $\rho = 1.069988$, fixed at this value for *all orders* of the perturbation. If we were to fix ρ by the initial conditions, we would need to update it in each order and would not be using the full power of MNF. In this procedure, ρ, as obtained from Equation (6.14), causes a slight mismatch in the relationship between ρ and $x(0)$ as given by the initial conditions [Equation (6.12)]. However, a detailed check shows that the mismatch is only in $O(\varepsilon^3)$.

Case a. The usual choice: $\alpha=\beta=0$. Equation (6.3) is written as

$$x_u(t) = \left[\rho_u + \tfrac{1}{32}\rho_u^3 + \tfrac{23}{6144}\rho_u^5\right]\cos\theta \\ - \left[\tfrac{1}{192}\rho_u^3 + \tfrac{13}{122,884}\rho_u^5\right]\cos 3\theta + \tfrac{1}{36,864}\rho_u^5 \cos 5\theta \quad (6.15)$$

The subscript "*u*" signifies that this is the *usual choice*.

We need to find the relationship between ρ_u and $x(0)=\pi/3$. A first-order correction yields $\rho_u = 1.01729$; a second-order correction, $\rho_u = 1.01644$. Thus, the radius of the zero-order term needs to be updated as one includes higher-order terms in the computation.

Case b. Canonical transformation: $\alpha=0$, $\beta=\tfrac{27}{2048}$. Here the subscript *c* stands for *canonical*. Here we find $O(\varepsilon)$: $\rho_c=1.01729$; $O(\varepsilon^2)$: $\rho_c=1.01598$; $O(\varepsilon^3)$: $\rho_c=1.01600$.

Case c. (0–0) initial conditions: $\alpha=\tfrac{5}{32}$, $\beta=-\tfrac{25}{1024}$. The radius is fixed by the initial conditions: $\rho_{0-0}=\pi/3=1.0472$.

Case d. Minimal normal forms: $\alpha=\tfrac{17}{64}$, $\beta=\tfrac{131}{18,192}$, and $\rho_{mnf}=1.06998$. The MNF choice of coefficients leads to a different way of relating the maximum displacement and the radius, ρ, of the fundamental. In cases a and b, as we go to higher orders, we

174   NORMAL FORMS: CONSERVATIVE PLANAR SYSTEMS

need to recompute this relationship. (If we wish to improve on our expression for the frequency in case c, we need to continue the expansion.) Here we use the free functions to "kill" all corrections to the frequency. Thus, the value of the radius is uniquely determined, not by the initial condition, but from the relationship between the exact frequency and the radius [Equation (6.14)]:

$$\omega = 1 - (\tfrac{1}{6})(\tfrac{3}{8})\rho_{mnf}^2, \quad \Rightarrow \quad \rho_{mnf} = 1.06997570$$

Note that this result yields a slight mismatch in the initial condition. Rather than being exactly $(\pi/3)$ through $O(\varepsilon)$, Equation (6.13) yields

$$x(0) = \tfrac{1}{3}\pi + 0.00047094$$

Also, as we have truncated the Fourier expansion in this approximation to

$$x(t) = 1.0540\cos\theta - 0.00638\cos 3\theta$$

energy is not constant as the approximate solution traverses the orbit. For example, it yields $E(t=0)=0.4985637$; $E(t=\pi/2)=0.48768002$. These are all higher-order effects, which are neglected once we have decided that the $O(\varepsilon)$ approximation satisfies our requirements.

This poblem is not particular to MNF. It occurs in any other choice of the free functions, once we have specified the precision level of the approximation. In particular, aapproximating the solution within ε precision, allows for a variable $O(\varepsilon^2)$ error in the energy.

Case e. Killing the fundamental: $\alpha = \tfrac{3}{16}$, $\beta = -\tfrac{15}{512}$. Here we find $O(\varepsilon)$: $\rho_{kf}=1.0532$; $O(\varepsilon^2)$: $\rho_{kf}=1.0540$.

6.2.6 Quantitative analysis

At this point, consider Equation (6.3) through $O(\varepsilon^2)$. In Table 6.1, we compare the exact values of the coefficients of the harmonic expansion for the solution of the Duffing equation with values obtained in perturbation calculations, for different choices of the free functions. From the table, it is clear that in the calculations carried out through $O(\varepsilon^2)$, the minimal normal forms choice yields the best agreement. The fundamental component is approximated in an excellent manner already in the first order, leaving the higher orders to improve the quality of the harmonic content of the solution.

THE DUFFING OSCILLATOR

Table 6.1 Comparison of coefficients of terms through $O(\varepsilon^2)$ in $x(t)$ as given by Equation (6.3)

	$\cos(\theta)$	$\cos(3\theta)$	$\cos(5\theta)$
Exact result	1.05409	−0.00694	0.000046
Usual			
1st order	1.05019	−0.00548	-----
2nd order	1.05332	−0.00662	0.000029
Canonical			
1st order	1.05019	−0.00548	-----
2nd order	1.05320	−0.00661	0.000029
Zero-zero			
1st order	1.05318	−0.00598	-----
2nd order	1.05396	−0.00680	0.000034
Minimal normal form			
1st order	1.05403	−0.00638	-----
2nd order	1.05408	−0.00689	0.000038
Killing the fundamental			
1st order	1.05318	−0.00608	-----
2nd order	1.05396	−0.00684	0.000035

176 NORMAL FORMS: CONSERVATIVE PLANAR SYSTEMS

Further discussion. A simple test of the quality of a perturbation expansion, is the degree to which the approximation deviates from the full solution. Our "mismatch test" studies the difference between the exact solution and the approximation through a certain order in ε:

$$m \equiv x_{exact}(t) - x_{approx}(t)$$

In Fig. 6.1 we plot the "mismatch" m for the case of the Duffing equation and approximations through first, second, and third order that use five different choices of the free functions. (For details, see Ref. 6). The minimal normal form approximation is much better than all other choices already in first order. It is also a better approximation in second order. Only in third order do some of the other choices begin to be comparable in quality to the minimal normal form approximation.

Comment. What have we accomplished? Given Equation (6.1b), we have introduced the near-identity transformation and constructed the T and U functions. We have an eigenvalue update. Thus, the normal form u equation *must* contain terms beyond U_0 (at least U_1) to have the capbility to generate the frequency update. Exploiting the free functions, one is able to terminate the U expansion and obtain a compact expression, given by Equation (6.8). Of course, all choices of the free functions lead to valid approximations for the displacement $x(t)$. The MNF expansion is very good, yielding an excellent approximation for the fundamental component early on.

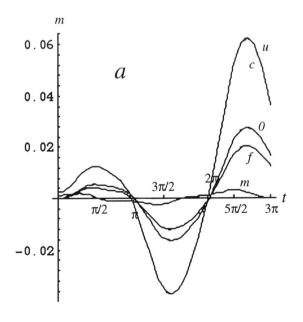

THE DUFFING OSCILLATOR

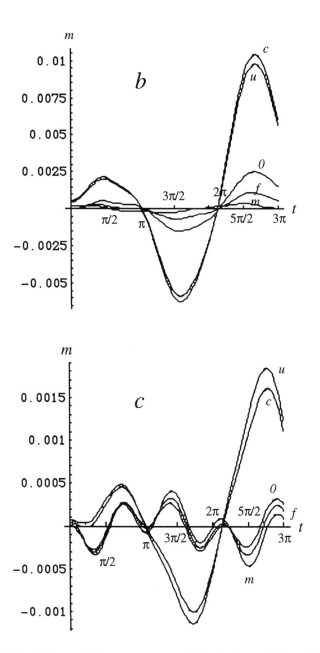

Figure 6.1 Mismatch between numerical solution of Equation (6.1a) and (a) first-; (b) second-, and (c) thrid-order approximation $[x(0)=\pi/3, dx/dt(0)=0,$ and $\varepsilon=-(1/6)]$. u - usual choice, no free functions; c - canonical-near identity transformation; 0 - "(0–0)" initial conditions; f - elimination of fundamental component; m - MNF

6.3 PRECISION AND DURATION OF THE VALIDITY OF THE EXPANSION

6.3.1 Dependence on accuracy with which the updated frequency is known

In conservative planar problems there is no need to compute the frequency through a perturbation expansion. Instead, one can calculate it from the integral of Equation (6.5a) [Equation (6.5b) in the case of the Duffing oscillator] which can be evaluated on a computer to any desired accuracy. The errors incurred by the use of an imprecise value of the frequency can be minimized and postponed to times much longer than $O(1/\varepsilon)$. For example, let us assume $\varepsilon=0.1$ and that the frequency has been computed with p significant decimal figures, namely

$$\omega = \omega_{approx} + O\left(10^{-(p+1)}\right)$$

Then, using the approximation ω_{approx} instead of the correct ω in, say, $\cos\omega t$, yields:

$$\cos\omega_{approx} t = \cos\left\{\left(\omega - O\left(10^{-(p+1)}\right)\right)t\right\}$$
$$= \cos\omega t \, \cos\left(O\left(10^{-(p+1)}\right)t\right) + \sin\omega t \, \sin\left(O\left(10^{-(p+1)}\right)t\right)$$

For $t \approx O(1/\varepsilon)$, this quantity is close to $\cos\omega t$ with an $O(\varepsilon^p)$ error. An $O(\varepsilon)$ error is generated only for $t \approx O(1/\varepsilon^p)$, that is, $t \approx O(10^p)$.

To stress this point, we show in Fig. 6.2 a comparison of the exact solution with the first-order approximation (i.e., with $u + \varepsilon T_1$), with ω computed with 16 ($\omega=0.9284471890586603$), 3 ($\omega=.928$), and 2 ($\omega=0.93$) significant figures.

Computing ω to any desired accuracy on a computer is a very simple task today, while finding T_n is a major job. Thus, the calculation with 2 or 3 significant figures for ω is only meant as an illustration. The example is given for two cases: the "worst case" of the *usual choice* (no free functions), and the "best case" (MNF).

With ω computed to a large number of digits, the advantage of MNF is visible. Of no lesser importance is the fact that, with high precision ω, both choicess yield a mismatch, m, that is seemingly periodic in time and does not exhibit a secular behavior, at least not within the timespan considered. Clearly, if the computation were carried out for much longer times, the inaccuracy in the numerical value of ω would have shown up as secular behavior.

PRECISION, DURATION OF VALIDITY OF EXPANSION 179

As ω becomes less accurate, the secular growth of m becomes more pronounced at earlier time. At sufficiently long times, the advantage of MNF is totally masked by the effect of the inaccuracy in ω.

6.3.2 Precision and duration of validity of MNF expansion

The error analysis of Chapter 4 has shown that, when the approximate solution is developed through $O(\varepsilon^n)$ and the frequency is computed through $O(\varepsilon^{n+1})$

$$z_a = u + \sum_{k=1}^{n} \varepsilon^k T_k \qquad u = \rho_a \exp(-i\,\varphi) \qquad \varphi = \omega_a t + \varphi_0 \quad (6.16a)$$

$$\omega_a = 1 + \sum_{k=1}^{n+1} \varepsilon^k \omega_k = \omega + O(\varepsilon^{n+2}) \qquad (6.16b)$$

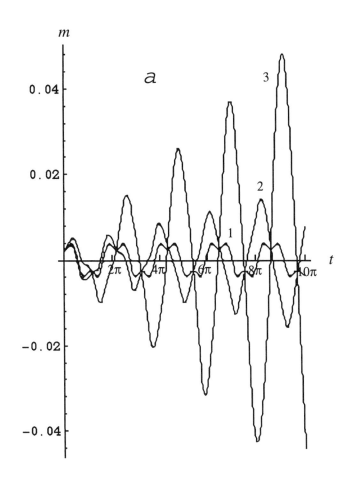

a

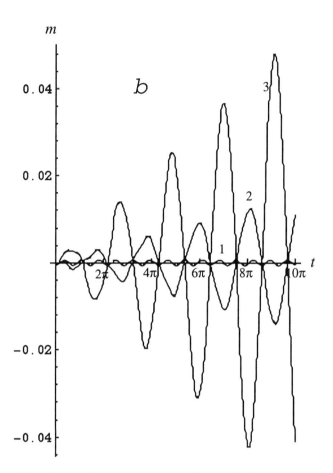

Figure 6.2 Mismatch between numerical solution of Equation (6.1a) [$x(0)=\pi/3$, $dx/dt(0)=0$, $\varepsilon=-\frac{1}{6}$] and first-order approximations: (a) "Usual choice"; (b) Minimal normal form choice; (1) High precision ω; (2) ω to 3 digits; (3) ω to 2 digits.

the deviation of $z_a(t)$ from the exact solution is fed by two sources:

$$O(\varepsilon^{n+1}) + O(\varepsilon^{n+2})t$$

Subscript a signifies approximations to the exact quantities. The ω_k are computed from the normal form and ω denotes the exact frequency.

The sources are a bounded term, arising from the error in the initial condition, and a secular term, resulting from the error in the frequency. For the bounded term, we invert Equation (6.16a) at $t=0$ (without loss of generality, we assume that φ_0 vanishes):

$$\rho_a = x(0) + \sum_{k=1}^{n} \varepsilon^k \rho_k = \rho + O(\varepsilon^{n+1}) \tag{6.17}$$

where ρ is the exact the amplitude. Thus, using ρ_a instead of ρ in Equation (6.16a) generates an $O(\varepsilon^{n+1})$ error in $[z_a(t) - z(t)]$.

For the secular term, consider the phase factor $\exp(\pm i\omega_a t)$:

$$\begin{aligned}\exp(\pm i\,\omega_a t) &= \exp(\pm i\,\omega t \pm i\,(\omega_a - \omega)t) \\ &\approx \exp(\pm i\,\omega t)(1 + O(\varepsilon^{n+2})t) \quad \text{for } |(\omega_a - \omega)t| \ll 1\end{aligned} \tag{6.18}$$

For $|(\omega_a - \omega)|t \ll 1$, the error in $\exp(\pm i\omega_a t)$ and, hence, in $[z_a(t) - z(t)]$, grows linearly in time from $O(\varepsilon^{n+2})$ at $t=O(1)$, to $O(\varepsilon^{n+1})$ at $t=O(1/\varepsilon)$.

Can one extend the validity of the $O(\varepsilon^{n+1})$ approximation to longer times at a given order of the expansion? The results of Section 6.2.6 indicate that, compared to other choices of the free functions, MNF have the capacity to do this, by reducing the growth rate of the secular term. We will now show how this reduction is attained for the Duffing oscillator, for $n=1$.

The quality of the approximation is controlled by $|(\omega_a - \omega)|$. The expansion of ω in terms of ε and $x(0)=A$ (Equation (6.6)) does not depend on the free functions. On the other hand, the choice of these functions does affect the value of ω_a. To see this, we express ω_a in terms of ε and A. Equation (6.11) yields through $O(\varepsilon)$:

$$\rho_a = A + \varepsilon\left(\tfrac{5}{32} - \alpha\right)A^3$$

Substituting for ρ_a the expression obtained in Equation (6.4c), we derive through second order

$$\begin{aligned}\omega_a &= 1 + \tfrac{3}{8}\varepsilon\rho_a^2 - \tfrac{3}{4}\varepsilon^2\left(\tfrac{17}{64} - \alpha\right)\rho_a^4 \\ &= 1 + \tfrac{3}{8}\varepsilon A^2 - \tfrac{21}{256}\varepsilon^2 A^4 + \varepsilon^3\tfrac{21}{8}\left(\alpha - \tfrac{5}{32}\right)\left(\tfrac{9}{32} - \alpha\right)A^6 + \cdots\end{aligned} \tag{6.19}$$

Expressed in terms of ε and A, the expansion of ω_a through $O(\varepsilon^2)$ generates higher powers of ε. The correct ε^2 term ($\omega_2 = -\tfrac{21}{256}A^4$) is obtained. However, as ω_a is not computed consistently through ε^3, $\omega_{a,3}$, the ε^3 term in Equation (6.19) is not equal to the correct ε^3 term in Equation (6.6) ($\omega_3 = \tfrac{81}{2048}A^6$) for any α. The resulting secular

182 NORMAL FORMS: CONSERVATIVE PLANAR SYSTEMS

behavior can be reduced if $|\omega_3 - \omega_{a,3}|$ is minimized. For example

$$\omega_3 - \omega_{3,a} = \begin{cases} \frac{15}{512} = 0.0293 & \alpha = \frac{7}{32} \quad \text{minimal deviation} \\ \frac{1149}{32,768} = 0.0351 & \alpha = \frac{17}{64} \quad \text{MNF choice} \\ \frac{1269}{8196} = 0.1548 & \alpha = 0 \quad \text{no free function choice} \end{cases}$$

The MNF deviation is very close to the minimum. The secular term it generates grows more slowly in time than in other common choices of the free functions. The same conclusion is reached in the next order in ε and for other types of single-oscillator systems.

In single-oscillator systems ω is known. Therefore, the best value of α can be found here. However, as ω is known, the preceding analysis is not necessary. If ω were given with infinite precision, the $O(\varepsilon^{n+1})$ error in z_a of equation (6.16a) would remain valid *forever*. In actual calculations, we can extend the validity of the $O(\varepsilon^{n+1})$ error far beyond $t=O(1/\varepsilon)$ by computing ω with a sufficiently high precision.

The single-oscillator problem should be viewed as a "laboratory," indicating what may be expected in systems with several degrees of freedom, where there is no analog of Equation (6.5a). The updated frequencies are not known and can be evaluated only by a perturbative expansion, containing free functions. While the "best" free functions cannot be found, MNF can be applied, as they do not depend on knowledge of the frequencies. In Chapter 10 we give an example of two coupled oscillators where MNF reduce the secular growth rate.

6.4 OSCILLATOR WITH QUADRATIC PERTURBATION

Consider a system with a quadratic nonlinearity

$$\frac{dx}{dt} = y \qquad \frac{dy}{dt} = -x - \varepsilon x^2 \qquad x(0) = A \qquad \frac{dx}{dt}(0) = 0 \quad (6.20)$$

We convert the equations to complex form, introduce the near-identity transformation and find through second order:

$$T_1 = \tfrac{1}{4}u^2 - \tfrac{1}{2}uu^* - \tfrac{1}{12}u^{*2} + f_1(uu^*)u$$

(6.21a)

$$U_1 = 0$$

$$T_2 = \tfrac{1}{24}u^3 + +\tfrac{5}{24}uu^{*2} - \tfrac{1}{48}u^{*3}$$

$$+ \tfrac{1}{2}f_1 u^2 - \tfrac{1}{2}(f_1 + f_1^*)uu^* - \tfrac{1}{6}f_1^* u^{*2} + f_2(uu^*)u \quad (6.21b)$$

$$U_2 = \tfrac{5}{12} i\, u^2 u^*$$

As this is a conservative system, the radius, ρ, of the zero-order term will be constant if the free functions are chosen to be real.

MNF. The development of the expansion is straightforward. One obtains the minimal normal form:

$$\frac{du}{dt} = -i\,u + \varepsilon^2 \tfrac{5}{12} i\, u^2 u^*$$

In obtaining this equation, we have made appropriate choices of the free functions. In this regard, for example, the first four T functions have been chosen as follows:

$$T_1 = \tfrac{1}{4}u^2 - \tfrac{1}{2}uu^* - \tfrac{1}{12}u^{*2}$$

$$T_2 = \tfrac{1}{24}u^3 - \tfrac{157}{288}u^2 u^* + \tfrac{5}{24}uu^{*2} - \tfrac{1}{48}u^{*3}$$

$$T_3 = \tfrac{5}{864}u^4 - \tfrac{19}{576}u^3 u^* + \tfrac{7}{96}u^2 u^{*2} + \tfrac{11}{192}uu^{*3} - \tfrac{1}{288}u^{*4}$$

$$T_4 = \tfrac{5}{6912}u^5 + \tfrac{31}{2304}u^4 u^* + \tfrac{15{,}355}{165{,}888}u^3 u^{*2}$$
$$- \tfrac{485}{6912}u^2 u^{*3} - \tfrac{1}{4608}uu^{*4} - \tfrac{5}{10{,}368}u^{*5}$$

Usual choice. In contrast to MNF, in the "usual" choice (no free functions) the normal form is updated in each even order:

$$\frac{du}{dt} = -i\,u + \varepsilon^2 \frac{5i}{12} u^2 u^* + \varepsilon^4 \frac{785 i}{1728} u^3 u^{*2}$$
$$+ \varepsilon^6 \frac{65{,}495\, i}{82{,}944} u^4 u^{*3} + \varepsilon^8 \frac{59{,}370{,}835\, i}{35{,}831{,}808} u^5 u^{*4} + O(\varepsilon^{10})$$

6.4.1 Surprise with "zero–zero" choice

Let us try to implement the initial conditions fully by the zero-order term, which we write as $u = \rho \cdot \exp(-i\varphi)$:

$$\rho_{0-0} = A, \qquad \varphi(0) = 0$$

Requiring that at $t= 0$, all T_n must vanish. In particular, in T_1, we find

$$f_1(uu^*) = \tfrac{1}{3}\rho_{0-0} = \tfrac{1}{3}\sqrt{uu^*}$$

Comment. A resonant term appears in the normal form for the first time in second order. Also, in T_1 (and in all T s of odd orders), all the terms are even-powered, while the (resonant) free function is odd-powered. Hence, mixing of powers will occur in T_n of odd order whenever we decide to include free functions. That is why f_1 is not a monomial or a polynomial in uu^*.

These observations are characteristic of the normal form expansion of equations with even-powered perturbations. An even-powered perturbation modifies the symmetry of the potential by adding an aymmetric term [cubic in the case of Equation (6.20)] to the symmetric unperturbed harmonic oscillator potential:

$$V(x) = \tfrac{1}{2}x^2 + \tfrac{1}{3}\varepsilon x^3$$

Because of the asymmetry of the added term, the oscillations are not centered around the origin ($x=y=0$) but around a shifted origin. In T_n of odd order the computed terms are all even-powered of the form

$$u^p u^{*2n-p}$$

Among these, the neutral term, $u^n \cdot u^{*n}$, does not contribute to the oscillations (it has no variable phase), but rather, adds a constant shift to the origin of oscillations. For instance, in first order, the origin is shifted by

$$-\tfrac{1}{2}\varepsilon uu^* = -\tfrac{1}{2}\varepsilon\rho^2$$

Numerical solutions of Equation (6.20) are displayed in Fig. 6.3. In all the graphs, $\varepsilon=-0.2$ and $y(0)=dx/dt(0)= 0$. The values used for $x(0)$ are 0.5, 1.0, and 1.5. As the amplitude grows, two features of the solution become evident. The distortion of the circle (which is the solution of the unperturbed equation) becomes greater due to the contribution of higher harmonics, and the shift of the origin grows.

OSCILLATOR WITH QUADRATIC PERTURBATION 185

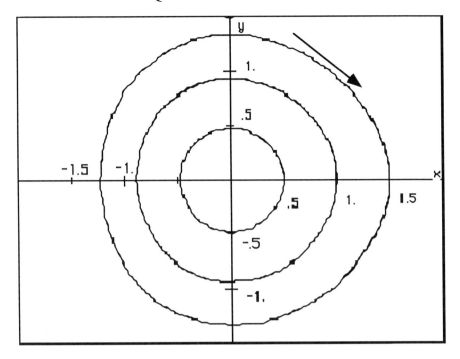

Figure 6.3 Solutions of Equation (6.20). Arrow indicates direction of motion along curves.

6.5 THE PENDULUM

Consider a pendulum of unit mass and length l (Figure 6.4), performing oscillations with a maximum angular displacement θ_m. (It is a system of one degree of freedom.) Its equation of motion is

$$l\frac{d^2\theta}{dt^2} + g\sin\theta = 0 \qquad (6.22)$$

We choose the unperturbed angular frequency $\omega_0=(g/l)^{1/2}$ to be equal to 1 and concentrate on small amplitudes. Thus, the nonlinear part of $\sin\theta$ constitutes a small perturbation relative to the linear (harmonic) part. We will show that a MNF can be found in this case. Namely, a choice of the free functions in T_i exists that makes $U_i=0$ for all $i>1$. Furthermore, it will become clear that, except for a multiplicative factor in U_1, the u equation is the same as for the Duffing problem. Hence, the two have the same zero-order approximation. They differ only in their harmonic structure.

Figure 6.4 The pendulum.

For small-amplitude oscillations, we scale the angular displacement as $x \to \sqrt{\varepsilon} \cdot x$ and expand in a Taylor series to obtain

$$\frac{d^2}{dt^2}x + x + \sum_{n \geq 1}(-)^{2n+1}\varepsilon^n \frac{x^{2n+1}}{(2n+1)!} = 0 \qquad (6.23a)$$

Introducing the z variables, $z = x + i\,(dx/dt)$, we have

$$\frac{dz}{dt} = -i\,z - i\sum_{n \geq 1}(-)^{2n+1}\varepsilon^n \frac{(z+z^*)^{2n+1}}{2^{2n+1}(2n+1)!} \qquad (6.23b)$$

Z_0 and Z_1 have the forms

$$Z_0(z) = -i\,z \qquad Z_1(z) = +i\left(\frac{1}{3!}\right)\frac{(z+z^*)^3}{2^3} \qquad (6.24)$$

Comparing Equation (6.24) with the relevant equations for the Duffing oscillator, Equations (6.1a,b), we see that, except for the factor $=(1/3!)$ they are the same. Thus, T_1 and U_1 should be multiples of the Duffing results. We now consider the structure of U_2. First, we observe that

$$Z_2(z) = -i \left(\frac{1}{5!}\right) \frac{(z+z^*)^5}{2^5} \qquad (6.25)$$

a contributor to U_2, is a quintic polynomial. All other terms in U_2, generated by the contributions of Z_1, T_1, T_1^*, U_1, and U_1^*, are also quintic polynomials. Therefore, T_2 will also be such a polynomial, and U_2 will be a quintic monomial, proportional to $u^3 u^{*2}$. We now note, from Equation (6.2b), that T_2 and U_2 for the Duffing problem are also quintic. Thus, the equations for the pendulum have the same general structure in this order, and a proper choice of the free function will enable one to make $U_2 = 0$, just as in the case of the Duffing Equation. The pattern is maintained in higher orders. We thus conclude that, with a proper choice of the free functions, we can have $U_i = 0$; $i > 1$. Therefore, the two equations have essentially the same minimal normal form and differ only in the composition of their higher harmonics.

6.6 THE "ALTERED" DUFFING OSCILLATOR

In conservative systems, the nonlinear perturbation modifies the potential by adding terms that are not quadratic in the coordinate. This results in a frequency update. The time variable of the unperturbed harmonic oscillator t, is not the natural one for the perturbed motion. The latter has a different natural time variable, $\tau = \omega t$, where ω is the updated frequency. The normal formal form expansion provides the frequency update. The simplest functional form of this update is given by the MNF choice; all U_n have been eliminated, except for a small number of low-order ones, e.g., U_1 in the case of the Duffing equation. We again focus our attention on the latter.

Clearly, as long as one uses t as the time variable, one *cannot* eliminate those U_k that appear in the MNF (U_1 in the case of the Duffing equation), as this is the minimum required for generating the frequency update. The question that arises is: What happens if we reformulate the perturbed problem in terms of *its own natural time variable*, τ? Clearly, this question will be easier to handle if we are given a simple expression for the updated frequency, ω. Thus, it is natural to choose for ω the MNF expression [Equation (6.8)].

We will show, using the Duffing equation as an example, that the

188 NORMAL FORMS: CONSERVATIVE PLANAR SYSTEMS

introduction of τ as the time variable alters the original equation, so that the MNF expansion for the altered oscillator has no frequency update at all, i.e., all $U_n=0$, *including* U_1. Pretending that we have not found MNF, we assume for the updated frequency a simple form: $\omega=1+\varepsilon\sigma$. (We anticipate that the MNF relation $\sigma=\frac{3}{8}\rho^2$ will emerge naturally. Note that in a conservative problem ρ is constant.) We now modify Equation (6.1a) by replacing the unperturbed frequency ($\omega_0=1$) by the updated value. Thus, the l.h.s. of the equation should be modified into

$$\frac{d^2x}{dt^2} + (1+\varepsilon\sigma)^2 x$$

Essentially, we have introduced a new "unperturbed" oscillator with an altered spring constant. Technically, this amounts to adding terms on the l.h.s. of Equation (6.1a). We therefore add the same terms on the r.h.s., obtaining

$$\frac{d^2x}{dt^2} + (1+\varepsilon\sigma)^2 x = -\varepsilon(x^3 - 2\sigma x) + \varepsilon^2 \sigma^2 x$$

$$x(0) = A, \quad \frac{dx}{dt}(0) = 0$$

(6.26)

The first step is to introduce the "scaled" time variable by writing $\tau=\omega t$. Then Equation (6.26) becomes

$$\frac{d^2x}{d\tau^2} + x = -\varepsilon\frac{(x^3 - 2\sigma x)}{\omega^2} + \frac{\varepsilon^2 \sigma^2 x}{\omega^2}$$

$$x(0) = A \quad \frac{dx}{d\tau}(0) = 0$$

(6.27)

When diagonalized, Equation (6.27) becomes

$$\frac{dz}{d\tau} = -i\,z + \frac{i\,\varepsilon}{\omega^2}\left(-\tfrac{1}{8}(z+z^*)^3 + \sigma(z+z^*)\right)$$

$$+ \frac{i\,\varepsilon^2}{\omega^2}\sigma^2 \tfrac{1}{2}(z+z^*)$$

(6.28)

$$\left(z \equiv x + i\,\frac{dx}{d\tau}\right)$$

THE "ALTERED" DUFFING OSCILLATOR

We note the following points.

1. Although we use our usual notation, the new complex variable z is *different* from the old one, as defined in Equation (6.1b). More specifically,

$$z_{new} = x + i\frac{dx}{d\tau} = x + i\frac{1}{1+\varepsilon\sigma}\frac{dx}{dt} =$$

$$x + i\frac{dx}{dt}\{1 - \varepsilon\sigma + (\varepsilon\sigma)^2 + \cdots\}$$

$$= z_{old} + \frac{(z_{old} - z_{old}^*)}{2}\{-\varepsilon\sigma + (\varepsilon\sigma)^2 + \cdots\}$$

Thus, they differ from one another by an $O(\varepsilon)$ correction in the imaginary part. As a result, the zero-order approximations u_{new} and u_{old} (with obvious notation) are also not identical.

2. In $O(\varepsilon)$, there are two competing terms in Equation (6.28): a new linear term and the traditional cubic one [see Equation (6.1b)]. Thus, an appropriate choice of σ can eliminate U_1.

3. A linear perturbation term, such as in Equation (6.28), introduces frequency updates in all orders in ε, which, in most problems, cannot be eliminated.

4. As $\omega^2 = (1+\varepsilon\sigma)^2$, we expand $1/\omega^2$ in Equation (6.28) in a power series in ε. This additional source of linear as well as cubic perturbation terms contributes in all orders.

The expansion yields

$$\frac{dz}{d\tau} = -i\,z + \varepsilon Z_1 + \varepsilon^2 Z_2 + \varepsilon^3 Z_3 + \cdots \quad (6.29a)$$

with

$$Z_1 = -\tfrac{1}{8}(z+z^*)^3 + \sigma(z+z^*)$$

$$Z_2 = \tfrac{1}{4}\sigma(z+z^*)^3 - \tfrac{3}{2}\sigma^2(z+z^*) \quad (6.29b)$$

$$Z_3 = -\tfrac{3}{8}\sigma^2(z+z^*)^3 + 2\sigma^3(z+z^*)$$

Through second order, one finds from the normal form expansion:

$$U_1 = -\tfrac{3}{8} i\, u^2 u^* + i\, \sigma u,$$

$$T_1 = \tfrac{1}{16} u^3 - \tfrac{3}{16} u u^{*2} - \tfrac{1}{32} u^{*3} + \tfrac{1}{2}\sigma u^* + \alpha u^2 u^* \qquad (6.30)$$

$$U_2 = \left\{ \tfrac{51}{256} - \tfrac{3}{8}(\alpha + \alpha^*) \right\} i\, u^3 u^{*2} - i\, \sigma^2 u$$

Note that a free term appears in T_1. We wish to have no frequency update, namely, that all U_n vanish. To achieve $U_1=0$, we need

$$\sigma = \tfrac{3}{8}\rho^2 = \tfrac{3}{8} u u^* \qquad (6.31)$$

Thus, the MNF result emerges naturally. One may now ask whether all the U_n for $n>1$ vanish automatically. The answer is that this is not the case. The freedom in the choice of the free functions that enabled us originally to derive a MNF expansion *still persists*. Thus, to make U_2 vanish, we must set $\alpha = \tfrac{5}{64}$. Introducing the final values of σ and α in the expression for T_1, we obtain

$$T_1 = \tfrac{1}{16} u^3 - \tfrac{1}{32} u^{*3} + \tfrac{5}{64} u^2 u^* \qquad (6.32)$$

Comment. Note that this expression for T_1 is different from the one Equation (6.2a). This is a manifestation of point 1 above.

While it was straightforward to make of U_1 and U_2 vanish, we need to uncover the pattern that makes the elimination of higher-order U_n possible. The structure of the terms in the interaction [Equation (6.29)] and in the normal form [Equation (6.30)], will provide the clue. Z_1 contains a linear term and a cubic polynomial. U_1 contains a linear term and a cubic monomial. In general, these terms cannot cancel each other (except at a particular point). However, with the choice of Equation (6.31) for σ and the fact that $\rho^2 = u u^*$ is constant in a conservative system, the linear terms in Z_1 and in U_1 effectively become cubic and can, therefore, compete with the original cubic terms. This is how U_1 is eliminated. In $O(\varepsilon^2)$ this pattern is repeated. Z_2 and U_2 are each composed of a linear term and a quintic one. With the chosen value of σ inserted, the linear term becomes quintic as well. However, as σ has been predetermined in first order, it is the free function in T_1 (the coefficient α) that enables us to eliminate U_2. The pattern that emerges is clear — in $O(\varepsilon^n)$, Z_n and U_n will each be composed of a linear term and a term of degree $(2n+1)$. The added powers of σ convert all the polynomials in Z_n and all the monomials in U_n into homogeneous ones of degree $(2n+1)$. The free function from

THE "ALTERED" DUFFING OSCILLATOR

the previous order enables one to eliminate U_n. In summary, we have introduced the natural time variable of the altered oscillator, so that the updated frequency is included in the formulation of the problem. Therefore, no frequency update occurs.

6.7 ADDITIONAL EXAMPLES

We list a few additional examples of perturbed harmonic oscillator equations and the associated minimal normal forms. Observe that perturbations that are even powers of the velocity do not lead to dissipation.

SYSTEM	MNF
1. $\ddot{x} + x + \varepsilon \dot{x}^2 = 0$	$\dot{u} = -i\,u + \varepsilon^2\,i\,\frac{1}{6}u^2 u^*$
2. $\ddot{x} + x + \varepsilon x^2 + \varepsilon^2 x^3 = 0$	$\dot{u} = -i\,u + \varepsilon^2\,i\,\frac{1}{24}u^2 u^*$
3. $\ddot{x} + x + \varepsilon \dot{x}^2 + \varepsilon^2 x^3 = 0$	$\dot{u} = -i\,u - \varepsilon^2\,i\,\frac{5}{24}u^2 u^*$
4. $\ddot{x} + x + \varepsilon \dot{x}^4 = 0$	$\dot{u} = -i\,u + \varepsilon^2\,i\,\frac{7}{40}u^4 u^{*3}$
5. $\ddot{x} + x + \varepsilon x^3 + \varepsilon^2 x^5 = 0$	$\dot{u} = -i\,u - \varepsilon\,i\,\frac{3}{8}u^2 u^*$
6. $\ddot{x} + x + \varepsilon x^5 = 0$	$\dot{u} = -i\,u - \varepsilon\,i\,\frac{5}{16}u^3 u^{*2}$
7. $\ddot{x} + x + \varepsilon x^5 + \varepsilon^2 x^3 = 0$	$\dot{u} = -i\,u - \varepsilon\,i\,\frac{5}{16}u^3 u^{*2} - \varepsilon^2\,i\,\frac{3}{8}u^2 u^*$
8. $\ddot{x} + x + \varepsilon^2 x^3 + \varepsilon^3 x^4 = 0$	$\dot{u} = -i\,u - \varepsilon^2\,i\,\frac{3}{8}u^2 u^*$
9. $\ddot{x} + x + \varepsilon x^2 + \varepsilon^2 x^3 + \varepsilon^3 x^4 + \varepsilon^4 x^5 = 0$	$\dot{u} = -i\,u + \varepsilon^2\,i\,\frac{1}{24}u^2 u^*$

6.7.1 Mixing of scales

We now concentrate on one particular case, system 7 in the list of examples given above. In diagonalized form, the equation becomes

$$\dot{z} = -i\,z - \varepsilon\,i\,\left(\tfrac{1}{32}\right)(z+z^*)^5 - \varepsilon^2\,i\,\left(\tfrac{1}{8}\right)(z+z^*)^3 \qquad (6.33)$$

The normal form expansion yields

$$U_1 = -\tfrac{5}{16} i\,u^3 u^{*2}$$

$$T_1 = \tfrac{1}{128} u^5 + \tfrac{5}{64} u^4 u^* - \tfrac{5}{32} u^2 u^{*3} - \tfrac{5}{128} u u^{*4} - \tfrac{1}{192} u^{*5} + F_1 u \qquad (6.34\text{a})$$

$$U_2 = -\tfrac{3}{8} i\,u^2 u^* + \tfrac{655}{3072} i\,u^5 u^{*4} - \tfrac{5}{8} i\,u^3 u^{*2}(F_1 + F_1^*) \qquad (6.34\text{b})$$

There are two known terms in U_2: a cubic and a quintic one. It is clear that by including in F_1 a term of the form

$$\tfrac{131}{768} u^2 u^{*2}$$

one can eliminate the quintic term. However, elimination of the cubic term can be achieved only if F_1 includes a singular term:

$$-\tfrac{3}{10}(u u^*)^{-1}$$

This means that T_1 will diverge for small amplitudes – certainly an undesired feature. Therefore, it is best to retain the cubic term, so that $U_2 \neq 0$. A minimal normal form can be still obtained, with $U_n = 0$ for $n \geq 3$ only, as given in the list of examples.

To see the pattern, it suffices to examine the qualitative structure of U_3. The latter will have the form

$$U_3 = i\left\{a u^4 u^{*3} + b u^7 u^{*6}\right\} - \tfrac{5}{8} i\,(F_2 + F_2^*) u^3 u^{*2} \qquad (6.34\text{c})$$

where a and b are calculable coefficients. All the monomials with known coefficients are of degrees not lower than the monomial multiplying the free function. Therefore, they can be canceled by including appropriate nonsingular monomials in F_2.

This pattern is repeated in higher orders. Namely, the known terms in U_n have degrees that are greater than or equal to the degree of the monomial multiplying the free-function contribution, so that U_n can be eliminated for all $n \geq 3$. The resulting normal form

ADDITIONAL EXAMPLES 193

$$\dot{u} = -iu - \varepsilon i \tfrac{5}{16} u^3 u^{*2} - \varepsilon^2 i \tfrac{3}{8} u^2 u^* \qquad (6.35)$$

is easily solved, yielding

$$\rho = \text{const} \qquad \omega = 1 + \tfrac{5}{16}\varepsilon\rho^4 + \tfrac{3}{8}\varepsilon^2\rho^2 \qquad (6.36)$$

From Equation (6.36), we see the significance of the result that the normal form cannot be terminated in first order. In simpler problems, such as the Duffing oscillator, there is only one quantity in which the expansion is carried out ($\varepsilon\rho^2$, in the case of the Duffing oscillator).

Here there are two expansion parameters: $\varepsilon\rho^4$ and $\varepsilon^2\rho^2$. For small amplitudes, $\rho < O(\varepsilon^{1/2})$, the $\varepsilon\rho^4$ term is smaller than the $\varepsilon^2\rho^2$ term. It is, therefore, the latter that determines the quality and time validity of a given approximation for the solution. The roles are interchanged when $\rho > O(\varepsilon^{1/2})$, in particular, when $\rho = O(1)$. Thus, the problem generates two independent timescales. The method of normal forms *identifies* them for us by preventing us from eliminating U_2. It also points out naturally that, when $\rho = O(\varepsilon^{1/2})$, there is mixing of the two timescales.

We mention in passing that, in other methods, e.g., the method of multiple timescales, a straightforward application of the expansion procedure leads to inconsistencies. These are resolved only after one extracts from the unknown variable a specific scale: $z = \varepsilon^{1/2} w$ in the present problem. Thus, formally, the expansion is then valid only for $z = O(\varepsilon^{1/2})$. Moreover, finding the correct scale requires some experience with the method, while the method of normal forms indicates the existence of the scale in a natural manner.

Comment. It is noteworthy that when the powers of x grow in concert with powers of ε in the perturbation (e.g., example 5 in the list provided at the beginning of Section 6.7 or the pendulum), then no competition of scales occurs, and a minimal normal form terminating at $n=1$ can be derived. In contrast, when the powers of ε and of x do not grow in concert, as in example 7 in the list, then competition of scales occurs and the minimal normal form must contain more terms.

6.7.2 Example where the minimal normal form is not useful

Not always is the minimal normal form useful. In many problems, to obtain the minimal normal form, one must resort to free functions that are singular just where some interesting phenomena occur. We demonstrate this with one particular example. Consider the equation

194 NORMAL FORMS: CONSERVATIVE PLANAR SYSTEMS

$$\ddot{x} + x + \varepsilon(x^3 \pm \alpha x^5) = 0 \tag{6.37}$$

The two signs are included in the quintic term because both suffer from the same problem, when one attempts to construct a minimal normal form. The coefficient $\alpha > 0$ is included to enable us to trace the effect of the quintic term. Without loss of generality, we also assume $\varepsilon > 0$.

In complex notation, Equation (6.37) becomes

$$\dot{z} = -i\, z - i\, \varepsilon\left\{\tfrac{1}{8}(z+z^*)^3 \pm \alpha \tfrac{1}{32}(z+z^*)^5\right\} \tag{6.38}$$

Through first order, the normal form procedure yields

$$U_1 = -\tfrac{3}{8} i\, u^2 u^* \mp \alpha \tfrac{5}{16} i\, u^3 u^{*2} \tag{6.39a}$$

$$T_1 = \begin{bmatrix} \tfrac{1}{16}u^3 - \tfrac{3}{16}uu^{*2} - \tfrac{1}{32}u^{*3} \\ \\ \mp \alpha \left\{ \begin{array}{l} \tfrac{1}{128}u^5 + \tfrac{5}{64}u^4 u^* - \tfrac{5}{32}u^2 u^{*3} \\ -\tfrac{5}{128}uu^{*4} - \tfrac{1}{192}u^{*5} \end{array} \right\} \end{bmatrix} + F_1 u \tag{6.39b}$$

There is no need to calculate U_2 or U_{n+1} for $n \geq 1$ explicitly. From the formalism of Chapter 4, we find that for $n \geq 1$, U_{n+1} will have the generic form

$$U_{n+1} = \tilde{U}_{n+1} - i\,(F_n + F_n^*)\left\{\tfrac{3}{8}u^2 u^* \pm \tfrac{5}{8}\alpha u^3 u^{*2}\right\} \tag{6.40}$$

where $\tilde{U}_{n+1}$ is the known part, calculated from lower-order contributions. In principle, one can eliminate U_{n+1}, with a free function of the form

$$F_n = -\tfrac{1}{2} i\, \frac{\tilde{U}_{n+1}}{\left(\tfrac{3}{8}u^2 u^* \pm \tfrac{5}{8}\alpha u^3 u^{*2}\right)} \tag{6.41}$$

Having eliminated all $U_{n \geq 2}$, the minimal normal form becomes

$$\dot{u} = -i\, u + \varepsilon U_1 = -i\, u + \varepsilon\left\{-\tfrac{3}{8} i\, u^2 u^* \mp \alpha \tfrac{5}{16} i\, u^3 u^{*2}\right\}$$

ADDITIONAL EXAMPLES

and the updated frequency is then given by

$$\omega = 1 + \varepsilon \tfrac{5}{16} \alpha \rho^2 \left\{ \frac{6}{5\alpha} \pm \rho^2 \right\} \qquad (6.42)$$

For the case of the (−) sign, Equation (6.37) exhibits an interesting behavior. For

$$\rho^2 < \frac{6}{5\alpha} \equiv \rho_1^2$$

Equation (6.37) yields $\omega>1$. For $\rho^2>\rho_1^2$, it yields $\omega<1$ and at $\rho^2=\rho_1^2$, $\omega=1$. The physical interpretation of this behavior is the following. In Equation (6.37), the perturbation, which is a sum of a cubic and a quintic term, modifies the linear force of the unperturbed harmonic oscillator. The corresponding potential is a double-hump potential. For $|x|<(1/\alpha)$, the cubic term dominates and the force (as well as the potential) is slightly higher than that of the unperturbed oscillator. For $|x|>(1/\sqrt{\alpha})$, the quintic term is larger, so that the force (in the case of a (−) sign) is slightly below that of a harmonic oscillator. [At much larger $|x|$ ($|x|\approx(\alpha\varepsilon)^{-1/4}$) the force vanishes: the potential has a maximum − a hump.] Thus, there should be a qualitative difference between the nature of solutions with amplitudes less than $\sim(1/\sqrt{\alpha})$ and those with amplitudes larger than $\sim(1/\sqrt{\alpha})$. What Equation (6.42) yields is the amplitude, ρ_1, of the zero-order term, which serves as the crossing point between the two domains.

However, there is an inherent difficulty with this result. If we insist on expanding $F_{n\geq 1}(uu^*)$ in positive powers of uu^*, for both (±) signs, the occurrence of the denominator in Equation (6.41) requires an infinite power series that converges only for

$$\rho^2 = uu^* < \frac{3}{5\alpha} = \tfrac{1}{2}\rho_1^2 \equiv \rho_0^2$$

The minimal normal form is thus obtained at the cost of a complicated near identity transformation.

In the case of the (−) sign, things are even worse. The denominator in Equation (6.41) vanishes at $\rho=\rho_0$. This is an undesirable feature, because it means that just near the region in which interesting physics occurs, the near-identity transformation is singular.

The following is an interesting possibility that retains the characteristic behavior of the frequency, while avoiding the singularity.

We rewrite Equation (6.40) in the following form

$$U_{n+1} = \tilde{U}_{n+1} + i\left(F_n + F_n^*\right)\{\tfrac{3}{8}u^2 u^*\} + 2\left(F_n + F_n^*\right)U_1 \quad (6.40\text{a})$$

and require that only a part of Equation (6.40a) vanish

$$\tilde{U}_{n+1} + i\left(F_n + F_n^*\right)\{\tfrac{3}{8}u^2 u^*\} = 0$$

This requirement is possible and does not yield a singular free function. One can show by induction that $\tilde{U}_{n+1}$ is a polynomial with resonant terms of degrees no lower than $u^2 u^*$, with imaginary coefficients, and that F_n are polynomials in uu* with real coefficients. The result is that, for all n, U_n do not vanish identically — they are polynomials with imaginary coefficients. (Obviously, in a conservative system, only the frequency is updated.) More interesting is the fact that one now has

$$U_{n+1} = 2U_1\left(F_n + F_n^*\right) \quad (6.43)$$

so that the normal form becomes

$$\dot{u} = -i\,u + \varepsilon U_1 \left\{ 1 + 2\sum_{n=1}^{\infty} \varepsilon^n \left(F_n + F_n^*\right) \right\} \quad (6.44)$$

Owing to the structure of U_1 [see Equation (6.39a)], for small ε, this normal form preserves the qualitative dependence of ω on ρ as the latter varies from below to above ρ_1, as concluded from Equation (6.42).

6.8 BOUNDARY-VALUE PROBLEMS: BIFURCATIONS AND NORMAL FORMS

The literature on boundary-value problems is vast. This section is not intended to cover this subject at great length, except for one particular aspect, related to the issue of the freedom in perturbation expansions.

Consider the following one-dimensional nonlinear boundary-value problem:

$$\frac{d^2 w}{dx^2} + w + f\left(w, \frac{dw}{dx}\right) = 0 \quad w(x=0) = a \quad w(x=L) = b \quad (6.45)$$

BOUNDARY-VALUE PROBLEMS

Instead of two initial conditions [$x(0)$ and $dx/dt(0)$] which we usually employ in this book, we now impose the values of the unknown function, w, at the two endpoints, $x=0$ and $x=L$, of a finite interval. This difference, in the type of constraints imposed on the solution, makes boundary-value problems more difficult to solve than initial-value problems. In initial-value problems, once a general solution to the equation is found, its value and derivative at a point are fixed by the initial values, and it then evolves according to the equation without further constraints. In boundary-value problems, starting from one of the boundaries, only the value of the solution is fixed there. The slope remains undetermined. Now one has to seek for the right value of the slope so that the solution indeed reaches the value imposed at the second boundary. In some problems this may be impossible.

The nonlinearity in Equation (6.45) is introduced through the function $f(w,(dw/dx))$, which is a nonlinear function of w and dw/dx (i.e., is at least quadratic in the two). Instead of using time as the independent variable, we have used in Equation (6.45) a position variable, x, since that equation is usually the stationary version of a partial-differential equation of the form

$$\frac{\partial w}{\partial t} = \frac{\partial^2 w}{\partial x^2} + w + f\left(w, \frac{dw}{dx}\right)$$

(6.46)

$$w(x=0,t) = \alpha(t) \quad w(x=L,t) = \beta(t) \quad w(t=0,x) = w_0(x)$$

Equation (6.46) is a nonlinear diffusion equation. One often studies versions of the equation with x replaced by the three-dimensional position vector. It could describe, for instance, temperature diffusion in a medium that has thermal properties that depend on the temperature itself in a significant, nonlinear manner. More complicated versions of Equation (6.46), where the unknown function is a vector of unknown entities, are abundant in the physical sciences. For instance, when the vector **w** contains the velocity components and the temperature of a fluid, equations of the generic form of Equation (6.46) are the basis of the theory of fluid dynamics. The diffusion term then represents the effect of viscosity in the equations. Equation (6.46) is also a prototype of dynamical equations describing the temporal evolution of chemical reactions in solutions. In this case, **w** is a vector of the concentrations of chemicals in the solution, a function $f(\mathbf{w})$ describes the reaction rates among the chemicals and the diffusion term describes their diffusion throughout the solution.

One is interested in the study of stationary solutions of such equations (i.e., in the limit $\partial w/\partial t=0$). The reasons for this are many. For instance, the stationary solutions may have particularly interesting

198 NORMAL FORMS: CONSERVATIVE PLANAR SYSTEMS

characteristics, or one may be interested in the transition from any given state of the system to a stationary solution. In particular, one studies the stability characteristics of the stationary solutions of the equation as some control parameter is varied, in the analysis of *bifurcations*. As this is not a text on bifurcation theory, we will demonstrate the concept of a bifurcation in one specific example: a Duffing oscillator in space (rather than time). The boundary-value problem is

$$\frac{d^2w}{dx^2} + w - w^3 = 0, \qquad w(x=0) = w(x=L) = 0 \quad (6.47)$$

The vanishing boundary conditions were chosen so as to simplify the problem.

We will be dealing with situations when the amplitude of w is small, so that the problem is amenable to a perturbation analysis. This being the case, we begin by studying the unperturbed equation:

$$\frac{d^2w}{dx^2} + w = 0, \qquad w(x=0) = w(x=L) = 0 \quad (6.48)$$

The general solution of Equation (6.28) is

$$w(x) = A \sin x + B \cos x \quad (6.49)$$

However, imposing the boundary conditions of Equation (6.48), the solution turns out to be

$$w = \begin{cases} 0 & L \neq \pi, 2\pi, 3\pi, \dots \\ A \sin x & L = \pi, 2\pi, 3\pi, \dots \end{cases} \quad (6.50)$$

In Equation (6.50), the amplitude A can obtain any value at $L=\pi$, 2π,... and must vanish for any other value of L. A plot of the amplitude of the solution against L is given in Fig. 6.5. This diagram provides a trivial example of the concept of bifurcation – the nature of a solution changes in a discontinuous manner as a control parameter, in this case the length of the interval, is varied. As the pattern repeats itself, let us concentrate on the vicinity of $L=\pi$. For $L<\pi$, there is one solution: $A \equiv 0$. At $L=\pi$, the character of the solution changes, and the amplitude splits off into two allowed branches. When L grows beyond π, the zero solution is again the only solution.

BOUNDARY-VALUE PROBLEMS

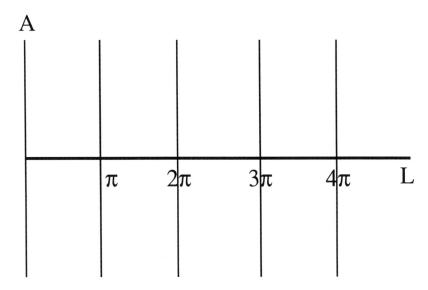

Figure 6.5 Bifurcation diagram for Equation (6.50).

The question that now arises is what happens to the bifurcation diagram when we introduce the nonlinear term. It turns out that one can obtain rigorous results only in the vicinity of the bifurcation point, $L=\pi$. We begin by guessing that the effect of the nonlinear perturbation will be to distort the straight perpendicular bifurcation lines. This would mean that, close to $L=\pi$, the amplitude will not be arbitrary any more, but will become a function of L, more specifically, of $L-\pi$, slightly distorted away from the unperturbed straight line. Therefore, for small $L-\pi$, we expect the amplitude to be small.

One can get a clue to what is expected, by resorting to Equation (6.5a) for the period T of the oscillations in the usual Duffing oscillator. With the substitutions

$$x \to w \qquad t \to x \qquad T \to 2L$$

Equation (6.5b) is converted into an expression for the length of the interval:

$$L = 2 \int_0^A \frac{dw}{\sqrt{(A^2 - w^2) - \frac{1}{2}(A^4 - w^4)}} \qquad (6.51)$$

Here we do not have a small parameter. However, for small amplitudes, the integral can be estimated by a first-order approximation

NORMAL FORMS: CONSERVATIVE PLANAR SYSTEMS

$$L = \pi\left(1 + \tfrac{3}{8}A^2\right)$$

yielding

$$A = \pm\sqrt{\tfrac{8}{3}\frac{(L-\pi)}{\pi}} \tag{6.52}$$

Equation (6.52) is valid only for $L \geq \pi$, where the amplitude grows along two branches, and is proportional to the small parameter $\sqrt{(L-\pi)}$. Thus, in the equation, one must scale the amplitude by the latter.

For $L < \pi$, only $A=0$ is allowed. Thus, the solution changes character at the critical value $L=\pi$, and we have a bifurcation. The resulting bifurcation diagram is shown in Fig. 6.6, where the broken line is the unperturbed bifurcation line.

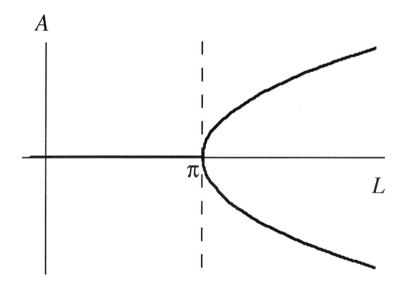

Figure 6.6 Bifurcation diagram for Equation (6.47) in the vicinity of $L=\pi$.

We now want to show the significance of such a bifurcation within the framework of a normal form expansion. One may immediately raise a valid question. The method of normal forms is one of several perturbation schemes aimed at eliminating secular behavior, that is, for extending the validity of the perturbation expansion for long times

BOUNDARY-VALUE PROBLEMS

(here, for long distances in x). However, the problem we are studying now is to be solved over a *finite* interval. Thus, secular terms are no threat to the validity of a perturbation expansion, and any method, including the naive expansion, would do. Indeed, for this reason, boundary-value problems are studied in the literature by a variety of expansion procedures, some of which are not concerned with the elimination of secular terms. Why, then, should we bother at all to use the elaborate procedure of a normal expansion?

Analyzing Equation (6.47) by the method of normal forms will elucidate the concept of bifurcations from a particular point of view. We will show that the occurrence of a bifurcation and the type of required rescaling are both related to the notion of frequency update, discussed extensively in this chapter. In the present problem, time is replaced by a spatial coordinate, x. Therefore, instead of a frequency update, we will talk about a wavenumber update. Before the boundary conditions are imposed, the method of normal forms yields a zero-order approximation to the full solution of the form

$$w_0(x) = A \sin k x + B \cos k x \qquad (6.53)$$

The boundary conditions imply that in Equation (6.53) we must have

$$B = 0 \qquad k = \frac{\pi}{L}$$

Notice that this value of k is different from the unperturbed value of 1, used in Equations (6.48) and (6.49). On the other hand, in a normal form analysis, we expect that the nonlinear perturbation will generate a modification of the wavenumber, k, by a small- amplitude-dependent correction:

$$k = 1 + O(A, B)$$

With $B=0$, the updated wavenumber, as computed in the normal form expansion, will be a function of the amplitude A. In general, it will be different from π/L. The latter value will be obtained only for a specific value of A. That particular value will provide a solution to Equation (6.47). The amplitude will turn out to depend on a small parameter: $(L-\pi)$. In the resulting bifurcation curve, the degeneracy exhibited by the solution of the unperturbed equation, Equation (6.48) (that at $L=\pi$, the amplitude A may take on any value; see Fig. 6.5), is removed.

To see how the results stated above are obtained in a perturbation scheme, we first change variables and define a small parameter:

$$x = \frac{Lt}{\pi} \qquad \left(\frac{L}{\pi}\right)^2 = 1 + \varepsilon \quad |\varepsilon| \ll 1$$

Equation (6.47) becomes

$$\frac{d^2w}{dt^2} + w + \varepsilon w - (1+\varepsilon)w^3 = 0 \qquad (6.54)$$

$$w(0) = w(\pi) = 0$$

Comment. Before embarking on a normal form analysis of Equation (6.54), we note that, with the boundary conditions taken into account, we can write the solution as a Fourier series:

$$w(t) = \sum_{n=1}^{\infty} A_n(\varepsilon) \sin nt$$

Inserting this expansion in Equation (6.54), we can solve recursively for the coefficients, A_n. Thus, for the equation to be satisfied, consistency requires that

$$A_1 = \sqrt{\tfrac{4}{3}\varepsilon} + O(\varepsilon^{3/2})$$

Hence, $w(t) = O(\varepsilon^{1/2})$. The ε dependence of each Fourier coefficient is updated as one proceeds with the expansion.

Now we turn to the normal form analysis of Equation (6.54). Anticipating that the amplitude will be small, we extract an explicit small size parameter in w:

$$w = \delta v$$

The equation now becomes

$$\frac{d^2v}{dt^2} + v + \varepsilon v - (1+\varepsilon)\delta^2 v^3 = 0 \qquad v(0) = v(\pi) = 0$$

The last preparatory step is to diagonalize the equation by writing $z = v + i\, dv/dt$, to obtain

BOUNDARY-VALUE PROBLEMS

$$\dot{z} = -i\,z - \tfrac{1}{2} i\,\varepsilon(z+z^*) + \tfrac{1}{8} i\,(1+\varepsilon)\delta^2(z+z^*)^3$$

$$\mathrm{Re}\,z(0) = \mathrm{Re}\,z(\pi) = 0$$

(6.55)

Comments

1. With the change of variables, the wavevector k should be replaced by $q = (L/\pi)\cdot k$. Thus, the boundary conditions require that $q=1$.
2. Although we expect the small size parameter, δ, to be related to ε, we leave it as independent, so as to demonstrate how scaling emerges as a natural consequence in the normal form approach.
3. By identifying the small parameter ε, the problem has obtained a richer structure than that of the original equation. There is a competition between a cubic and a linear term. This competition will enable us to find a solution with a small but nonvanishing amplitude obeying the constraint $q=1$.

We now perform a normal form expansion in the small parameters, ε and $\mu = \delta^2$. (Note that δ appears squared to begin with.) We write

$$z = u + \varepsilon T_{10} + \mu T_{01} + \cdots \qquad (6.56)$$

The notation for T_{mn} is evident; the first index denotes the power of ε and the second one, powers of μ. The normal form analysis yields through first order

$$U_{10} = -\tfrac{1}{2} i\, u \qquad T_{10} = -\tfrac{1}{4} u^* + f_{10}(u u^*) u$$

$$U_{01} = \tfrac{3}{8} i\, u^2 u^* \qquad (6.57)$$

$$T_{01} = \tfrac{1}{16} u^3 - \tfrac{3}{16} u u^{*2} - \tfrac{1}{32} u^{*3} + f_{01}(u u^*) u$$

Therefore, the normal form in first order is

$$\dot{u} = -i\,u - \tfrac{1}{2} i\,\varepsilon u + \tfrac{3}{8} i\,\mu u^2 u^* \qquad (6.58)$$

Writing

$$u = \rho \cdot \exp(-i\,\varphi) \qquad \varphi = q\cdot t + \varphi_0$$

204 NORMAL FORMS: CONSERVATIVE PLANAR SYSTEMS

we obtain for the updated wavenumber, q, the following expression:

$$q = 1 + \tfrac{1}{2}\varepsilon - \tfrac{3}{8}\mu\rho^2 \tag{6.59}$$

Imposing the boundary conditions on the first-order approximation for z at both $t=0$ and $t=\pi$ ($x=0$ and $x=L$, respectively), we find that one must have

$$q = 1$$
$$\varphi_0 = \pm \frac{\pi}{2} \mp \varepsilon \operatorname{Im} f_{10} \mp \mu \operatorname{Im} f_{01} \tag{6.60}$$

(The ± signs indicates that there are two branches.) We set the free functions equal to zero, as they are not useful at the level of approximation we have chosen here. The requirement that $q=1$ means that in this order of the approximation we must have

$$\tfrac{1}{2}\varepsilon = \tfrac{3}{8}\mu\rho^2 = \tfrac{3}{8}\delta^2\rho^2 \tag{6.61}$$

We first note that Equation (6.61) requires that $\varepsilon>0$, that is, $L>\pi$. Therefore, the amplitude A of the oscillations must satisfy a relation identical to Equation (6.52):

$$A = \begin{cases} 0 & L<\pi \\ \pm\delta\rho = \pm\left(\tfrac{4}{3}\varepsilon\right)^{1/2} = \pm\left(\tfrac{4}{3}\dfrac{L^2-\pi^2}{\pi^2}\right)^{1/2} & L\geq\pi \\ \cong \pm\left(\tfrac{8}{3}\dfrac{L-\pi}{\pi}\right)^{1/2} \end{cases} \tag{6.62}$$

In summary, the modification of the trivial bifurcation diagram of the unperturbed problem [Equation (6.48)] into an interesting diagram can be viewed as a consequence of the "conflict" between two effects: the tendency of the nonlinear perturbation to modify the wavenumber and the requirement that q remains at 1 (k remains at π/L in the original variable, x) imposed by the boundary conditions.

We now give another example of an oscillatory system that is perturbed by a nonlinear term such that no bifurcation occurs. We will show that this result goes hand in hand with the fact that the solution to that particular equation, when analyzed by the method of normal forms, does not have a wavenumber update. Consider the equation

$$\frac{d^2w}{dx^2} + w + \varepsilon\left(-2w + 6w\frac{dw}{dx}\right)$$
$$+ \varepsilon^2\left(w - 6w\frac{dw}{dx} + 4w^3\right) - 8\varepsilon^3 w^3 + 4\varepsilon^4 w^3 = 0 \qquad (6.63a)$$

$$w(0) = w(L) = 0, \qquad \varepsilon = 1 - \frac{\pi}{L}, \qquad |\varepsilon| \ll 1$$

Clearly, this equation has been concocted to achieve our goal. In fact, it can be recast in the form

$$\frac{d^2w}{dx^2} + \left(\frac{\pi}{L}\right)^2 w + 6\varepsilon\left(\frac{\pi}{L}\right)w\frac{dw}{dx} + 4\varepsilon^2\left(\frac{\pi}{L}\right)^2 w^3 = 0$$
$$(6.63b)$$
$$w(0) = w(L) = 0$$

We first note that the unperturbed linear part of Equation (6.63b) together with the boundary conditions is solved, for *any* L, by

$$w_0(x) = A \sin k_0 x \qquad k_0 = \frac{\pi}{L}$$

with no constraints on A.

The full Equation (6.63a) is solved, for *any* L, by

$$w(x) = \frac{A \sin\left(\frac{x\pi}{L}\right)}{1 - 2\varepsilon A \cos\left(\frac{x\pi}{L}\right)} \qquad (6.64)$$

with A arbitrary. [$A = \pm 1/(2\varepsilon)$ included!] The solution obeys the boundary conditions. It is periodic in x with period $2L$. An expansion of the solution in a Fourier expansion would, therefore, be of the form

$$w(x) = \sum_n a_n \sin\left(n\frac{x\pi}{L}\right)$$

Thus, the fundamental wavenumber is still $k_0 = \pi/L$. The nonlinear perturbation has not modified the wavenumber.

206 NORMAL FORMS: CONSERVATIVE PLANAR SYSTEMS

Returning to Equation (6.63a), we note that the perturbation added to the (unperturbed) linear part is such that it eliminates the bifurcation phenomenon completely. Let us see how this feature is understood from the point of view of a normal form expansion. For the sake of convenience, we again change variables form x to t, given by

$$t = \frac{\pi x}{L}$$

Equation (6.63b) now becomes

$$\frac{d^2w}{dt^2} + w + 6\varepsilon w \frac{dw}{dt} + 4\varepsilon^2 w^3 = 0 \tag{6.65}$$

$$w(0) = w(\pi) = 0 \qquad \varepsilon = 1 - \frac{\pi}{L} \qquad |\varepsilon| \ll 1$$

Equation (6.65) is obviously solved by

$$w(x) = \frac{A \sin t}{1 - 2\varepsilon A \cos t} \tag{6.66}$$

with arbitrary A.

For the normal form analysis, we diagonalize Equation (6.65) by using $z = w + i\, dw/dt$. The equation for z is

$$\dot{z} = -i\,z - \tfrac{3}{2}\varepsilon\!\left(z^2 - z^{*2}\right) - \tfrac{1}{2} i\, \varepsilon^2 \!\left(z + z^*\right)^3 \tag{6.67}$$

$$\operatorname{Re} z(0) = \operatorname{Re} z(\pi) = 0$$

Computing the normal form through second order, we find

$$U_1 = U_2 = 0 \qquad T_1 = -\tfrac{3}{2} i\, u^2 - \tfrac{1}{2} i\, u^{*2} + f_1(uu^*)\cdot u \tag{6.68}$$

As expected, there is no frequency update (at least through second order). Therefore, u, the zero-order approximation, can be written as

$$u = \rho \cdot \exp(-i\,t - i\,\varphi_0) \tag{6.69}$$

To impose the boundary conditions, we write the real part of z through first order:

BOUNDARY-VALUE PROBLEMS

$$w = \operatorname{Re} z = \rho\cos(t + \varphi_0) + \varepsilon \begin{pmatrix} -\rho^2 \sin(2t + 2\varphi_0) \\ + \operatorname{Re} f_1\, \rho\cos(t + \varphi_0) \\ + \operatorname{Im} f_1\, \rho\sin(t + \varphi_0) \end{pmatrix} \quad (6.70)$$

The boundary conditions of Equation (6.67) require

$$\varphi_0 = \pm \frac{\pi}{2} \qquad \operatorname{Im} f_1 = 0$$

converting Equation (6.70) into

$$w = \mp\rho\sin t\{1 + \varepsilon(\mp 2\rho\cos t + \operatorname{Re} f_1)\} \quad (6.71)$$

We note that with the choice

$$\operatorname{Re} f_1 = 0$$

Equation (6.71) is the first-order approximation in the expansion of

$$w = \frac{(\mp\rho)\sin t}{1 - 2\varepsilon(\mp\rho)\cos t}$$

which is just in the form of Equation (6.66).

In summary, Equation (6.63) constitutes an example in which the perturbation eliminates the bifurcation structure from the solution of the unperturbed equation. Within the framework of the method of normal forms, this phenomenon is associated with the *lack of wavenumber* or *frequency update*.

REFERENCES

1. Perko, L.. M., "Higher Order Averaging and Related Methods for Perturbed Periodic and Quasi-Periodic Systems," *SIAM J. Appl. Math.* **17**, 698 (1968).
2. Kahn, P. B., and Y. Zarmi, "Minimal Normal Forms in Harmonic Oscillations with Small Nonlinear Perturbations," *Physica D* **54**, 65 (1991).
3. Landau, L. D., and E. M. Lifshitz, *Mechanics*, Pergamon Press, Bristol (1960).
4. Davis, H. T., *Introduction to Nonlinear Differential and Integral Equations*, Dover, New York (1962).

5. Hellemann, R. H. G., and E. W. Montroll, "On a Nonlinear Perturbation Theory without Secular Terms: 1. Classical Coupled Anharmonic Oscillators," *Physica* **74**, 22 (1974).
6. Kahn, P. B., D. Murray, and Y. Zarmi, "Freedom in Small Parameter Expansion for Nonlinear Perturbations," *Proc. Roy. Soc.* (London) **A443**, 83 (1993).

7

DISSIPATIVE PLANAR SYSTEMS

In this chapter we consider dissipative systems that are derived from perturbed harmonic oscillators. In these systems, both frequency and amplitude updating may occur. For this reason, one may choose the free function to have a complex coefficient.

The systems fall into two classes:

Systems with a fixed point. In simple dissipative systems, motion is characterized by attraction to, or repulsion from, a fixed point. If the fixed point is attractive, we call this *positive damping*. Such systems execute damped oscillations. The freedom associated with the T functions can be exploited to obtain equations that are more compact than in the traditional NF method, so that the characteristics of the flow are captured early on and the computation is fairly easy.

Limit cycle systems. The next level of intricacy is attraction to, or repulsion from, a closed periodic orbit. Such an orbit is called a *limit cycle*. When the trajectory is attracted to an asymptotic closed periodic orbit, the latter is called a *stable limit cycle* (Figure 7.1a). When the trajectory is repelled away from a closed periodic orbit, the latter is called an *unstable limit cycle* (Figure 7.1b). (A limit cycle that is unstable as $t \to \infty$, can be viewed as a stable one for $t \to -\infty$.) Rayleigh introduced the notion of limit cycles in [1,2], where he developed an approximate model for the description of maintained vibrations that result from the interplay of negative and positive damping terms. He mentions that this situation occurs in various systems: organ pipes, violin strings, electromagnetic tuning forks, etc. (Such phenomena were traditionally analyzed by the method of harmonic balance.) Van der Pol developed some seminal arguments that led to a systematic attack on limit cycle problems [3].

Limit cycles may occur in systems of any dimensionality. In the present chapter, we focus on limit cycle problems in the plane. Moreover, we address only problems of linear harmonic motion that is perturbed by a small dissipative nonlinearity. The unperturbed system then has a "center" as its equilibrium point. The eigenvalues of the

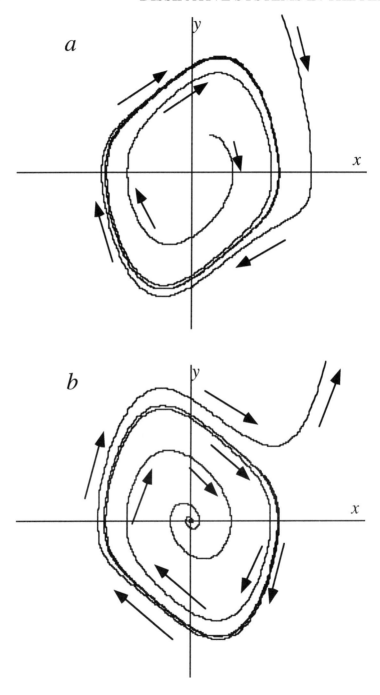

Figure 7.1 Examples of stable (a) and unstable (b) limit cycles. Arrows indicate direction of motion as time grows.

DISSIPATIVE SYSTEMS IN THE PLANE

unperturbed system are $\pm i$. This corresponds to a closed orbit that is a circle in the phase plane. The amplitude of that circle is a free parameter, determined by the total energy of the motion. As a parameter representing the dissipative perturbation is varied, stability of the circular motion is lost. Of the infinity of possible unperturbed circles, one particular circle is selected by the perturbation as the basis for the evolution of a limit cycle. The onset of a limit cycle phenomenon owing to the variation of a control parameter, is referred to as a *Hopf bifurcation*.

We develop the normal form expansion for limit cycle systems in the plane and find that the notion of minimal normal forms is not the best choice for such problems. Rather, it is better to use the free functions to fix the values of the dynamical parameters, associated with the onset of limit cycle behavior, in a low order of the calculation. Specifically, the u equation is constructed so that there is no updating of the radius of the *limit circle* (which is the asymptotic limit of the zero-order approximation), as the higher-order terms in ε are incorporated. We say that the system has been *renormalized*.

7.1 OSCILLATORS WITH CUBIC DAMPING

In this section we study three oscillators that are subjected to cubic damping: a linear oscillator, the pendulum, and the Duffing oscillator. In all cases, due to the damping, the origin is a stable attractor.

7.1.1 Linear oscillator with cubic damping

Consider the following equation for cubic damping:

$$\ddot{x} + x + \varepsilon \dot{x}^3 = 0, \qquad x(0) = a, \quad \dot{x}(0) = b, \quad |\varepsilon| \ll 1 \quad (7.1a)$$

In the diagonalized complex notation $z = x + iy$, Equation (7.1a) becomes

$$\dot{z} = -i\,z + \tfrac{1}{8}(z - z^*)^3 \qquad (7.1b)$$

In the near-identity transformation, the T functions through $O(\varepsilon^2)$ are:

$$T_1 = \tfrac{1}{16} i\, u^3 - \tfrac{3}{16} i\, u u^{*2} + \tfrac{1}{32} i\, u^{*3} + F_1 \qquad (7.2a)$$

$$T_2 = -\tfrac{9}{1024} u^5 + \tfrac{9}{256} u^4 u^* - \tfrac{27}{512} u^2 u^{*3} - \tfrac{27}{1024} u u^{*4} + \tfrac{3}{512} u^{*5}$$
$$+ \tfrac{3}{16} i\, F_1 u^2 - \tfrac{3}{16} i\, F_1 u^{*2} - \tfrac{3}{8} i\, F_1^* u u^* + \tfrac{3}{32} i\, F_1^* u^{*2} + F_2$$

Writing the free functions appearing in T_1 and T_2 as $F_1 = f_1 \cdot u$ and $F_2 = f_2 \cdot u$, respectively, the U functions are determined and yield through $O(\varepsilon^3)$ the following normal form:

$$\frac{du}{dt} = -i\,u - \varepsilon \tfrac{3}{8} u^2 u^*$$

$$+ \varepsilon^2 \begin{pmatrix} -\tfrac{3}{8}(f_1 + f_1^*) u^2 u^* + \tfrac{3}{4} f_1' u^3 u^{*2} \\ + \tfrac{27}{256} i\, u^3 u^{*2} \end{pmatrix}$$

$$+ \varepsilon^3 \begin{pmatrix} -\tfrac{3}{8}(2 f_1^2 + 3 f_1 f_1^*) u^2 u^* \\ + \left[\tfrac{27}{128} i\,(f_1 + f_1^*) - \tfrac{3}{4} f_1 f_1' \right] u^3 u^{*2} \\ - \tfrac{3}{4}(f_1' + f_1'^*) f_1' u^4 u^{*3} - \tfrac{3}{8}(f_2 + f_2^*) u^2 u^* \\ + \tfrac{3}{4} f_2' u^3 u^{*2} - \tfrac{567}{8192} u^4 u^{*3} \end{pmatrix} \tag{7.2b}$$

The normal form includes resonant contributions in each order. Through sixth order in ε, for example, one then obtains in the *usual choice*

Usual NF:
$$\frac{du}{dt} = -i\,u - \varepsilon \tfrac{3}{8} u^2 u^* + \varepsilon^2 \tfrac{27}{256} i\, u^3 u^{*2}$$
$$- \varepsilon^3 \tfrac{567}{8192} u^4 u^{*3} + \varepsilon^4 \tfrac{7965}{262144} i\, u^5 u^{*4} \tag{7.2c}$$
$$- \varepsilon^5 \tfrac{62721}{4194304} u^6 u^{*5} + \varepsilon^6 \tfrac{214083}{268435456} i\, u^7 u^{*6}$$

Equations (7.2c) is updated continually as one proceeds to higher orders.

Writing
$$u = \rho e^{-i\phi}$$

Equation (7.2c) gives

$$\frac{d\rho}{dt} = -\tfrac{3}{8} \varepsilon \rho^3 - \tfrac{567}{8192} \varepsilon^3 \rho^7 - \tfrac{62{,}721}{4{,}194{,}304} \varepsilon^5 \rho^{11} + O(\varepsilon^7) \tag{7.3a}$$

$$\frac{d\phi}{dt} = \omega = 1 - \tfrac{27}{256}\varepsilon^2 \rho^4 - \tfrac{7965}{262,144}\varepsilon^4 \rho^8$$
$$- \tfrac{214,083}{268,435,456}\varepsilon^6 \rho^{12} + O(\varepsilon^7) \quad (7.3b)$$

Equations (7.3a,b) are typical of dissipative systems; in general, both the frequency ω and the radius ρ are updated. In the *usual choice* (no free functions) the radius and phase equations quickly become calculationally unwieldy, as is indcated by Equations (7.3a,b). (For additional details and technical aspects of this and related calculations, see Ref. 4.)

However, the free functions can be chosen to eliminate the second- and third-order terms U_2 and U_3. For instance, requiring that U_2 vanishes constitutes an equation for f_1, which is solved by

$$f_1 = \left(A - \tfrac{9}{64}i\right)uu^* + B$$

Here A and B are arbitrary real constants. The simplest choice, $A=B=0$, yields

$$f_1 = -\tfrac{9}{64}i\, uu^*$$

In a similar manner, the simplest choice for f_2 that eliminates U_3 is

$$f_2 = \tfrac{675}{8192}(uu^*)^2$$

Higher-order resonant contributions to the normal form can be also eliminated to yield the following expression, correct to all orders:

MNF: $\qquad \dfrac{du}{dt} = -i\,u - \tfrac{3}{8}\varepsilon u^2 u^* \qquad (7.4a)$

$\dfrac{d\rho}{dt} = -\tfrac{3}{8}\varepsilon\rho^3 \qquad \dfrac{d\phi}{dt} = \omega = 1 \qquad (7.4b)$

Solving Equation (7.4b), one obtains

$$\phi = t + \phi_0, \quad \Rightarrow \omega = 1 \quad (7.5)$$

$$\rho = \frac{\rho_0}{\sqrt{1 + \tfrac{3}{4}\varepsilon\rho_0^2 t}} \quad (7.6)$$

Thus, the minimal normal form captures the behavior of the zero-order approximation in a simple manner. Both the fundamental frequency of oscillation and the radius are determined at this order and are not updated by higher-order corrections. Moreoever, both are given by extremely simple expressions. In particular, although the eigenvalues of the unperturbed problem lie in a Siegel domain, the frequency, ω, remains equal to 1. Higher-order corrections show up only in T_n, in the near-identity transformation.

In Figure 7.2 we compare the numerical solution of Equation (7.1) for the initial conditions $x(0)=1$, $y(0)=0$, and positive damping ($\varepsilon=+\frac{3}{4}$), with two second-order approximations (i.e., x is computed from $z=u+\varepsilon T_1+\varepsilon^2 T_2$): with (1) MNF, and (2) a *usual* normal form (no free functions). The results of both choices are in good agreement with the numerical solution and display long time validity. A large value of ε has been chosen to provide a stricter test for the quality of the approximation. (For small ε, the higher-order corrections are small, anyway, so that any zero-order approximation suffices.)

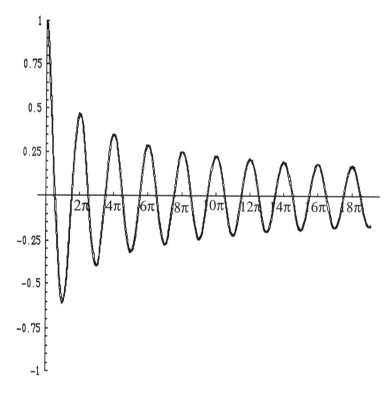

Figure 7.2 Solution and approximations to solution of Equation (7.1), $\varepsilon=+\frac{3}{4}$, $x(0)=1$, $dx(0)/dt=0$.

The two approximations are excellent and cannot be distinguished on the scale of the graph. Studying the numbers in greater detail, we find that the MNF choice still fares better. The *usual* normal form approximation misses the exact solution by an error that is just under 3% near $t=0$ and reaches 1.6% around $t=19\pi$. The MNF approximation errors are 0.4% and 0.02%, respectively. Clearly, for sufficiently long times, the zero-order terms in *both* choices become equally good approximations, as they approach the exact solution. All higher-order terms, being polynomials in u and u^* of degree at least 3, vanish more rapidly.

7.1.2 Pendulum with cubic damping

Consider the equation

$$\frac{d^2x}{dt^2} + \left(\frac{dx}{dt}\right)^3 + \sin x = 0 \tag{7.7a}$$

Converting Equation (7.7a) into a set of two first-order equations:

$$\frac{dx}{dt} = y \qquad\qquad \frac{dy}{dt} = -\sin x - y^3 \tag{7.7b}$$

one finds that the system has fixed points at $(x_n, y_0) = (n\pi, 0)$.

A linear stability analysis reveals that, for odd n, the point is a saddle node, while, for even n, it is a neutrally stable fixed point. To see what the actual nature of the fixed points is, one must take into account the effect of the nonlinearity. Figure 7.3 shows the first few fixed points and a sample of trajectories, obtained by the numerical solution of Equation (7.7a). The saddle nodes retain their nature, while the neutrally stable fixed points (the origin in Figure 7.3) become attractive. To understand this, we need to perform a detailed analysis. As we are interested in the vicinity of the origin, we can safely apply perturbation theory for small amplitudes. We introduce the scaling

$$x = \sqrt{\varepsilon}\,\tilde{x}$$

and expand the sine function in a Taylor series. The equation of motion for the rescaled variable becomes

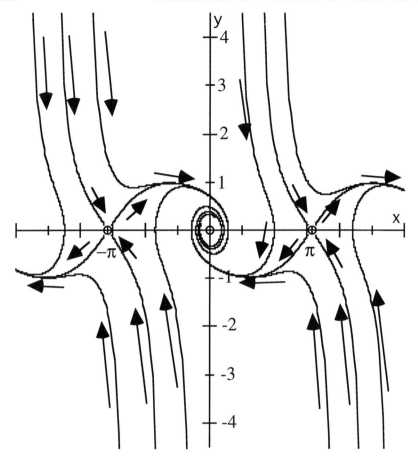

Figure 7.3 Fixed points and typical trajectories of pendulum with cubic damping [Equation (7.7a)]. Arrows indicate direction of motion in time.

$$\frac{d^2\tilde{x}}{dt^2} + \tilde{x} + \varepsilon\left\{\left(\frac{d\tilde{x}}{dt}\right)^3 - \tfrac{1}{6}\tilde{x}^3\right\} + \sum_{n=2}^{\infty} \varepsilon^n (-1)^n \frac{\tilde{x}^{2n+1}}{(2n+1)!} = 0 \quad (7.7c)$$

In the diagonalized (complex) form, this equation becomes

$$\dot{z} = -i\,z + \varepsilon\left\{i\,\tfrac{1}{48}(z+z^*)^3 + \tfrac{1}{8}(z-z^*)^3\right\} \\ -i\sum_{n=2}^{\infty} \varepsilon^n (-1)^n \frac{(z+z^*)^{2n+1}}{2^{2n+1}(2n+1)!} \quad (7.7d)$$

OSCILLATORS WITH CUBIC DAMPING

The normal form analysis yields (a prime denotes a derivative with respect to uu^*)

$$U_1 = \{-\tfrac{3}{8} + \tfrac{1}{16}i\}u^2 u^* \tag{7.8a}$$

$$T_1 = \left(\tfrac{1}{16}i - \tfrac{1}{96}\right)u^3 + \left(-\tfrac{3}{16}i + \tfrac{1}{32}\right)u^2 u^* + \left(\tfrac{1}{32}i + \tfrac{1}{192}\right)u^{*3} + f_1 u \tag{7.8b}$$

$$U_2 = \{\tfrac{1}{64} + \tfrac{111}{1024}i\}u^3 u^{*2}$$
$$+ \{-\tfrac{3}{8} + \tfrac{1}{16}i\}(f_1 + f_1^*)u^2 u^* + \tfrac{3}{4}f_1' u^3 u^{*2} \tag{7.8c}$$

Writing $u = \rho \cdot \exp(-i\varphi)$, the normal form equations for ρ and φ through $O(\varepsilon)$ are

$$\frac{d\rho}{dt} = -\varepsilon \tfrac{3}{8}\rho^3 \tag{7.9a}$$

$$\frac{d\varphi}{dt} = 1 - \varepsilon \tfrac{1}{16}\rho^2 \tag{7.9b}$$

Thus, the amplitude decays to zero (reflecting the fact that the origin is a stable attracting fixed point) and the frequency update is time-dependent.

The question that arises is whether a near-identity transformation exists so that Equations (7.9a,b) become a minimal normal form. To achieve this, we first need to find an f_1 that makes U_2 vanish identically. Requiring the latter, Equation (7.8c) becomes a differential equation for the uu^* dependence of f_1 that is solved by

$$f_1 = \alpha_1 uu^* - \tfrac{1}{48}uu^* \ln(uu^*)$$
$$+ i\left\{\left(-\tfrac{37}{256} - \tfrac{1}{6}\alpha_1\right)uu^* + \tfrac{1}{288}\left(uu^* \ln(uu^*) - uu^*\right) + \alpha_2\right\} \tag{7.10}$$

The constants α_1 and α_2 are undetermined in this order. The real part of f_1 is responsible for eliminating the real part of U_2 [which would have affected the Equation (7.9a) for ρ], and the imaginary part, for eliminating a second order frequency update of Equation (7.9b).

This form of the free function is not of the type we have been using throughout this text. It cannot be written as a polynomial or an entire function of uu^*. However, one need not worry about the singular behavior of the ln(uu^*) term in $f_1 \cdot u$ when the amplitude tends to zero, since it is multiplied by $u^2 u^*$; hence it decays as $\rho^3 \ln(\rho^2)$ when $\rho \to 0$. A detailed study of higher orders reveals that U_n can be also eliminated at the price of the recurrence of a ln(uu^*) term or powers of it. Again, each ln(uu^*) is multiplied by ρ^k, with $k \geq 3$. Thus, no singularity develops in the near-identity transformation at $\rho = 0$.

In summary, a minimal normal form can be obtained, given by Equations (7.9a,b), at the price of a near-identity transformation, the structure of which is different from what we have become used to.

Suppose that one insists on having a near-identity transformation of a simpler form, namely, that it can be written in terms of polynomials or entire functions of uu^*. Then one cannot eliminate ReU_2, but *can* eliminate ImU_2, by choosing for f_1

$$f_1 = i\left\{-\tfrac{37}{256} uu^* + \alpha_2\right\} \tag{7.11}$$

A detailed study shows that, once the requirement of eliminating ReU_2 (the part that affects the amplitude equation) is dropped, *all $U_{n \geq 3}$ can be eliminated* by polynomial f_n. The resulting near-identity transformation has a much simpler structure, while the MNF is modified only slightly. Equation (7.9b) for the phase remains unchanged, and Equation (7.9a) for the amplitude is replaced by

$$\frac{d\rho}{dt} = -\varepsilon \tfrac{3}{8}\rho^3 + \varepsilon^2 \tfrac{1}{64}\rho^5 \tag{7.12}$$

Thus, exploiting the nonuniqueness of the expansion, one derives compact expressions, exact to all orders - the MNF. By contrast, the *usual* case has a frequency and radius update in each order.

Note that Equation (7.12) has an unstable fixed point at $\rho^2 = 24/\varepsilon$. Clearly, at such large amplitudes, perturbation theory is not valid. In the original variables of Equation (7.7a), this corresponds to amplitudes of the order of $(24)^{1/2} = 2\sqrt{6}$. This may be a crude reminder of the saddle node fixed point at $x = \pi$, $y = 0$ of the original equation.

7.1.3 Duffing oscillator with cubic damping

The conclusions of the preceding section are not particular to the pendulum. Consider, for example, a Duffing oscillator with cubic damping:

$$\ddot{x} + x + \left(-bx^3 + a\dot{x}^3\right) = 0 \tag{7.13}$$

OSCILLATORS WITH CUBIC DAMPING

where a and b are positive real coefficients. The fixed points are at

$$(x,y) = \begin{cases} (0,0) \\ \left(\pm \dfrac{1}{\sqrt{b}}, 0\right) \end{cases} \quad (y = \dot{x})$$

As in the case of the damped pendulum, the origin is an attractor, while the other two fixed points are saddle points. Again, to analyze Equation (7.13) near the origin, we rescale the variable x by writing

$$x = \sqrt{\varepsilon}\, \tilde{x}$$

Equation (7.13) becomes

$$\ddot{\tilde{x}} + \tilde{x} + \varepsilon\left(-b\tilde{x}^3 + a\dot{\tilde{x}}^3\right) = 0 \qquad (7.14)$$

A minimal normal form can be obtained that terminates in $O(\varepsilon)$, at the price of the appearance of harmless logarithms in the near-identity transformation. Except for appropriate modifications in numerical coefficients, it will be identical to Equations (7.9a,b).

Alternatively, one may insist on having a near-identity transformation with a simpler structure, i.e., that is composed of polynomials only in u and u^*. The resulting MNF terminates in $O(\varepsilon^2)$:

$$\dot{u} = -i\,u + \tfrac{3}{8}\varepsilon(ib-a)u^2 u^* + \tfrac{3}{32}\varepsilon^2 a b u^3 u^{*2} \qquad (7.15a)$$

Equation (7.15a) yields an equation for ρ, ($u = \rho \cdot \exp(-i\varphi)$):

$$\dot{\rho} = -\tfrac{3}{8}\varepsilon a \rho^3 + \tfrac{3}{32}\varepsilon^2 a b \rho^5 \qquad (7.15b)$$

Equation (7.15b) has a stable fixed point at $\rho = 0$, and an unstable one at $\rho^2 = [4/(b\varepsilon)]$. In the original, unscaled variable, this corresponds to an amplitude of the order of $(4/b)^{1/2}$, which is of the order of the location of the saddle fixed point.

Apart from numerical constants, Equation (7.15b) has the same structure as Equation (7.12) for the damped pendulum. In both cases, $\rho = 0$ is a stable fixed point, and an unstable fixed point exists at a large value of ρ, (reminiscent of an unstable fixed point in the original problem), where perturbation theory is invalid.

Example 7.1. The preceding analysis also applies to a generalization of Equation (7.14):

$$\ddot{x} + x + \varepsilon\left(a_{30} x^3 + a_{21} x^2 \dot{x} + a_{12} x \dot{x}^2 + a_{03} \dot{x}^3\right) = 0 \quad (7.16a)$$

If we choose $a_{30}<0$, then Equation (7.16a) has fixed points at

$$(x,y) = \begin{cases} (0,0) \\ \left(\dfrac{\pm 1}{\sqrt{-a_{30}}}, 0\right), \end{cases} \quad (y = \dot{x})$$

In the complex (diagonalized) notation, Equation (7.16a) becomes

$$\dot{z} = i\,z + \varepsilon\left\{\alpha z^3 + \beta z^2 z^* + \gamma z z^{*2} + \delta z^{*3}\right\} \quad (7.16b)$$

The complex constants α, β, γ, and δ are combinations of a_{ij} of Equation (7.16a).

The $z^2 z^*$ term, the only resonant one in Equation (7.16b), is the only one to contribute to U_1. The expression for β is

$$\beta = \left\{-\tfrac{1}{8} a_{21} - \tfrac{3}{8} a_{03}\right\} - i\left\{\tfrac{3}{8} a_{30} + \tfrac{1}{8} a_{12}\right\} \quad (7.17)$$

Re β contributes to the amplitude equation and Im β, to the phase equation. If α, β, γ, and δ are such that a dissipation term is generated in $O(\varepsilon^2)$, then the problems encountered in the previous examples occur here as well. Namely, for a MNF that terminates in $O(\varepsilon)$, the near-identity transformation must include (harmless) $\ln(uu^*)$ terms. [Note that if $a_{21}=-3a_{03}$, so that Re β=0, then the amplitude equation is not affected by the dissipation in $O(\varepsilon)$ and one must go to, at least $O(\varepsilon^2)$, to see the dissipative effect.]

In conclusion, Equation (7.1) for the linear oscillator with cubic damping, Equation (7.7) for the pendulum with cubic damping, and Equation (7.14) for the Duffing oscillator with cubic damping and its generalization of Equation (7.16) have almost identical minimal normal forms. One can obtain for them minimal normal forms that terminate in $O(\varepsilon)$. The resulting first-order equations for the amplitude ρ are the same, except for possible differences in numerical coefficients. This is a reflection of the fact that the origin is a stable attractor in all cases. The MNFs differ only in the phase update.

In all cases, except for the damped linear oscillator, the near-identity

OSCILLATORS WITH CUBIC DAMPING

transformation could not be constructed merely from polynomials of u and u^*. Harmless $\ln(uu^*)$ terms were required to obtain a MNF that terminates in $O(\varepsilon)$. By "harmless" we mean that each $\ln(uu^*)$ term in the near-identity transformation was multiplied by at least u^2u^*, so that the singularity of the logarithm at the origin is eliminated.

Also, the nonlinear damped oscillators of Equations (7.7), (7.14), and (7.16) are equivalent. Near-identity transformations exist that generate for them the *same* minimal normal forms (except for possible differences in numerical coefficients). This is true for MNFs that terminate in either $O(\varepsilon)$ or $O(\varepsilon^2)$. The resulting MNF reflects the fact that in each case the origin is a stable fixed point. Also, the MNFs that terminate in $O(\varepsilon^2)$, in some crude sense, "remind" us of the fact that far away, in a domain where the perturbative analysis *around the origin* does not apply, the original equation has a saddle fixed point.

Finally, in a damped system, the effect of higher orders is diminished as the radius decreases. Hence, all perturbation expansions become comparable. However, the MNF expression is the most compact one.

7.2 LIMIT CYCLE SYSTEMS

In this section, we exploit the freedom in the normal form expansion to simplify the description of systems that have limit cycles.

7.2.1 Definition of a limit cycle

A stable limit cycle is a periodic closed curve in phase space that is an attractor for all trajectories in some domain in phase space. There are no other closed periodic trajectories arbitrarily close to the limit cycle. The occurrence of limit cycles in harmonic oscillations that are perturbed by a small nonlinear perturbation has been studied extensively in the literature [1–3, 5–16]. In this case, it is convenient to use the zero-order approximation as a basis, and to picture the approaching orbits as spirals and the limit cycle as a circle.

In general, the radius of the *limit circle* (the asymptotic limit of the zero-order approximation) is updated by the effect of the perturbation in all orders. We exploit the freedom in the near-identity transformation to obtain as our zero-order approximation a limit circle that is not updated as one computes higher-order terms in the expansion. Thus, to emphasize this point, the radius of the limit circle is determined by the lowest-order calculation. The higher-order terms only introduce the harmonic components of the limit cycle and do not affect the radius of the limit circle.

DISSIPATIVE SYSTEMS IN THE PLANE

Example 7.2. We begin with an example due to Poincaré:

$$\frac{dx}{dt} = (x+y) - \varepsilon(x-y)(x^2+y^2) \qquad (7.18a)$$

$$\frac{dy}{dt} = -(x-y) - \varepsilon(x+y)(x^2+y^2) \qquad \varepsilon > 0 \quad (7.18b)$$

We concentrate our attention on the fixed point at (0,0).

Analysis. Transform x and y to polar coordinates:

$$x = r\cos(\varphi) \qquad y = r\sin(\varphi)$$

Then, Equations (7.18a) and (7.18b) yield

$$x\frac{dx}{dt} + y\frac{dy}{dt} = r\frac{dr}{dt} = r^2 - \varepsilon r^4 \qquad (7.19a)$$

$$x\frac{dy}{dt} - y\frac{dx}{dt} = r^2\frac{d\varphi}{dt} = -r^2 - \varepsilon r^4 \qquad (7.19b)$$

The radial equation has a stationary value when

$$r = \hat{r} = \varepsilon^{-1/2} \qquad (7.20a)$$

The phase equation is of much less interest. It is usually sufficient to observe that when $r=\hat{r}$, one has

$$\frac{d\varphi}{dt} = -1 - \varepsilon\hat{r}^2 \qquad (7.20b)$$

Let us show that $r=\hat{r}$ is stable under small perturbations. Substituting

$$r = \hat{r} + u$$

in Equation (7.19a) and retaining only leading terms, we find

$$\frac{du}{dt} = -2u \qquad (7.21)$$

LIMIT CYCLE SYSTEMS

The solution for u in Equation (7.21) decays to zero for any initial condition, $u(0)>0$ or $u(0)<0$, corresponding to $r(0)>\hat{r}$ and $r(0)<\hat{r}$, respectively. Hence, if we start from an *interior* location (that is, $r(0)<\hat{r}$), then $dr/dt>0$. Thus, as time increases, the system spirals out toward a closed curve in the phase plane that acts as an attractor. Note that if our original coordinates are $(0,0)$ the system remains there. If the system is perturbed or has a different initial condition, the radius spirals out to the value $\hat{r}$. If we start from an *exterior* location ($r(0)>\hat{r}$), then $dr/dt<0$, and the system spirals in toward the closed curve. We call such a system a *limit cycle*.

We have only done a local stability analysis. It requires more effort to show that the limit cycle is a global attractor.

7.3 FIRST-ORDER NORMAL FORM ANALYSIS FOR LIMIT CYCLE SYSTEMS

We introduce the normal form expansion and compute the lowest-order terms that lead to limit cycle behavior. After this is done, we show how one can use the free functions to "renormalize" the u equation. The general equation is of the form

$$\frac{d^2x}{dt^2} + x = \varepsilon g_1\left(x, \frac{dx}{dt}\right) + \text{h.o.t.} \qquad (7.22a)$$

In diagonalized form, writing $z = x+iy$, the equation becomes

$$\frac{dz}{dt} = -i\,z + \varepsilon Z_1(z,z^*) + \text{h.o.t.} \qquad (7.22b)$$

We denote higher-order terms by the abbreviation "h.o.t.."

One now constructs the T functions associated with the normal form expansion. In pursuing the analysis, we come across new aspects of the perturbation expansion that may be traced to the fact that the fixed point at the origin is unstable and the orbit is repelled and asymptotically approaches the limit cycle orbit. The orbit itself consists of a zero-order approximation that is a *limit circle* and harmonic contributions that "fill out" the complete limit cycle curve. We introduce the standard transformations and obtain equations for U_n that we write in the form

$$U_n = f_n(uu^*) \cdot u \qquad (7.23)$$

Let us see how the limit circle appears in a few examples. (We compute only the leading terms.)

Example 7.3: Van der Pol oscillator

$$\frac{d^2x}{dt^2} + x = \varepsilon(1-x^2)\frac{dx}{dt}, \qquad \varepsilon > 0 \qquad (7.24)$$

In this case U_1 becomes

$$U_1 = \tfrac{1}{2}u\left(1 - \tfrac{1}{4}uu^*\right) \qquad (7.25a)$$

and in Equation (7.23)

$$f_1 = \tfrac{1}{2}\left(1 - \tfrac{1}{4}uu^*\right) \qquad (7.25b)$$

The Van der Pol oscillator is analyzed in detail in Section 7.4.4.

Example 7.4: Rayleigh oscillator. In physical media, one expects there to be friction. Usually, the friction is assumed to be linear, leading to the decay of vibrations. To explain how stable sound oscillations can occur in such media despite the friction, Rayleigh constructed an equation that includes a nonlinear friction term in addition to the linear one:

$$\frac{d^2x}{dt^2} + x = \varepsilon \frac{dx}{dt}\left[1 - \tfrac{1}{3}\left\{\frac{dx}{dt}\right\}^2\right] \qquad \varepsilon > 0 \qquad (7.26)$$

Note that the positive linear friction term competes with the negative cubic one. Because of this competition, stable oscillations exist. The Rayleigh oscillator is analyzed in detail in Section 7.4.5.

Equations (7.24) and (7.26) are equivalent. Differentiating Equation (7.26) with respect to time and defining dx/dt as a new variable, v, the Van der Pol oscillator Equation (7.24) is obtained for v. Hence, the two produce similar solutions. They have the same normal form (it is easy to see that they have the same U_1) and thus, have both limit cycles of radius close to 2. However, their harmonic structures are different. The harmonic content of the two oscillators will be developed in Sections 7.4.4 and 7.4.5.

QUALITATIVE ASPECTS OF THE EXPANSION

7.4 QUALITATIVE ASPECTS OF THE EXPANSION

In this section, we set up the expression for U_{n+1} in such a form that conditions required on the free functions, to have the final expansion in its most convenient form, become apparent.

7.4.1 Structure of the expansion

We need to introduce some notation that will be used in the analysis.

1. We write $U_n = f_n(uu^*) \cdot u$. (We will use parentheses to indicate functional dependence and not to group factors for multiplication.)
2. The free functions in the T_n are written as $F_n(uu^*)u$, where F_n is a function of (uu^*) associated with the order n..
3. It is convenient to take derivatives with respect to (uu^*):

$$\frac{\partial}{\partial u} = u^* \frac{d}{d(uu^*)} \qquad \frac{\partial}{\partial u^*} = u \frac{d}{d(uu^*)}$$

Thus, for example

$$\frac{\partial}{\partial u}\left[f_1(uu^*)u\right] = f_1' uu^* + f_1 \qquad \frac{\partial}{\partial u^*}\left[f_1(uu^*)u\right] = f_1' u^2$$

We have indicated the derivative with respect to (uu^*) by a prime. Hereafter, we generally suppress the argument and just write f_n and F_n, etc., except where it might be necessary for clarity.

Assume that we have carried the general expansion out through nth order. Following the formalism presented in Chapter 4, we write the equation in the $(n+1)$st order in the following form:

$$U_{n+1} = [T_{n+1}, Z_0]$$
$$+ T_n \frac{\partial Z_1}{\partial u} + T_n^* \frac{\partial Z_1}{\partial u^*} - U_1 \frac{\partial T_n}{\partial u} - U_1^* \frac{\partial T_n}{\partial u^*} + \{U_{n+1}\} \qquad (7.27a)$$

where $\{U_{n+1}\}$ is known. It is constructed from contributions of the perturbation itself in all orders or from terms involving T_i with $i \leq n-1$ and U_i with $i \leq n$. These have all been fully determined in lower-order calculations. The function T_n has been determined in the nth-order calculation, *except for its unknown part*, which we write as

$F_n(uu^*)u$. The expression for U_{n+1} in Equation (7.27a) contains resonant as well as nonresonant terms. The new, unknown, T_{n+1} has to be chosen so as to eliminate all the nonresonant contributions. Once this is achieved, U_{n+1} becomes a sum of resonant terms only.

Notice that the term

$$T_n \frac{\partial Z_1}{\partial u} + T_n^* \frac{\partial Z_1}{\partial u^*} - U_1 \frac{\partial T_n}{\partial u} - U_1^* \frac{\partial T_n}{\partial u^*}$$

of Equation (7.27a) contributes resonant terms in two ways. First, there is a contribution involving the free function $F_n u$. It is not known yet. We write it explicitly in the form of a Lie bracket:

$$F_n u \frac{\partial U_1}{\partial u} + F_n^* u^* \frac{\partial U_1}{\partial u^*} - U_1 \frac{\partial [F_n u]}{\partial u} - U_1^* \frac{\partial [F_n u]}{\partial u^*}$$

Second, a contribution of terms involving T_n occurs through its known component, computed in the previous order of the calculation. Using T_{n+1} in order to eliminate all nonresonant contributions in Equation (7.27a), we are left with the resonant contributions. These involve the known part, denoted by $<U_{n+1}>$. Thus, U_{n+1} is now written as

$$U_{n+1} = F_n u \frac{\partial U_1}{\partial u} + F_n^* u^* \frac{\partial U_1}{\partial u^*} \\ - U_1 \frac{\partial [F_n u]}{\partial u} - U_1^* \frac{\partial [F_n u]}{\partial u^*} + <U_{n+1}> \quad (7.27b)$$

With points 1–3 of the beginning of this Section in mind, we write

$$\frac{\partial U_1}{\partial u} = \frac{\partial [f_1 u]}{\partial u} = \frac{df_1}{d(uu^*)} uu^* + f_1 = f_1' \cdot (uu^*) + f_1$$

Inserting this result in Equation (7.27b), the latter takes the form

$$U_{n+1} = 2\{\operatorname{Re} F_n f_1' u^2 u^* - \operatorname{Re} f_1' F_n u^2 u^*\} + <U_{n+1}> \quad (7.28a)$$

Alternatively, writing

$$<U_{n+1}> = <f_{n+1}> \cdot u$$

we obtain the following expression for f_{n+1}:

QUALITATIVE ASPECTS OF THE EXPANSION

$$f_{n+1} = 2\{\operatorname{Re} F_n f_1' - \operatorname{Re} f_1 F_n'\}uu^* + <f_{n+1}> \qquad (7.28b)$$

It is convenient to separate real and imaginary parts:

$$\operatorname{Re} f_{n+1} = 2\{\operatorname{Re} F_n \operatorname{Re} f_1' - \operatorname{Re} f_1 \operatorname{Re} F_n'\}uu^* + \operatorname{Re}<f_{n+1}>$$
$$\operatorname{Im} f_{n+1} = 2\{\operatorname{Re} F_n \operatorname{Im} f_1' - \operatorname{Re} f_1 \operatorname{Im} F_n'\}uu^* + \operatorname{Im}<f_{n+1}> \qquad (7.28c)$$

The normal form, the equation for the zero-order approximation is

$$\dot{u} = -i\,u + \sum_{n\geq 1}\varepsilon^n U_n = -i\,u + \sum_{n\geq 1}\varepsilon^n f_n u \qquad (7.29a)$$

Introducing $u = \rho\cdot\exp(-i\varphi)$, the equations for the radius and the phase become

$$\frac{d\rho}{dt} = \rho\left\{\varepsilon \operatorname{Re} f_1(\rho^2) + \sum_{n\geq 2}\varepsilon^n \operatorname{Re} f_n(\rho^2)\right\} \qquad (7.29b)$$

$$\frac{d\varphi}{dt} = 1 - \sum_{n\geq 1}\varepsilon^n \operatorname{Im} f_n(\rho^2) \qquad (7.29c)$$

7.4.2 Observations

$\operatorname{Re} f_{n+1}$ contribute only to the radius equation and $\operatorname{Im} f_{n+1}$, to the phase equation.

Andronov et. al. [5] have shown that if $\operatorname{Re} f_1$ has a simple zero at some $\rho = \rho_0$, then: (1) the perturbed system has a limit cycle; (2) for $\varepsilon \to 0$, the limit cycle collapses into a circle of radius ρ_0. However, it is obvious from Equation (7.29b) that for $\varepsilon \neq 0$, when $t \to \infty$, if the limit cycle is stable, the zero-order term, u, spirals toward a circle (the stable *limit circle*), the radius of which is generally not equal to ρ_0. It is modified by higher orders in the perturbation calculation. This was the reason why we split the r.h.s. of Equation (7.29b) in two parts: the first order contribution and all higher-order contributions. While $\operatorname{Re} f_1$ vanishes at $\rho = \rho_0$, generally, this is not the case with $\operatorname{Re} f_n$ for $n \geq 2$. As a result, the zero in the r.h.s. of Equation (7.29b) is slightly shifted from ρ_0. For example, the traditional expansion of the limit circle radius for the Van der Pol oscillator is

$$\rho = 2 + \varepsilon^2 a + \varepsilon^4 b + \cdots \qquad (7.30)$$

with $a, b,...$ calculable coefficients. The need to update the radius of the limit circle in higher orders of the perturbation results in a cumbersome calculation of the basic frequency, both away from the limit cycle, where the frequency is time dependent, and also as one approaches the limit cycle, where it becomes constant. As is evident from Equation (7.29c), in any given order, the correction to ω includes corrections to ρ in all orders.

Naturally, one may ask whether minimal normal forms, as developed for conservative and damped systems, can be applied without modification to limit cycle systems. If this were so, then all U_n would vanish for $n > 1$, and the radial and phase equations would be written as

$$\frac{d\rho}{dt} = \varepsilon \rho \operatorname{Re} f_1(\rho^2) \qquad \frac{d\varphi}{dt} = 1 + \varepsilon \operatorname{Im} f_1(\rho^2) \qquad (7.31)$$

yielding equations for the radius and the frequency that are terminated at $O(\varepsilon)$.

It turns out that imposing the condition $U_n=0$ for $n>1$ is not a sound choice. This can be seen as follows. If we require that U_{n+1} vanishes, Equation (7.28b) serves as a differential equation for F_n. As the systems are assumed to have limit cycles, $\operatorname{Re} f_1$ has a simple zero at some amplitude $uu^* = (\rho_0)^2$. Therefore, an F_n that solves Equation (7.28b) has a singular point at $uu^* = (\rho_0)^2$. Owing to this singularity, F_n is analytic and has an expansion in powers of uu^* that converges only for $uu^* < (\rho_0)^2$. This feature is undesirable for $t \to \infty$, as $uu^* \approx (\rho_0)^2$ is just the vicinity of the limit cycle, the most interesting characteristic of the system.

7.4.3 Limit circle radius renormalization

As the attempt at simplifying the normal form equation by requiring that $U_n = 0$ for $n>1$ creates unwanted difficulties in the free functions F_n, we turn elsewhere. We note that the perturbation expansion has a characteristic that one may wish to avoid, namely, that the radius of the limit circle is modified by contributions in all orders in ε [see Equations (7.29b) and (7.30)]. To eliminate that undesired feature, we must require that all $\operatorname{Re} f_n$ vanish at one point (which will be the radius of the limit circle). That point should be the same point where $\operatorname{Re} f_1$

QUALITATIVE ASPECTS OF THE EXPANSION

vanishes, namely, $\rho=\rho_0$. Thus it suffices to require, that for all $n \geq 1$, $\operatorname{Re} F_n$ obeys the equation:

$$2 \operatorname{Re} F_n(\rho^2) \operatorname{Re} f_1'(\rho_0^2)\rho_0^2 + \operatorname{Re} <f_{n+1}(\rho^2)> = 0 \quad (7.32)$$

Note that in Equation (7.32) f_1' is computed at $\rho^2=(\rho_0)^2$. As $(\rho_0)^2$ is a simple zero of f_1, we have $f_1'(\rho_0^2) \neq 0$. As $<f_{n+1}>$ is a known polynomial or an analytic function in $uu^* = \rho^2$, Equation (7.32) determines Re F_n. Employing this choice in Equations (7.28a,b), we find that the part in U_{n+1} that contributes to the radius equation is

$$\operatorname{Re} f_{n+1} \cdot u = \begin{cases} 2\operatorname{Re} F_n \left[\operatorname{Re} f_1'(\rho^2) uu^* - f_1'(\rho_0^2)\rho_0^2 \right] \\ -2 \operatorname{Re} f_1' \operatorname{Re} F_n' uu^* \end{cases} \cdot u \quad (7.33)$$

As a result, $\operatorname{Re} f_{n+1}$ vanishes at $\rho=\rho_0$ for all $n \geq 1$. Hence we can write

$$\operatorname{Re} f_{n+1}(\rho^2) = (\rho^2 - \rho_0^2) a_n(\rho^2) \quad (7.34)$$

Equation (7.29b) now becomes

$$\frac{d\rho}{dt} = \rho(\rho^2 - \rho_0^2)\left\{ \varepsilon a_0(\rho^2) + \varepsilon^2 a_1(\rho^2) + \cdots \right\} \quad (7.35a)$$

$$\omega = \frac{d\varphi}{dt} = 1 - \varepsilon \operatorname{Im} f_1(\rho^2) - \varepsilon^2 \operatorname{Im} f_2(\rho^2) + \cdots \quad (7.35b)$$

The terms that might have potentially modified the radius away from its lowest-order value, ρ_0, do not do so. They only modify the rate of approach in time of ρ toward its asymptotic value, which now is ρ_0. We have organized the U equation so that the effect of higher-order terms is concentrated exclusively at $\rho=\rho_0$. It therefore seems appropriate to say that we have *renormalized* the radius of the *limit circle* so that it remains at ρ_0 despite the effect of the perturbation. The computation of the asymptotic, fully updated, frequency is much simpler than in the usual case. [See comment following Equation (7.30).] One substitutes $\rho=\rho_0$ in the expression for ω. The nth-order term in the expression for ω does not generate higher-order corrections in it.

7.4.4 The Van der Pol oscillator revisited

We will illustrate the results of the general scheme through the example of the Van der Pol oscillator. In terms of z variables, Equation (7.24) takes the form

$$\dot{z} = -i\,z + \tfrac{1}{8}\varepsilon\left\{1 - (z+z^*)^2\right\}(z-z^*) \qquad (7.36)$$

We compute the T and U functions, obtaining

$$T_1 = \tfrac{1}{4} i\, u^* - \tfrac{1}{16} i\, u^3 - \tfrac{1}{16} i\, uu^{*2} - \tfrac{1}{32} i\, u^{*3} + F_1(uu^*)u$$
$$U_1 = \tfrac{1}{2} u - \tfrac{1}{8} u^2 u^* \qquad f_1 = \tfrac{1}{2}\left(1 - \tfrac{1}{4} uu^*\right) \qquad (7.37a)$$

$$U_2 = \tfrac{11}{256} i\, u^3 u^{*2} - \tfrac{3}{16} i\, u^2 u^* + \tfrac{1}{8} i\, u$$
$$+ \left(-\tfrac{1}{4}\mathrm{Re}\, F_1 - F_1'\left(1 - \tfrac{1}{4} uu^*\right)\right) u^2 u^* \qquad (7.37b)$$

The MNF option, i.e., that U_2 of Equation (7.37b) vanishes identically, requires that F_1 have a pole at $uu^* = 4$ and is, therefore, not a good choice. Even worse, a nonvanishing $\mathrm{Re}\, F_1$ in Equation (7.37b) introduces a spurious time dependence in the radius. Therefore, it is best to choose for the free term in T_1, $F_1 = 0$. With this done, we have

$$T_2 = -\tfrac{5}{1024} u^5 - \tfrac{1}{256} u^4 u^* + \tfrac{5}{512} u^2 u^{*3} + \tfrac{1}{1024} uu^{*4} + \tfrac{5}{1536} u^{*5}$$
$$- \tfrac{5}{128} u^3 - \tfrac{1}{32} uu^{*2} + \tfrac{1}{128} u^{*3} \qquad\qquad + F_2(uu^*)u \qquad (7.38a)$$

$$U_3 = -\tfrac{13}{8192} u^4 u^{*3} + \tfrac{11}{1024} u^3 u^{*2} - \tfrac{3}{128} u^2 u^*$$
$$+ \left\{-\tfrac{1}{4}\mathrm{Re}\, F_2 - F_2'\left(1 - \tfrac{1}{4} uu^*\right)\right\} u^2 u^* \qquad (7.38b)$$

Employing Equation (7.29), we find the radius, ρ, and frequency, ω. The frequency is then found to be as follows:

$$\omega = 1 - \varepsilon^2\left(\tfrac{11}{256}\rho^4 - \tfrac{3}{16}\rho^2 + \tfrac{1}{8}\right) + O(\varepsilon^4) \qquad (7.39)$$

[Note that, through $O(\varepsilon^2)$, there is no free function contribution here, as we have chosen $F_1 = 0$.]

QUALITATIVE ASPECTS OF THE EXPANSION

Usual choice: *free function* $F_2 = 0$. The radial equation becomes

$$\frac{d\rho}{dt} = \tfrac{1}{2}\varepsilon\rho\left(1 - \tfrac{1}{4}\rho^2\right) + \varepsilon^3\left(-\tfrac{1}{8192}\rho^7 + \tfrac{11}{1024}\rho^5 - \tfrac{3}{128}\rho^3\right) +$$

$$\varepsilon^5\rho^3\left(\begin{array}{c}-\tfrac{19}{1024} + \tfrac{435}{16,384}\rho^2 - \tfrac{687}{65,536}\rho^4 \\ + \tfrac{13,843}{9,437,184}\rho^6 - \tfrac{2035}{37,748,736}\rho^8\end{array}\right) + O(\varepsilon^7) \quad (7.40)$$

Observe that $d\rho/dt$ does not vanish at $\rho^2=4$: the zero of the r.h.s. is shifted to

$$\rho^2 = 4 - \tfrac{3}{16}\varepsilon^2 + O(\varepsilon^4)$$

Thus, higher-order terms in the expansion generate higher-order corrections to the radius of the limit circle that cause a deviation from its value in the zero-order approximation [$(\rho_0)^2=4$].

"Renormalization" choice. We need an F_2 that will produce the renormalization choice leading to Equation (7.35a). Refer to Equations (7.28b) and (7.28c). To impose Equation (7.32), we must first identify the known terms in that equation for the case $n=2$. We note that with $\rho_0=2$

$$f_1'(\rho_0^2)\rho_0^2 = \frac{d}{d(uu^*)}\left[\tfrac{1}{2} - \tfrac{1}{8}uu^*\right]\rho_0^2 = -\tfrac{1}{2} \quad (7.41)$$

Equation (7.32) then becomes

$$2\operatorname{Re} F_2 \operatorname{Re} f_1'(\rho_0^2)\rho_0^2 = -\operatorname{Re} F_2 = -\operatorname{Re}\langle f_3(\rho^2)\rangle \quad (7.42)$$

With U_3 given by Equation (7.38), together with Equation (7.41), we obtain

$$\operatorname{Re} F_2 = -\tfrac{13}{8192}(uu^*)^3 + \tfrac{11}{1024}(uu^*)^2 - \tfrac{3}{128}(uu^*) \quad (7.43)$$

$$\operatorname{Im} F_2 = 0 \quad (7.44)$$

With the results for F_2, given by Equations (7.43) and (7.44), we compute the last quantity in Equation (7.38b):

$$-\tfrac{1}{4}\operatorname{Re} F_2 u^2 u^* - F_2'\left(1-\tfrac{1}{4}uu^*\right)u^2 u^*$$

The expression for U_3 becomes

$$U_3 = \left(\tfrac{13}{4096}u^4 u^{*3} - \tfrac{11}{1024}u^3 u^{*2}\right)\left(1-\tfrac{1}{4}uu^*\right) \qquad (7.45)$$

Performing the calculations through fifth order in a similar manner, the radial equation is modified, replacing Equation (7.40) by

$$\frac{d\rho}{dt} = \tfrac{1}{2}\varepsilon\rho\left(1-\tfrac{1}{4}\rho^2\right)\times \left\{ \begin{array}{l} 1 + \varepsilon^2\left(\tfrac{13}{2048}\rho^6 + \tfrac{11}{512}\rho^4\right) \\[4pt] + \varepsilon^4 \left(\begin{array}{l} -\tfrac{231}{4096}\rho^4 + \tfrac{441}{8192}\rho^6 \\ -\tfrac{33{,}913}{1{,}572{,}864}\rho^8 + \tfrac{15{,}467}{2{,}359{,}296}\rho^{10} \\ -\tfrac{46{,}356}{3{,}355{,}432}\rho^{12} + \tfrac{15{,}717}{134{,}217{,}728}\rho^{14} \end{array} \right) \\[4pt] + O(\varepsilon^6) \end{array} \right\} \qquad (7.46)$$

Unlike in Equation (7.40), the "remainder terms" are also inside the parentheses. Hence, the value of the limit circle radius is unaffected by the higher-order corrections; it remains fixed at $\rho=2$. Also, with the present choice of the free functions, U_3 contains a term of the form $(u^5 u^{*4})$. No such term exists in U_3 as given by Equation (7.38).

Finally, the expression for the asymptotic frequency becomes a power series in ε and ρ, but ρ is fixed at $\rho=2$:

$$\omega = 1 - \tfrac{1}{16}\varepsilon^2 + \tfrac{17}{3072}\varepsilon^4 + \tfrac{35}{884{,}736}\varepsilon^6 + O(\varepsilon^8) \qquad (7.47)$$

Hence, the second and higher-order terms do not generate higher order corrections in ω.

Comment. To make this last point clear, we need to compute U_4. We choose $\operatorname{Re} F_3 = 0$ for the same reason that we chose $\operatorname{Re} F_1 = 0$; namely, because any other choice would lead to a spurious time dependence in the radius. We employ Equation (7.43), which gives the various terms in $\operatorname{Re} F_2$. To make the resulting expression easier to read, we introduce the following constants: $a_1 = -\tfrac{3}{128}$, $a_2 = +\tfrac{11}{1024}$ and $a_3 = -\tfrac{13}{8192}$. We now compute T_3 and U_4 to find

QUALITATIVE ASPECTS OF THE EXPANSION 233

$$U_4 = i \left\{ \begin{array}{l} \left[\frac{1}{128} - \frac{11}{256}\rho^2\right] + \left[\frac{281}{4096} - \frac{3}{8}a_1\right]\rho^4 \\ + \left[-\frac{1019}{32,768} - \frac{3}{8}a_2 + \frac{11}{64}a_1\right]\rho^6 \\ + \left[\frac{3319}{786,432} - \frac{3}{8}a_3 + \frac{11}{64}a_2\right]\rho^8 + \frac{11}{64}a_3\rho^{10} \end{array} \right\} \quad (7.48)$$

Substituting $\rho=2$, we obtain

$$U_4 = -i\frac{17}{3072}$$

yielding a frequency correction of $+\frac{17}{3072}\varepsilon^4$ in Equation (7.47). Observe that, had we made the no free function choice, we would have

$$U_4 = +i\frac{1168}{49,152}$$

at $\rho=2$. This looks like a *different* correction to the frequency. However, with zero free functions, the limit circle radius is *not* $\rho=2$. Instead, it is

$$\rho = 2 - \frac{3}{64}\varepsilon^2$$

Using the latter in Equation (7.39), the same correction to the asymptotic frequency is obtained in $O(\varepsilon^2)$.

7.4.5 The Rayleigh oscillator revisited

As indicated in Example 7.4, the Rayleigh and Van der Pol oscillators are of similar structure. Therefore, it is worthwhile to study the Rayleigh oscillator, using the analysis of Section 7.4.4, and exploit the free functions to obtain "renormalized" equations. As in the case of the Van der Pol oscillator, it is best to choose the first-order free function $F_1=0$. (A nonzero F_1 introduces a spurious time dependence of the radius.) The generators through second order are given by

$$T_1 = i\tfrac{1}{48}u^3 - i\tfrac{1}{16}uu^{*2} + i\tfrac{1}{96}u^{*3} + i\tfrac{1}{4}u^* \quad (7.49a)$$

$$\begin{aligned}T_2 = &-\tfrac{1}{1024}u^5 + \tfrac{1}{256}u^4u^* - \tfrac{3}{512}u^2u^{*3} - \tfrac{3}{1024}uu^{*4} + \tfrac{1}{1536}u^{*5} \\ &- \tfrac{1}{128}u^3 + \tfrac{1}{32}uu^{*2} + \tfrac{1}{128}u^{*3} + F_2(uu^*)u \end{aligned} \quad (7.49b)$$

Implementing the "renormalization" procedure, we find that to obtain the desired structure, one needs an $F_2(uu^*)u$ of the following form:

$$F_2(uu^*)u = -\tfrac{3}{128}u^2 u^* + \tfrac{19}{1024}u^3 u^{*2} - \tfrac{21}{8192}u^4 u^{*3} \qquad (7.49c)$$

With this done, we obtain the U functions:

$$U_1 = \tfrac{1}{2}u(1 - \tfrac{1}{4}uu^*) \qquad (7.50a)$$

$$U_2 = i\tfrac{1}{8}u - i\tfrac{1}{16}u^2 u^* + i\tfrac{2}{256}u^3 u^{*2} \qquad (7.50b)$$

$$U_3 = u(1 - \tfrac{1}{4}uu^*)\{-\tfrac{19}{1024}(uu^*)^2 + \tfrac{21}{4096}(uu^*)^3\} \qquad (7.50c)$$

$$U_4 = i\tfrac{1}{128}u - i\tfrac{41}{122{,}88}u^2 u^* \qquad (7.50d)$$

Notice that the choice of the free functions has allowed us to achieve the "renormalized" form for U_3. The amplitude and phase equations become

$$\frac{d\rho}{dt} = \tfrac{1}{8}\varepsilon\rho(4 - \rho^2)\left[1 - \varepsilon^2\{\tfrac{19}{128}\rho^4 - \tfrac{21}{512}\rho^6\} + O(\varepsilon^4)\right] \qquad (7.51a)$$

$$\omega = 1 + \varepsilon^2\left[-\tfrac{1}{8} + \tfrac{1}{16}\rho^2 - \tfrac{3}{256}\rho^4\right]$$
$$+ \varepsilon^4\left[-\tfrac{1}{128} + \tfrac{41}{12{,}288}\rho^2\right] + O(\varepsilon^6) \qquad (7.51b)$$

7.5 MULTIPLE LIMIT CYCLES: IDENTIFICATION OF THE EXPANSION PARAMETER

On several occasions, we have seen that the actual expansion parameter in a perturbation scheme is not the small parameter that is formally extracted in front of the perturbation. For example, in both the Duffing and the Van der Pol equations, the perturbations include cubic terms. As a result, the actual expansion parameter is not ε but $\varepsilon\rho^2$, where ρ is the amplitude of the zero-order term. Issues such as the timespan of the validity of the approximation in a given order or the convergence of the expansion depend on the size of $\varepsilon\rho^2$ rather than of ε. For instance, a necessary condition for the validity of the expansion near the limit cycle in the case of the Van der Pol oscillator is $4|\varepsilon|<1$.

This observation becomes relevant when one wishes to analyze dynamical problems that have more than one limit cycle solution. In particular, in Section 7.4.3 we described a choice of the free functions

MULTIPLE LIMIT CYCLES

in the perturbation expansion that "renormalizes" the radius of the limit circle. To be able to apply this idea to limit cycles with large radii, one must first assure oneself that the perturbation expansion can be applied when the amplitudes are large.

In this section we study three problems that have similar structures. They all have an infinite number of limit cycles, but different nonlinear perturbations. In particular, the actual expansion parameters are different in the three problems, leading to different conclusions concerning the effect of the perturbations on the solutions within the framework of perturbation theory. The equations we study are

$$\ddot{x} + x + \varepsilon \sin \dot{x} = 0 \quad (7.52)$$

$$\ddot{x} + x + \varepsilon \dot{x} \cos x = 0 \quad (7.53)$$

$$\ddot{x} + x + \varepsilon \dot{x}^3 \cos x = 0 \quad (7.54)$$

Figures 7.4–7.6 show phase space plots of numerical solutions for the first five stable limit cycles of each equation, respectively. In each case, the graphs depict the solutions after they have evolved for a long time, so that the trajectories are very close to the limit cycles.

7.5.1 Qualitative considerations

Before starting a perturbative analysis of Equations (7.52–7.54), we make a few qualitative comments. First, Equation (7.52) is a modification of the Rayleigh oscillator, Equation (7.26). In particular, the first two terms in the Taylor expansion of the sine function in Equation (7.52) are identical in form to those of Equation (7.26):

$$\dot{x} - \tfrac{1}{6}\dot{x}^3$$

instead of

$$\dot{x} - \tfrac{1}{3}\dot{x}^3$$

Hence, for small amplitudes, solutions of the two equations should behave in a similar manner, executing an outgoing spiral motion in phase plane. However, already close to the first limit cycle, the solutions are quite different. The amplitude of the (only) limit cycle in Equation (7.26) is close to 2. If we use the cubic approximation for the sine function, in Equation (7.52), the factor of $\tfrac{1}{6}$, replacing $\tfrac{1}{3}$ in the Rayleigh oscillator, the value 2 for a limit cycle radius is replaced by $2\sqrt{2} \approx 2.8$. The amplitude of the true limit cycle is at $\rho=3.8$.

Next, we note that Equation (7.53) is a modification of the Van der Pol oscillator, Equation (7.24). Also, just as Equations (7.26) and (7.24) are equivalent, in the sense that Equation (7.24) can be obtained by taking a derivative of Equation (7.26) with respect to time, so are Equations (7.52) and (7.53). By taking one additional time derivative of Equation (7.52), the substitution $\dot{x} \Rightarrow x$ yields Equation (7.53). We therefore expect Equations (7.52) and (7.53) to have the same normal form equation for the zero-order terms of their solutions, but different near-identity transformations for the construction of the full solutions. Finally, Equation (7.54) demonstrates the effect of additional powers in the nonlinearity on the perturbation expansion.

We observe from Fig. 7.4 [displaying limit cycles for Equation (7.52)], that, when $\dot{x}$ is small (i.e., motion is close to the x axis), the perturbation is small and its effect on the motion is negligible. As the system moves away from the x axis, $\dot{x}$ becomes larger and $\sin \dot{x}$ oscillates rapidly. Within less than a quarter of a cycle of the full motion, $\dot{x}$ oscillates a number of times. This number increases with the amplitude of the limit cycle. For example, in one-quarter of the period of motion in the second limit cycle (for which the amplitude is approximately 10.2) in Fig. 7.4, dx/dt varies from 0 to about 10. Thus, $\sin \dot{x}$ goes through slightly over $\frac{3}{2}$ of a cycle ($3\pi \approx 9.4$). In the fifth limit cycle (for which the amplitude is approximately 29.0) $\sin \dot{x}$ goes through almost five full cycles ($10\pi \approx 31.4$).

Since the period of oscillation of the full system is much longer than that of the oscillations of $\sin \dot{x}$, the system cannot follow the oscillations in the latter in an efficient manner. Therefore, averaged over time, the perturbation is expected to have a small effect on the motion also away from the x axis. This effect is expected to become less pronounced as one moves to limit cycles with progressively larger amplitudes. Thus, for large amplitude limit cycles, we expect the solutions to be progressively less affected by the perturbation. This qualitative expectation is borne out by the numerical solutions shown in Fig. 7.4. The outer limit cycles are very close to pure circles. We have chosen a large value of ε so that some distortion would be still visible in the lowest limit cycle (for which the amplitude is not small, either). With small values of ε, even the lowest limit cycle is very close to a circle.

We now turn to Equation (7.53). In a similar manner, we observe in Fig. 7.5 that when $\dot{x}$ is small, x is large. Therefore, $\cos(x)$ oscillates rapidly, and its effect on the motion is averaged (almost) to zero. Thus, the motion there is only weakly affected by the perturbation.

MULTIPLE LIMIT CYCLES

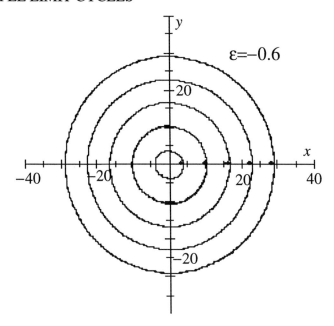

Figure 7.4 First five stable limit cycles of Equation (7.52). Dots represent zeros of $J_1(\rho)$.

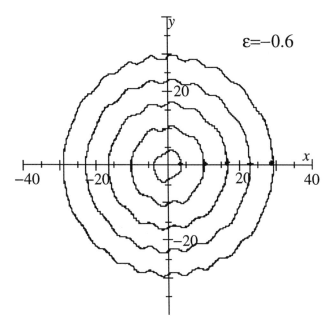

Figure 7.5 First five stable limit cycles of Equation (7.53). Dots represent zeros of $J_1(\rho)$.

On the other hand, when x is small (i.e., near the y axis), $\cos(x)$ is close to unity and $\dot{x}$ is large. Hence, we expect the perturbation to have a significant effect the motion. These expectations are borne out by the solutions depicted in Fig 7.5. Near the x axis, the trajectories are smooth, while around the y axis, they develop wiggles – the effect of the perturbation.

The qualitative analysis of Equation (7.54) is similar, with one difference. Near the axis, $\dot{x}$ is larger than unity, and hence $(\dot{x})^3$ should have a more pronounced effect on the motion than in the case of Equation (7.53). This effect is amplified as we move to the outer limit cycles, since they have larger amplitudes. These observations are borne out by the solutions shown in Fig. 7.6. A small value of ε was picked so as to avoid excessively large oscillations of the solutions, so that, while they are visibly different from the solutions to the first two equations, the amplitudes are of similar magnitude.

Finally, from inspection of Equations (7.52–7.54), we expect the actual expansion parameters in the perturbation expansion for the solutions of the three equations to be different. Namely, in each case, the actual expansion parameter should be ε times a different power of ρ. This expectation will be demonstrated in the following section.

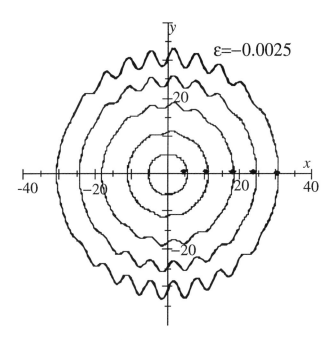

Figure 7.6 First five stable limit cycles of Equation (7.54). Dots represent zeros of $J_2(\rho)$.

7.5.2 Perturbation analysis

For a more detailed picture of the effect of the perturbation in each of Equations (7.52)–(7.54), we turn to the method of normal forms. We study Equation (7.52) first. In the complex notation, $z = x + i\,\dot{x}$, Equation (7.52) becomes

$$\dot{z} = -i\,z - i\,\varepsilon \sin\left(\frac{z + z^*}{2i}\right) \quad (7.52a)$$

Denoting the zero-order approximation by u, the first order equation, relating U_1 and T_1, is

$$U_1 = [Z_0, T_1] - i\,\sin\left(\frac{u - u^*}{2i}\right) \quad (7.55)$$

U_1 must contain only the resonant contribution on the r.h.s. of Equation (7.55). The latter can be found by expanding the sine function in a power series in u and u^* and collecting all resonant monomials (i.e., all monomials of the form $u^{k+1}u^{*k}$, $k \geq 0$). However, it can be also found by a somewhat simpler procedure. Writing $u = \rho \cdot \exp(-i\varphi)$, Equation (7.55) then becomes

$$U_1 = -i\,T_1 - \frac{\partial T_1}{\partial \varphi} + i\,\sin(\rho \sin \varphi) \quad (7.55a)$$

Viewing the perturbation as a periodic function in φ, we expand it in a Fourier series in trigonometric functions of φ. U_1, the resonant part of the interaction, will be the term in the Fourier expansion that is proportional to $\exp(-i\varphi)$, found by computing the integral

$$U_1 = \exp(-i\,\varphi)\frac{1}{2\pi}\int_0^{2\pi} i\,\sin(\rho \sin \varphi)\exp(i\,\varphi) \quad (7.56)$$

This integral can be written in terms of the Bessel function J_1 [20]:

$$U_1 = -J_1(\rho)\exp(-i\,\varphi) \quad (7.57)$$

Thus, through first order, the normal form is

240   DISSIPATIVE SYSTEMS IN THE PLANE

$$\dot{u} = -i\,u - \varepsilon J_1(\rho)\exp(-i\,\varphi) \qquad (7.58)$$

The resulting equations for the radius and the phase are

$$\dot{\varphi} = 1 \qquad \dot{\rho} = -\varepsilon J_1(\rho) \qquad (7.59)$$

Thus, in this order, there is no frequency update. More important, since $J_1(\rho)$ has an infinite number of *simple* zeros, we expect the system to have limit cycles with amplitudes that are close to the zeros of J_1 [5]. For $\varepsilon<0$, the limit cycles that correspond to odd (even) zeros are stable (unstable). The roles are reversed for $\varepsilon>0$.

Having found U_1, we can compute T_1. Equation (7.55a) becomes

$$i\,T_1 + \frac{\partial T_1}{\partial \varphi} = i\,\sin(\rho\sin\varphi) + J_1(\rho)\exp(-i\,\varphi) \qquad (7.60)$$

Writing

$$T_1 = \exp(-i\,\varphi)S \qquad (7.61)$$

S obeys the following equation:

$$\frac{\partial S}{\partial \varphi} = i\,\sin(\rho\sin\varphi)\exp(i\,\varphi) + J_1(\rho) \qquad (7.62)$$

which is solved by

$$S = i\int_0^\varphi \sin(\rho\sin\eta)\exp(i\,\eta)\,d\eta + J_1(\rho)\varphi + A(\rho) =$$

$$\int_0^\varphi \sin(\rho\sin\eta)(\,i\,\cos\eta - \sin\eta)\,d\eta + J_1(\rho)\varphi + A(\rho)$$

(7.63)

where $A(\rho)$ is an arbitrary free function, for which we assume $A(\rho)=0$. (The second-order analysis shows this to be the simplest choice.) This implies that $T_1(\theta=0)=0$, namely, that the first-order term vanishes when the trajectory crosses the x axis (the turning point).

Superficially, it seems that the solution for S includes a secular term owing to the explicit $J_1(\rho)$ term in Equation (7.62). However, we must keep in mind [see Equations (7.55a) and (7.57)] that this term has a counterpart, $-J_1(\rho)$, which cancels it, hidden in the first term on the

MULTIPLE LIMIT CYCLES

r.h.s. of Equation (7.62). Thus, our worry is unjustified.

To identify the actual expansion parameter in the problem, we need to find the large ρ behavior of T_1. Defining

$$N = \left[\frac{\varphi}{\pi/2}\right] \qquad \theta = \varphi - N\left(\frac{\pi}{2}\right) \qquad \theta < (\pi/2)$$

(the square brackets denote the integral part) and using the identities

$$\int_{k\pi}^{(k+1/2)\pi} \sin(\rho \sin\eta)\sin\eta \, d\eta = \int_{k\pi}^{(k+1/2)\pi} \sin(\rho \cos\eta)\cos\eta \, d\eta = \frac{\pi}{2} J_1(\rho)$$

$$\int_{k\pi}^{(k+1)\pi} \sin(\rho \sin\eta)\cos\eta \, d\eta = 0$$

which hold for any integer k, we obtain for S:

$$S = i\frac{(1 - \cos(\rho \sin\varphi))}{\rho} + \theta J_1(\rho)$$

$$- \int_0^\theta \sin\left(\rho\begin{Bmatrix}\sin\eta\\\cos\eta\end{Bmatrix}\right)\begin{Bmatrix}\sin\eta\\\cos\eta\end{Bmatrix} d\eta \qquad \begin{Bmatrix}N \text{ even}\\N \text{ odd}\end{Bmatrix}$$

(7.64)

The first term on the r.h.s. of Equation (7.64) decreases asymptotically as $(1/\rho)$. The J_1 term behaves for large ρ like $1/\rho^{1/2}$. Again, it should be kept in mind that this term is canceled by an identical one, hidden in the integral. If $\theta=0$ [φ is an exact multiple of $(\pi/2)$], then S decreases like $1/\rho$. Otherwise, $0<\theta<\pi/2$, and $S\approx 1/\rho^{1/2}$.

The two cases (N even amd odd) in the integral in Equation (7.64) behave in a similar manner. We present the analysis for the case of even N. Integrating by parts we find

$$\int_0^\theta \sin(\rho\sin\eta)\sin\eta \, d\eta = -\frac{\cos(\rho\sin\theta)}{\rho}\tan\theta + \frac{1}{\rho}\int_0^\theta \frac{\cos(\rho\sin\eta)}{\cos^2\eta} d\eta$$

By repeated integration by parts, we similarly find that the new integral contributes a term of $O(1/\rho^2)$ and another bounded integral. Thus, for *fixed* θ<π/2, the leading asymptotic behavior of the l.h.s. is $O(1/\rho)$. Consequently, the large ρ dependence of T_1 is

$$T_1 = S\exp(-i\,\varphi) = \begin{cases} O(\rho^{-1}) & \theta = 0,\ \pi/2 \\ O(\rho^{-1/2}) & 0 < \theta < \pi/2 \end{cases} \quad (7.65)$$

A detailed analysis of U_2 shows that it constitutes a mere phase update and that there is no gain in including a free function in T_1. This is why we have chosen $A(\rho)=0$ in Equation (7.64). We also find that, for large ρ

$$U_2 = O(\rho^{-3/2})$$

From this analysis we conclude:

1. The phase update is small, especially for limit cycles with large amplitudes. Thus, the frequency of oscillations of the limit cycles (especially those with large amplitudes) is updated by a very small correction, becoming progressively closer to the unperturbed value of unity for larger-amplitude limit cycles.
2. For large amplitudes, the size of the first-order correction decreases as $|\varepsilon T_1|=|\varepsilon|O(\rho^{-1/2})$. Thus, it is small compared to the zero-order term, u, and decreases for larger amplitudes. Hence, we expect perturbation theory to work well in this case, especially for large-amplitude limit cycles. As the corrections decrease with amplitude, we expect that, for larger amplitudes, the limit cycles get progressively closer to the unperturbed pure circles, with radii that are very close to the positions of the zeros of $J_1(\rho)$. This is so, in particular, because near the x axis of Figure 7.4, the perturbation drops even faster as a function of ρ [see Equation (7.65)].
3. T_1 vanishes near the x axis, where θ=0. As the second-order term in the normal form is only a phase update, at least through $O(\varepsilon^2)$, the intersections of the actual phase-space trajectories with the x axis should be very close to the zeros of J_1 and should tend to the latter for larger-amplitude limit cycles.
4. Point 3 is of interest from another aspect. Andronov et al. [5] showed that if U_1 has a simple zero at some $\rho=\rho_0$, then as ε→0, the limit cycle degenerates to a circle of radius ρ_0. In the present problem, although ε is kept fixed, the actual expansion parameter, $\varepsilon/\rho^{1/2}$, decreases as the amplitudes increases and again, the limit cycles tend to circles with radii that are very close

MULTIPLE LIMIT CYCLES

to the zeros of J_1. From Figure 7.4 we see that the solutions behave as expected.

We now turn to Equation (7.53). As stated earlier, the latter is equivalent to Equation (7.52). Therefore, one expects the normal forms of the two equations to be identical. An analysis, similar to the one carried out for Equation (7.52), shows that U_1 is the same as in Equation (7.57). Thus, again, one expects to have limit cycles with amplitudes that are close in values to the zeros of J_1. Because of the small effect of the perturbation close to the x axis (as discussed in Section 7.6.1), we expect the phase-space trajectories to intersect the x axis very close to these zeros. This is found to be the case, as indicated by Figure 7.5. However, the different structure of the perturbations implies that T_1 should be different in the present problem. Writing T_1 as in Equation (7.61), we obtain

$$S = i\left(\sin\rho - \sin(\rho\cos\varphi)\cos\varphi\right)$$

$$+ i\,\frac{1-\cos(\rho\cos\varphi)}{\rho} + \theta J_1(\rho) \qquad (7.66)$$

$$-\rho\int_0^\theta \cos\left(\rho\begin{Bmatrix}\cos\eta\\\sin\eta\end{Bmatrix}\right)\begin{Bmatrix}\sin^2\eta\\\cos^2\eta\end{Bmatrix}d\eta \qquad \begin{Bmatrix}N\text{ even}\\N\text{ odd}\end{Bmatrix}$$

Again, for $\theta=0, \pi/2$, the integral cancels the J_1 term exactly, so that the remainder of S behaves asymptotically like $O(\rho^0)$ as ρ increases. The detailed analysis of the integral indicates that the same is true for $0<\theta<\pi/2$. Thus, we conclude that, for large ρ

$$T_1 = S\exp(-i\,\varphi) = O(\rho^0) \qquad (7.67)$$

In summary, the limit cycles of Equation (7.53) should have the same frequency as those of Equation (7.52). Note also that near the x axis, where $\theta=0$, T_1 vanishes. As the second-order term in the normal form is only a phase update, at least through $O(\varepsilon^2)$, the intersections of the actual phase-space trajectories with the x axis should coincide with that of the zero-order term, so that they should be very close to the zeros of J_1. However, away from the x axis, the effect of T_1 on the solution is felt. It is $O(\rho^0)$. This observation implies that the solutions should exhibit two distinctive features:

1. Unlike the case of Equation (7.52), the effect of the perturbation should not diminish as one goes to large amplitude limit cycles.
2. For large radii, the perturbation is $O(\rho^0)$. Hence, the amplitude of the modulation that is superimposed on the pure circle by T_1 should be similar for all large radius limit cycles. Only the number of "wiggles" should grow with the radius of the limit cycle. The reason is quite obvious. For instance, the argument of the term $\sin(\rho \cos \varphi)$ in Equation (7.67) varies between ρ, when $\varphi=0$, and 0, when $\varphi=\pi/2$. For large ρ, this argument will go several times through 2π cycles. Therefore, T_1 will oscillate a number of times, the number increasing with ρ. Since the zeros of J_1 are spaced by approximately π and since only every alternate zero is a radius of a stable limit cycle, T_1 performs one additional full oscillation as one goes from one stable limit cycle to the next.

All these qualitative statements are borne out by the solutions depicted in Fig. 7.5.

We now turn to Equation (7.54). In complex notation, it becomes

$$\dot{z} = -i\, z + \tfrac{1}{8}(z - z^*)^3 \cos\left(\frac{z + z^*}{2}\right) \tag{7.54a}$$

Replacing z by $u=\rho\cdot\exp(-i\varphi)$, the zero-order term, the perturbation becomes

$$Z_1 = i\, \rho^3 \sin^3\varphi \cos(\rho\cos\varphi) \tag{7.68}$$

Rather than identifying the resonant part in Z_1, by an integral of the type used in Equation (7.54), we exploit the Fourier expansion [20]:

$$\cos(\rho\cos\varphi) = J_0(\rho) + 2\sum_{k=1}^{\infty}(-1)^k J_{2k}(\rho)\cos 2k\varphi$$

Inserting this expansion in Equation (7.68), we readily identify the resonant part in Z_1 [the term proportional to $\exp(-i\varphi)$]:

$$U_1 = \left(\tfrac{1}{8}\right)\rho^3\left(3J_0(\rho) + 4J_2(\rho) + J_4(\rho)\right)\exp(-i\,\varphi)$$
$$= 3\rho J_2(\rho)\exp(-i\,\varphi) \tag{7.69}$$

The final expression for U_1 is derived through identities obeyed by the Bessel functions [20]. A detailed analysis of T_1 leads to the conclusion that in this problem, the large-amplitude behavior of T_1 is

$$T_1 = O(\rho^2) \tag{7.70}$$

Equations (7.69) and (7.70) lead to the following conclusions:

1. The amplitudes of the limit cycles are expected to be close in magnitude to the zeros of $J_2(\rho)$.
2. With the no-free-function option, T_1 vanishes as one approaches the x axis ($\theta=0$), so that the intersection of the phase-space trajectory with the x axis should be at the zero order-value of the amplitude, namely, at the zero of J_2.
3. T_1 should generate a modulation of the full solution around the zero-order circle approximation. The number of "wiggles" should increase as one goes to larger amplitude limit cycles.
4. The amplitude of this modulation should be approximately proportional to ρ^2. Since ρ is equal (to a good approximation) to a zero of J_2, the modulation amplitudes should grow from one limit cycle to the next in proportion to the squares of the corresponding zeros.

These expectations are borne out by the limit cycles depicted in Figure 7.6. In particular, point 4 was checked in detail. The modulation amplitudes grow approximately in proportion to the square of the magnitudes of the zeros of J_2. However, the growth is not strictly in proportion to the squares of the zeros. The reason is most probably the effect of higher-order contributions that may be sizable in this problem. The actual expansion parameter here is not ε, but $\varepsilon\rho^2$. This means that perturbation theory is not applicable to the analysis of limit cycles with sufficiently large amplitudes.

In summary, we have analyzed three similar dynamical systems, all having an infinite number of limit cycles. However, because the effective expansion parameter is different in each case, the effect of the perturbation as well as the efficacy of the perturbation expansion are different. In the case of Equation (7.52), the effective expansion parameter is ε/ρ, so that the higher limit cycles are hardly affected by the perturbation. The zero-order term constitutes an excellent approximation. In the case of Equation (7.53), the first-order correction is, to a good approximation, independent of the amplitude for large amplitudes. Hence the quality of the perturbation expansion depends on the size of ε. In the case of Equation (7.54), the perturbation expansion is not valid for limit cycles of large amplitudes.

REFERENCES

1. Rayleigh (J. W. Strutt), *The Theory of Sound*, Vols. 1,2, London (1877–8); (American Edition, Dover, New York, 1954).
2. Rayleigh (J. W. Strutt), "On Maintained Vibrations," *Phil. Mag.* **XV**, 1 (1883).
3. Van der Pol, B., "On Oscillation Hysteresis in a Simple Triode Generator," *Phil. Mag.* **43**, 700 (1922).
4. Murray, D., "Normal Form Investigations for Dissipative Systems," *Mech. Res. Commun.* **21**, 231 (1994).
5. Andronov, A. A., E. A. Leontovich, I. I. Gordon, and A. G. Maier, *Theory of Bifurcations of Dynamic Systems on a Plane*, Israel Program of Scientific Translations: Jerusalem (1971).
6. Coddington, E. A., and N. Levinson, *Theory of Ordinary Differential Equations*, McGraw-Hill, New York (1955).
7. Melnikov, V. K., "On the Stability of the Center for Time Periodic Perturbations," *Trans. Moscow Math. Soc.*, **12**, 1 (1963).
8. Bender, C. M., and S. A. Orszag, *Advanced Mathematical Methods for Scientists and Engineers*, McGraw-Hill, New York, (1978).
9. Nayfeh, A. H., *Perturbation Methods*, Wiley, New York, 1973).
10. Sanders, J. A., and F. Verhulst, *Averaging Methods in Nonlinear Dynamical Systems*, Springer-Verlag, New York (1985).
11. Jordan, D. W., and P. Smith, *Nonlinear Ordinary Differential Equations*, Oxford University Press, Oxford (1983).
12. Ye Yan-Qian et. al., "Theory of Limit Cycles", *Am. Math. Soc. Transl. Math. Monogr.* **66** (1986).
13. Lloyd, N. G., "Limit Cycles of Polynomial Systems – Some Recent Developments," *London Math. Soc. Lect. Notes Ser.* **127**, 192 (1987).
14. Verhulst, F., *Nonlinear Differential Equations and Dynamical Systems*, Springer-Verlag, Berlin (1990).
15. Kahn, P. B., *Mathematical Methods for Scientists and Engineers* Wiley, New York (1990).
16. Kahn, P. B., and Y. Zarmi, "Radius Renormalization in Limit Cycles," *Proc. Roy. Soc.* (London) **A440**, 189 (1993).
17. Gradshteyn, I. S., and I. M. Ryzhik, *Tables of Integrals, Series and Products*, Academic Press, New York (1965).

8

NONAUTONOMOUS OSCILLATORY SYSTEMS

8.1 INTRODUCTION

In previous chapters we have studied oscillations of autonomous systems in the plane (i.e., conservative or dissipative systems with one degree of freedom). In every example, the basic system was a harmonic oscillator and the perturbation either maintained the conservative nature of the system, or introduced dissipation. For conservative systems, the motion remained periodic, following a closed curve in the planar phase space. For a small perturbation, the phase plane curve was a slightly distorted circle. Compared to the unperturbed motion, it was characterized by two modifications. First, the basic frequency was slightly modified away from its unperturbed value. Second, small components with higher harmonics of the basic frequency were added to the fundamental circle. For dissipative systems the motion was either spiraling down to the origin, or to infinity, or exhibited a limit cycle (again, a slightly distorted circle). In both cases, the phase-space portrait of the system never crossed itself. (A direct consequence of the uniqueness of the solutions of differential equations, guaranteed when a condition such as the Lipschitz condition is obeyed.)

The interest in nonautonomous oscillatory systems stems from the enormous richness of phenomena they exhibit. In this chapter, we study harmonic oscillatory systems that are perturbed by small (linear or nonlinear) perturbations with explicit time-dependent terms. A nonautonomous oscillatory system is described by the general equation

$$\ddot{x} + \omega^2 x + \varepsilon F(x, \dot{x}, t; \varepsilon) = 0 \qquad (8.1)$$

with F depending explicitly on time.

Here, the phase portrait may cross itself in the x–(dx/dt) plane. Let the system pass through a point (a,b) in the planar phase space at time

$t=t_1$. Because of its intrinsic time dependence, the "force" F acquires different values at different times. As a result, the trajectory may pass through the same point at a different time, $t=t_2$. Time is no more a mere parameter but plays the role of a third coordinate. In the literature, such systems are often called "1.5 degrees of freedom" ones. The reason is that in classical mechanics every degree of freedom has two coordinates: a position coordinate and its canonically conjugate momentum. This can be exposed by converting the system described by Equation (8.1) into an autonomous three dimensional one. Introducing a spurious variable τ, Equation (8.1) is rewritten as

$$\ddot{x} + \omega^2 x + \varepsilon F(x, \dot{x}, \tau; \varepsilon) = 0 \qquad (8.2a)$$

$$\frac{d\tau}{dt} = 1 \qquad (8.2b)$$

It is customary to plot in the three-dimensional phase space of x, (dx/dt) and τ (or t). A typical plot is shown in Figure 8.1.

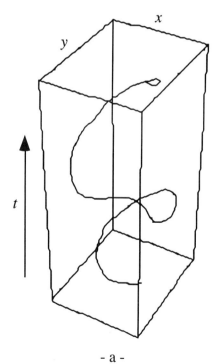

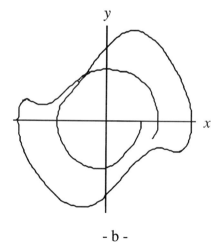

Figure 8.1 Phase-space portrait of a nonautonomous system with one degree of freedom: (a) 3D plot; (b) projection on 2D phase plane. Lines that do not cross in the 3D plot may do so in the 2D plot.

INTRODUCTION

We see that lines that do not cross in the three-dimensional phase space may do so in the two-dimensional projection. Equation (8.2) is a special case of a more general "genuine" three-dimensional problem where Equation (8.2b) is replaced by a more complicated equation:

$$\frac{d\tau}{dt} = G(x, \dot{x}, \tau : \varepsilon) \qquad (8.3)$$

Many specific examples of Equation (8.1), or, equivalently, Equation (8.2), exhibit the wealth of characteristics of high- (at least three-) dimensional systems, such as period doubling bifurcations and chaos. For example, the solution $x(t)$ of the equation of the periodically forced nonlinear oscillator

$$\ddot{x} + x + \mu \dot{x} + \varepsilon x^3 = f \cos \omega t \qquad (8.4)$$

exhibits that type of complex behavior for a certain range of values of f and ω [1].

8.2 AN EXAMPLE FROM THE THEORY OF ACCELERATOR DESIGN

In the design of circular particle accelerators, one studies the time dependence of the motion of charged particles through the accelerator. In the lowest-order approximation, the horizontal component of motion obeys an equation of the type [2]

$$\ddot{x} + \omega_h^2 x = \varepsilon x^p \cos mt \qquad (8.5)$$

(Higher orders in the perturbation expansion lead to coupling between the horizontal and vertical components of the motion. Such problems are discussed in Chapter 9.) In Equation (8.5), p and m are usually integers and ω_h is the natural frequency of horizontal oscillations. The force [r.h.s. of Equation (8.5)] has an explicit periodic time dependence. The nonlinear nature of the force is revealed when a multipole expansion is applied to the magnetic field.

We study examples of Equation (8.5) in the first-order approximation of a perturbation expansion. Analysis of specific cases ($p=1$ – the linear Mathieu equation and $p=3$ – the nonlinear Mathieu equation) through second order is carried out in the following sections. We search for situations in which the r.h.s. of

Equation (8.5) resonates with the natural oscillations of the unperturbed equation.

In the first-order approximation, one easily finds resonance combinations by noting that in the unperturbed limit, $x(t)$, behaves like a linear combination of $\exp(\pm i\, \omega_h t)$:

$$x(t) = x_0(t) = a\exp(i\,\omega_h t) + a^*\exp(-i\,\omega_h t) \tag{8.6}$$

Hence, resonance is expected whenever $x^p \cdot \cos(mt)$ includes a component behaving in the unperturbed limit also like $\exp(\pm i\,\omega_h t)$. Using the a binomial expansion of x^p, with x given by Equation (8.6), we find that the condition for resonance is

$$q\omega_h \pm m = \pm \omega_h \qquad \Rightarrow \omega_h = \frac{m}{|q \pm 1|}$$

$$(q = -p, -(p-2), \ldots, (p-2), p) \qquad q \pm 1 \neq 0 \tag{8.7}$$

In Equation (8.7), q is an integer. Note that we keep only positive values of ω_h.

For the study of the behavior of the solution near the resonant frequency, we write

$$\omega_h^2 = \left(\frac{m}{n}\right)^2 + \delta \qquad n = |q \pm 1| \tag{8.8}$$

The small parameter δ measures the distance from the resonance point. In Table 8.1, a list of the resonance points is provided for $0 \leq p \leq 4$.

Table 8.1 Resonance values of the frequency ω_h in Equation (8.7)

p	0	1	2	3	4
Resonant ω_h	m	$m/2$	m $m/3$	$m/2$ $m/4$	m $m/3$ $m/5$

THEORY OF ACCELERATOR DESIGN

8.2.1 The case $p=1$

For $p=1$, Equation (8.5) becomes

$$\ddot{x} + \left(\frac{m}{2}\right)^2 x + \delta x = \varepsilon x \cos mt \qquad (8.9)$$

We now change to complex notation, writing

$$z = x + i\left(\frac{2}{m}\right)\dot{x} \qquad v = e^{-imt} \qquad (8.10)$$

The equations for z and v are

$$\dot{z} = -i\left(\frac{m}{2}\right)z - i\left(\frac{\delta}{m}\right)(z+z^*)$$
$$+ i\left(\frac{\varepsilon}{2m}\right)(z+z^*)(v+v^*) \qquad (8.11a)$$

$$\dot{v} = -i\,mv \qquad (8.11b)$$

Thus, the nonautonomous linear second-order differential Equation (8.9) has been converted into an equivalent nonlinear autonomous system of two degrees of freedom, described by two first-order differential equations [Equations (8.11)]. The variable z is affected by v in a nonlinear manner, whereas v is known. Now we can use the normal form expansion for autonomous systems.

Equation (8.11) is a special case of more general systems, where v is coupled to z in a nontrivial manner.

We now expand z in a near-identity transformation:

$$z = u + \varepsilon T_1 + \cdots$$

Through first order, the normal form for u is obtained by retaining only the resonant terms in Equation (8.11a):

$$\dot{u} = -i\left(\frac{m}{2}\right)u - i\left(\frac{\delta}{m}\right)u + i\left(\frac{\varepsilon}{2m}\right)u^* v \qquad (8.12)$$

Writing

$$u = f(t)e^{-i\left(\frac{m}{2}\right)t} \tag{8.13}$$

(Note that we have explicitly taken out the fast part of the time dependence of the solution.) Equation (8.12) becomes

$$\dot{f} = -i\left(\frac{\delta}{m}\right)f + i\left(\frac{\varepsilon}{2m}\right)f^* \tag{8.14}$$

There are two ways to proceed at this point. One way is to write the equation for f^*:

$$\dot{f}^* = -i\left(\frac{\varepsilon}{2m}\right)f + i\left(\frac{\delta}{m}\right)f^* \tag{8.15}$$

Equations (8.14) and (8.15) are coupled linear equations for f and f^* that are solved by exponentials, $\exp(\lambda t)$. The exponent λ is obtained from the eigenvalue equation

$$\begin{vmatrix} -\lambda - i\left(\frac{\delta}{m}\right) & i\left(\frac{\varepsilon}{2m}\right) \\ -i\left(\frac{\varepsilon}{2m}\right) & -\lambda + i\left(\frac{\delta}{m}\right) \end{vmatrix} = \lambda^2 + \left(\frac{\delta}{m}\right)^2 - \left(\frac{\varepsilon}{2m}\right)^2 = 0 \tag{8.16}$$

Clearly, the solution exhibits stable oscillations for

$$|\delta| > \left|\frac{\varepsilon}{2}\right| \tag{8.17a}$$

and is unstable for

$$|\delta| < \left|\frac{\varepsilon}{2}\right| \tag{8.17b}$$

The case $|\delta|=|\varepsilon/2|$ requires special treatment.

A second way to analyze Equation (8.14) is in polar coordinates by writing:

THEORY OF ACCELERATOR DESIGN

$$f = \rho \exp(-i\varphi) \tag{8.18}$$

where φ is the slowly time-dependent part of the phase. Equations for ρ and φ are obtained from Equation (8.14):

$$\dot{\rho} = -\left(\frac{\varepsilon}{2m}\right)\rho \sin 2\varphi \tag{8.19}$$

$$\dot{\varphi} = \left(\frac{\delta}{m}\right) - \left(\frac{\varepsilon}{2m}\right)\cos 2\varphi \tag{8.20}$$

From Equation (8.20), we see that when Equation (8.17b) is obeyed, φ has a fixed point at

$$\cos 2\varphi^* = \frac{2\delta}{\varepsilon} \tag{8.21}$$

independently of ρ. If we now write

$$2\varphi = 2\varphi^* + \eta \tag{8.22}$$

we find that η obeys the following equation:

$$\dot{\eta} = \left(\frac{\delta}{m}\right)(1 - \cos\eta) + \left(\frac{\varepsilon}{2m}\right)\sin 2\varphi^* \sin\eta \tag{8.23}$$

Retaining only terms linear in η, Equation (8.23) yields

$$\dot{\eta} \cong \left(\frac{\varepsilon}{2m}\right)\sin 2\varphi^* \eta \qquad \left(\sin 2\varphi^* = \pm\sqrt{1 - \left(\frac{2\delta}{\varepsilon}\right)^2}\right) \tag{8.24}$$

Thus, for $\varepsilon \cdot \sin 2\varphi^* < 0$, φ^* is a stable fixed point:

$$\varphi \xrightarrow[t \to \infty]{} \varphi^*$$

As a result, asymptotically in time, Equation (8.19) for ρ becomes

$$\dot{\rho} \cong -\left(\frac{\varepsilon}{2m}\right)\sin 2\varphi^* \rho \qquad (8.25)$$

The constant coefficient in Equation (8.25) is positive (at the fixed point $\varepsilon \cdot \sin 2\varphi^* < 0$), so that ρ grows exponentially in time:

$$\rho \to \rho_0 \exp\left(\left|\frac{\varepsilon}{2m}\sin 2\varphi^*\right|t\right) \qquad (8.26)$$

To summarize, in the domain defined by Equation (8.17b), φ has a stable fixed point. As a result, the solution diverges: ρ diverges exponentially in time.

In the domain defined by Equation (8.17a), $(d\varphi/dt)$ has a unique sign (either always positive or always negative) and never vanishes. Consequently, φ grows monotonically in time and does not go to a fixed point. One can then solve for ρ in terms of φ by dividing Equation (8.19) by Equation (8.20) to obtain

$$\frac{1}{\rho}\frac{d\rho}{d\varphi} = \frac{-(\varepsilon/(2\delta))\sin 2\varphi}{1-(\varepsilon/(2\delta))\cos 2\varphi} \qquad (8.27)$$

The solution of Equation (8.27) oscillates with a finite amplitude:

$$\rho = \rho_0 \left\{\frac{1-(\varepsilon/(2\delta))\cos 2\varphi_0}{1-(\varepsilon/(2\delta))\cos 2\varphi}\right\}^{1/2} \qquad (8.28)$$

Thus, in first order, both methods of analysis of Equation (8.9) lead to the same conclusions regarding the domain of stability of its solutions in the $\varepsilon - \delta$ plane. The case

$$|\delta| = \left|\frac{\varepsilon}{2}\right| \qquad (8.29)$$

requires special treatment. From Equation (8.29) it follows that, for any ρ, the r.h.s. of Equations (8.20) and (8.21) vanish at $\varphi = \varphi^*$ given by

$$\cos 2\varphi^* = \frac{2\delta}{\varepsilon} = (-1)^n \quad \Rightarrow \quad 2\varphi^* = n\pi \qquad (8.30)$$

THEORY OF ACCELERATOR DESIGN

Writing

$$2\varphi = n\pi + \eta \tag{8.31}$$

we obtain

$$\frac{d}{dt}\rho^2 = (-1)^{n+1}\left(\frac{\varepsilon}{2m}\right)\rho^2 \sin\eta \tag{8.32}$$

$$\frac{d}{dt}\eta = (-1)^n\left(\frac{\varepsilon}{m}\right)(1 - \cos\eta) \tag{8.33}$$

Equations (8.32) and (8.33) have a first integral:

$$(1 - \cos\eta)\rho^2 = C \tag{8.34}$$

(The constant C is determined by the initial conditions.) For $|\eta| \ll 1$, Equation (8.34) yields

$$\eta\rho \cong \pm\sqrt{2C} \tag{8.35}$$

Thus, near the fixed points $2\varphi^* = n\pi$, the lines of flow are hyperbolas. In this order, the stability properties of the point $(\rho,\eta)=(0,0)$ cannot be determined. They are revealed in a second-order analysis.

8.2.2 The case $p=2$

For $p=2$, there are two resonance combinations: $\omega_h \approx m$, $m/3$. Through first order, the normal forms for both cases are given by

$$\dot{u} = \begin{cases} -i\,mu - i\left(\dfrac{3\delta}{2m}\right)u + i\left(\dfrac{3\varepsilon}{8m}\right)u^{*2}v & \omega_h^2 = \left(\dfrac{m}{3}\right)^2 + \delta \\[2ex] -i\,mu - i\left(\dfrac{\delta}{2m}\right)u + i\left(\dfrac{\varepsilon}{4m}\right)uu^*v & \omega_h^2 = m^2 + \delta \end{cases} \tag{8.36}$$

Again, writing explicitly the fast time dependence of the solution in the form

$$u = \begin{cases} \exp\left(-i\left(\dfrac{m}{3}\right)t\right)f(t) & \omega_h^2 = \left(\dfrac{m}{3}\right)^2 + \delta \\ \exp(-imt)f(t) & \omega_h^2 = m^2 + \delta \end{cases} \quad (8.37)$$

we obtain for $f(t)$ the equations

$$\dot{f} = \begin{cases} -i\left(\dfrac{3\delta}{2m}\right)f + i\left(\dfrac{3\varepsilon}{8m}\right)f^{*2} & \omega_h^2 = \left(\dfrac{m}{3}\right)^2 + \delta \\ -i\left(\dfrac{\delta}{2m}\right)f + i\left(\dfrac{\varepsilon}{4m}\right)f^{*} & \omega_h^2 = m^2 + \delta \end{cases} \quad (8.38)$$

In polar coordinates [as in Equation (8.18)], the equations for both cases have similar structures:

$$\dot{\rho} = \begin{cases} -\left(\dfrac{3\varepsilon}{8m}\right)\rho^2 \sin 3\varphi & \omega_h^2 = \left(\dfrac{m}{3}\right)^2 + \delta \\ -\left(\dfrac{\varepsilon}{4m}\right)\rho^2 \sin\varphi & \omega_h^2 = m^2 + \delta \end{cases} \quad (8.39)$$

$$\dot{\varphi} = \begin{cases} \left(\dfrac{3\delta}{2m}\right) - \left(\dfrac{3\varepsilon}{8m}\right)\rho\cos 3\varphi & \omega_h^2 = \left(\dfrac{m}{3}\right)^2 + \delta \\ \left(\dfrac{\delta}{2m}\right) - \left(\dfrac{\varepsilon}{4m}\right)\rho\cos\varphi & \omega_h^2 = m^2 + \delta \end{cases} \quad (8.40)$$

As a result, the qualitative behavior of the solutions is similar in both cases. We therefore, analyze $\omega_h \approx (m/3)$ as an example. For the latter, the r.h.s. of Equations (8.39) and (8.40) vanish at

$$\left.\begin{array}{l} 3\varphi = 3\tilde{\varphi} = n\pi \\ \rho = \tilde{\rho} = \left(\dfrac{4\delta}{\varepsilon}\right)(-1)^n \end{array}\right\} \quad \omega_h^2 = \left(\dfrac{m}{3}\right)^2 + \delta \quad (8.41)$$

The stability characteristics of the fixed points are found as follows. We first note that since $\rho \geq 0$, fixed points can only occur for the following combinations:

THEORY OF ACCELERATOR DESIGN

$$\left(\frac{\delta}{\varepsilon}\right) > 0 \quad \Rightarrow \quad n = 2k \qquad 3\tilde{\varphi} = 0, 2\pi, 4\pi, \ldots$$

$$\left(\frac{\delta}{\varepsilon}\right) < 0 \quad \Rightarrow \quad n = 2k+1 \qquad 3\tilde{\varphi} = \pi, 3\pi, 5\pi, \ldots \tag{8.42}$$

We note that as δ varies from negative to positive values, ω_h, the frequency of oscillations moves from below to above the resonant point. (Experimentally, ω_h is the control parameter that is varied for a given coupling to the magnetic field, i.e., given value of ε.)

Let us now write

$$\rho = \tilde{\rho} + \xi \qquad 3\varphi = 3\tilde{\varphi} + \eta \tag{8.43}$$

Insert Equation (8.43) in Equations (8.39) and (8.40) for the case $\omega_h \approx (m/3)$ and linearize the resulting equations around the fixed point. We find

$$\dot{\xi} \cong -\left(\frac{3\delta}{2m}\right)\eta \qquad \dot{\eta} \cong (-1)^{n+1}\left(\frac{9\varepsilon}{8m}\right)\xi \tag{8.44}$$

Equations (8.44) are solved by exponentials $\exp(\lambda t)$, with

$$\lambda = \pm\sqrt{(-1)^{n+1}\frac{27\varepsilon\delta}{16m^2}} = \pm\sqrt{-\frac{27\tilde{\rho}\varepsilon^2}{64m^2}} \tag{8.45}$$

At the fixed points, λ is a pure imaginary quantity. Thus, small perturbations generate small amplitude oscillations around the fixed point that are neutrally stable.

We wish to obtain a qualitative description of the motion followed by the solution of the first order normal form. Depending on the initial conditions, the solution may be either bounded or unbounded. We assume $(\delta/\varepsilon) > 0$. [The case $(\delta/\varepsilon) < 0$ can be analyzed in a similar manner.] For $\varepsilon > 0$ this corresponds to $\delta > 0$, i.e., the unperturbed frequency is above resonance. We next define

$$w = \cos 3\varphi \tag{8.46}$$

Using Equations (8.39) and (8.40) for the case $\omega_h \approx (m/3)$, we find

$$\frac{dw}{d\rho} + \frac{3w}{\rho} = \frac{12}{\rho^2}\left(\frac{\delta}{\varepsilon}\right) \qquad (8.47)$$

The phase plots of allowed trajectories are given by the solution of Equation (8.47):

$$w \equiv \cos 3\varphi = \frac{A}{\rho^3} + \frac{6(\delta/\varepsilon)}{\rho} \qquad (8.48)$$

The first term on the r.h.s. of Equation (8.48) is the general solution of the homogeneous part of Equation (8.47). The constant A is determined by the initial conditions. We define

$$g(\rho) \equiv -\rho^3 - 6\left(\frac{\delta}{\varepsilon}\right)\rho^2, \qquad f(\rho) \equiv \rho^3 - 6\left(\frac{\delta}{\varepsilon}\right)\rho^2$$

For the detailed analysis, we will need plots of $f(\rho)$ and $g(\rho)$. They are given in Figure 8.2 for $(\delta/\varepsilon) > 0$.

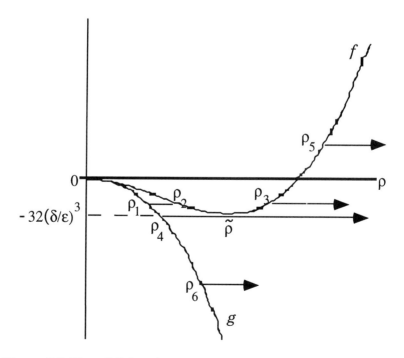

Figure 8.2 Plot of $f(\rho)$ and $g(\rho)$ of Equation (8.49). The lines denote allowed ranges of ρ.

THEORY OF ACCELERATOR DESIGN   259

For physical motion, $-1 \leq w = \cos 3\varphi \leq 1$, implying that A must obey

$$g(\rho) \leq A \leq^2 f(\rho) \tag{8.49}$$

The minimum of $f(\rho)$ is at the radius of the fixed point $\tilde{\rho} = 4(\delta/\varepsilon)$ and is given by

$$f(\tilde{\rho}) = -32\left(\frac{\delta}{\varepsilon}\right)^3 \equiv A_0 \tag{8.50}$$

We now distinguish four cases. For ease of the analysis, we first cast the equation for $d\varphi/dt$ in a new form. Inserting Equation (8.48) in Equation (8.40), we obtain

$$\dot{\varphi} = -\left(\frac{3\delta}{4m}\right) - \left(\frac{3\varepsilon}{8m}\right)\left(\frac{A}{\rho^2}\right) \tag{8.51}$$

Plots of w versus ρ are given for various values of A in Figure 8.3.

1. $A \geq 0$. The constraint $g(\rho) \leq A$ is always obeyed, as, for $(\delta/\varepsilon) > 0$, $g(\rho) \leq 0$. For any of $A \geq 0$, there is a single solution, $\rho = \rho_5$, to the equation $A = f(\rho)$, so that for any $\rho \geq \rho_5$, the inequality $A \leq f(\rho)$ is obeyed. Hence, motion is allowed for all $\rho \geq \rho_5$. In this case we note

   a. From Equation (8.48) we see that, for $A > 0$, $0 \leq \cos 3\varphi \leq 1$, corresponding to $-\pi/2 \leq 3\varphi \leq \pi/2$.
   b. As ρ increases, $\cos 3\varphi$ decreases (see Figure 8.3); hence $d(3\varphi)/d\rho > 0$.
   c. For $A > 0$, we deduce from Equation (8.51)

$$\frac{d\varphi}{dt} < -\left(\frac{3\delta}{4m}\right) < 0 \tag{8.52}$$

As a result, for any initial value in the range $(-\pi/2, +\pi/2)$, 3φ will decrease monotonically and reach its lowest allowed value $(-\pi/2)$ in *a finite time*!

To obtain the evolution of ρ in time, insert Equation (8.48) in Equation (8.39), yielding

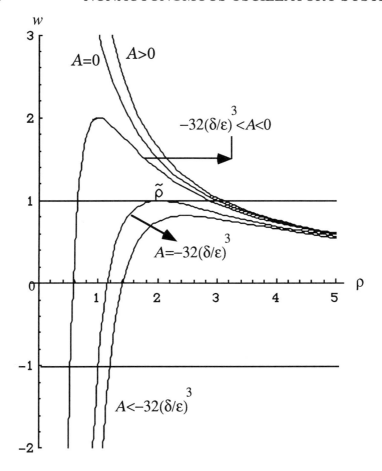

Figure 8.3 Plot of w versus ρ for different values of A.

$$\dot{\rho} = \pm \left(\frac{3\varepsilon}{8m}\right)\rho^2 \sqrt{1 - \left((A/\rho^3) + 6\left(\left(\frac{\delta}{\varepsilon}\right)/\rho\right)\right)^2} \qquad (8.53)$$

where the (−) sign corresponds to $0 \leq 3\varphi \leq (\pi/2)$ and the (+) sign, to $-(\pi/2) \leq 3\varphi \leq 0$. Now start with initial values obeying $\rho_0 > \rho_5$ and $3\varphi_0 > 0$ (corresponding to sin $3\varphi_0 > 0$). From Equation (8.39) we see that here $d\rho/dt < 0$ [corresponding to the (−) sign in Equation (8.53)]. Thus, ρ and 3φ both decrease until ρ reaches ρ_5 from above and (necessarily) 3φ reaches 0. At $\rho = \rho_5$, the r.h.s. of Equation (8.53) vanishes, so that ρ_5 is a fixed point of that equation. However, the coupling to the φ equation causes a drift of ρ from that point. As 3φ

THEORY OF ACCELERATOR DESIGN

continues to decrease, sin 3φ changes sign so that $d\rho/dt$ becomes positive and ρ grows away from ρ_5. We are now on the branch that will lead to a finite-time blowup of the radius: 3φ goes to $-(\pi/2)$ so that $\cos 3\varphi$ goes to zero in a finite time. Therefore, ρ must diverge at the same time.

We can get an approximate idea about the divergence of ρ by noting that ρ grows indefinitely on the second branch [(+) sign]. Hence, we can approximate Equation (8.53) by its large ρ limit:

$$\dot{\rho} \cong +\left(\frac{3\varepsilon}{8m}\right)\rho^2 \tag{8.54}$$

which is solved by

$$\rho = \frac{\rho_0}{1 - (3\varepsilon/(8m))\rho_0 t}$$

The radius diverges when $t \to (8m/3\varepsilon\rho_0) = O(1/\varepsilon)$. As we will see in Section 8.3, this *does not mean* that the full solution necessarily blows up. The blowup may be a problem of the first-order approximation. To sort the problem out, one must study the effect of the next orders in the perturbation expansion.

2. $-32(\delta/\varepsilon)^3 < A < 0$. This is the interesting case. From Figure 8.2, we see that, for a given value of A, there are three intersection points solving the equality $A = f(\rho)$. Of these, the pair ρ_1 and ρ_2 corresponds to bounded oscillations with $\rho_1 \leq \rho \leq \rho_2$. The single point ρ_3 is the lower boundary for a domain in which, again, unbounded motion is obtained (at least in first order).

For $\rho_1 \leq \rho \leq \rho_2$, Equation (8.39) yields

$$\dot{\varphi} \geq \left(\frac{3\delta}{2m}\right) - \left(\frac{3\varepsilon}{8m}\right)\rho \geq \left(\frac{3\delta}{2m}\right) - \left(\frac{3\varepsilon}{8m}\right)\rho_2$$

$$> \left(\frac{3\delta}{2m}\right) - \left(\frac{3\varepsilon}{8m}\right)\tilde{\rho} = 0 \tag{8.55}$$

Thus, φ grows monotonically in time. At $\rho = \rho_1$ and at $\rho = \rho_2$, $d\rho/dt = 0$. However, at both points, φ continues to grow, so that each time ρ reaches either ρ_1 or ρ_2, $d\rho/dt$ changes sign:

$$\dot{\rho} = \begin{cases} -\left(\dfrac{3\varepsilon}{8m}\right)\rho^2\sqrt{1-\cos^2 3\varphi} & 2n\pi \leq 3\varphi \leq (2n+1)\pi \\ +\left(\dfrac{3\varepsilon}{8m}\right)\rho^2\sqrt{1-\cos^2 3\varphi} & (2n+1)\pi \leq 3\varphi \leq (2n+2)\pi \end{cases} \quad (8.56)$$

Hence, the motion is oscillatory, with ρ oscillating slowly [over timescales of $O(1/\varepsilon)$] between the turning points ρ_1 and ρ_2, while $\cos 3\varphi$ oscillates between ± 1.

For $\rho \geq \rho_3$, we note from Figure 8.3 that $0 \leq \cos 3\varphi \leq 1$ [hence $-(\pi/2) \leq 3\varphi \leq +(\pi/2)$]. Moreover, from Equation (8.40), using the constraint $-32(\delta/\varepsilon)^3 < A < 0$ and the fact that $\rho_3 > \tilde{\rho}$, we obtain

$$\dot{\varphi} < -\left(\frac{3\delta}{4m}\right) + \left(\frac{3\varepsilon}{8m}\right)\frac{32(\delta/\varepsilon)^3}{\rho^2} \leq -\left(\frac{3\delta}{4m}\right) + \left(\frac{3\varepsilon}{8m}\right)\frac{32(\delta/\varepsilon)^3}{\rho_3^2}$$
$$< -\left(\frac{3\delta}{4m}\right) + \left(\frac{3\varepsilon}{8m}\right)\frac{32(\delta/\varepsilon)^3}{\tilde{\rho}^2} = 0 \quad (8.57)$$

Thus, φ decreases monotonically in time: 3φ will reach $-(\pi/2)$ in a finite time, and similar to the analysis for $A>0$, ρ will diverge.

3. $A=-32(\delta/\varepsilon)^3$: *The separatrix*. The motion is bounded between the intersection points: $\rho_4 \leq \rho \leq \tilde{\rho}$ and is unbounded for $\rho > \tilde{\rho}$. It takes an infinite amount of time to reach $\tilde{\rho}$ from either side. This is the *separatrix*. It separates between the domain in phase space where bounded motion is possible and the domain in which only unbounded motion is allowed. Equation (8.48) yields the phase-space portrait of the separatrix with $A=-32(\delta/\varepsilon)^3$.

It is interesting to see how the phase portrait looks like near two special points.

Near the fixed point ($\rho=\tilde{\rho}$, $3\varphi=0$). Near that point, we write

$$\rho = \tilde{\rho} + \xi \qquad |\xi| \ll \tilde{\rho} \quad (8.58)$$

(Noe that at the fixed point, $\cos 3\varphi = +1$.) Now insert Equation (8.58) in Equation (8.48) and expand the result. In lowest order one finds that the approach to the fixed point along the phase portrait of the separatrix is approximated by two intersecting straight lines:

THEORY OF ACCELERATOR DESIGN

$$\xi = \rho - \tilde{\rho} \cong \pm\sqrt{3}\tilde{\rho}\varphi \qquad (8.59)$$

Near $3\varphi = \pi$. There $\cos 3\varphi = -1$, Equation (8.48) yields $\rho = \rho_4 = 2(\delta/\varepsilon) = \tilde{\rho}/2$. Inserting in Equation (8.48)

$$\rho = \rho_4 + \xi \qquad \varphi = \pi/3 + \eta \qquad (8.60)$$

With both ξ and η small, we obtain in lowest order a parabola:

$$\xi = \rho - \rho_4 \cong \left(\frac{\delta}{\varepsilon}\right)\eta^2 = \left(\frac{\delta}{\varepsilon}\right)\left(\varphi - \frac{\pi}{3}\right)^2 \qquad (8.61)$$

4. $A < -32(\delta/\varepsilon)^3$. From Figure 8.3 we see that now $\cos 3\varphi$ varies between -1 ($3\varphi = \pi$) and

$$\cos 3\varphi_{max} = \left(32\frac{(\delta/\varepsilon)^3}{-A}\right)^{\frac{1}{2}} < 1 \qquad (8.62)$$

occurring at

$$\rho = \rho_{max} = \left(\frac{-A}{2(\delta/\varepsilon)}\right)^{\frac{1}{2}} > 4\left(\frac{\delta}{\varepsilon}\right) = \tilde{\rho} \qquad (8.63)$$

Thus, the range of 3φ is

$$3\varphi_{max} \leq 3\varphi \leq \pi \qquad (8.64)$$

In this case the equality $A = f(\rho)$ has no solution since one always has $A < f(\rho)$. However, for a given value of A, now the equality $A = g(\rho)$ has one solution, denoted by ρ_6. All values of $\rho \geq \rho_6$ are permitted. One can show that here, again, ρ diverges in a finite time.

Typical trajectories are depicted in Figure 8.4.

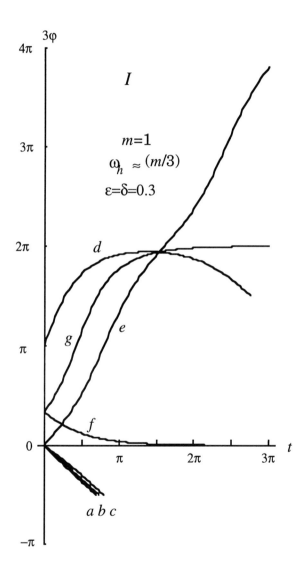

THEORY OF ACCELERATOR DESIGN

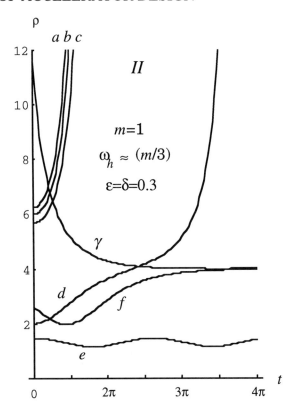

Figure 8.4 Plots of 3φ (panel *I*) and ρ (panel*II*) versus time [Equations (8.39)and (8.40)]: <u>Unbounded solutions</u>: (a) *A*=10; (b) *A*=0; (c) *A*=-1 ; (d) *A*=-33; <u>Bounded periodic solution</u>: (e) *A*=−10; <u>Separatrix</u>: (f) ,(g) *A*=−32: both solutions are attracted to fixed point.

8.3 THE MATHIEU EQUATION

The Mathieu equation (see, e.g., Ref. 3) describes a harmonic oscillator with a frequency that is modulated by a small periodic perturbation:

$$\ddot{x} + x(\omega^2 + \varepsilon \cos 2t) = 0 \qquad |\varepsilon| \ll 1 \qquad (8.65)$$

It appears in many areas of science, for instance, in solid state and plasma physics. The solution $x(t)$ may exhibit either stable periodic oscillations or be exponentially unstable, depending on the values of ω and ε. Usually, the analysis of the equation requires numerical methods. However, when ω is close to an integer n:

$$\omega^2 = n^2 + \mu \qquad |\mu| \ll 1 \qquad (8.66)$$

perturbation methods can be employed [3,4]. Normal forms are particularly suitable for this purpose. (When ω is close to an integer, the periodic term resonates with the natural frequency at some order of the expansion.) Stability of the oscillations depends on the values of ε and μ. We will find when the solution is unstable.

Substituting Equation (8.66) and using the complex notation $(z=x+i(1/n)(dx/dt))$, Equation (8.65) becomes

$$\dot{z} = -i\,nz - i\left(\frac{\mu}{2n}\right)(z+z^*) - i\left(\frac{\varepsilon}{2n}\right)(z+z^*)\cos 2t \qquad (8.67)$$

As in Section 8.2, we now convert the one-degree-of-freedom nonautonomous system of Equation (8.67) into an autonomous one with two degrees of freedom:

$$\dot{z} = -i\,n\,z - i\left(\frac{\mu}{2n}\right)(z+z^*) - i\left(\frac{\varepsilon}{4n}\right)(z+z^*)(v^2 + v^{*2})$$

$$\dot{v} = -i\,v \qquad v = e^{-it} \qquad (8.68)$$

8.3.1 The case $n=1$

We follow the formalism of Chapter 4. First, expand z in a near-identity transformation in powers of the two small parameters:

$$z = u + \varepsilon T_{10} + \mu T_{01} + \varepsilon^2 T_{20} + \varepsilon\mu T_{11} + \mu^2 T_{02} + \cdots \qquad (8.69)$$

Then insert this expansion in Equation (8.68) to find

$$T_{10} = \left(\tfrac{1}{8}\right) v^2 u - \left(\tfrac{1}{8}\right) v^{*2} u - \left(\tfrac{1}{16}\right) v^{*2} u^* + F_{10} \qquad (8.70a)$$

$$T_{01} = -\left(\tfrac{1}{4}\right) u^* + F_{01} \qquad (8.70b)$$

$$U_{10} = -i\left(\tfrac{1}{4}\right) v^2 u^* \qquad (8.71a)$$

$$U_{01} = -i\left(\tfrac{1}{2}\right) u \qquad (8.71b)$$

$$U_{20} = i\left(\tfrac{1}{64}\right)u + i\left(\tfrac{1}{4}\right)\left(-v^2 F_{10}^* + v^2 u^* \frac{\partial F_{10}}{\partial u} - v^{*2} u \frac{\partial F_{10}}{\partial u^*}\right) \quad (8.72a)$$

$$U_{11} = i\left(\tfrac{1}{8}\right)v^2 u^* + i\left(\tfrac{1}{2}\right)\left(-F_{10} + u\frac{\partial F_{10}}{\partial u} - u^*\frac{\partial F_{10}}{\partial u^*}\right)$$
$$+ i\left(\tfrac{1}{4}\right)\left(-v^2 F_{01}^* + v^2 u^*\frac{\partial F_{01}}{\partial u} - v^{*2} u\frac{\partial F_{01}}{\partial u^*}\right) \quad (8.72b)$$

$$U_{02} = i\left(\tfrac{1}{8}\right)u + i\left(\tfrac{1}{2}\right)\left(-F_{01} + u\frac{\partial F_{01}}{\partial u} - u^*\frac{\partial F_{01}}{\partial u^*}\right) \quad (8.72c)$$

I. No free functions. Without free functions, and noting that $v \cdot v^* = 1$, the normal form becomes

$$\dot{u} = -i\left\{1 + \left(\tfrac{1}{2}\right)\mu - \left(\tfrac{1}{64}\right)\varepsilon^2 - \left(\tfrac{1}{8}\right)\mu^2\right\}u - i\left\{\left(\tfrac{1}{4}\right)\varepsilon - \left(\tfrac{1}{8}\right)\varepsilon\mu\right\}v^2 u^* \quad (8.73)$$

Noting that $v = \exp(-it)$, we write

$$u = f e^{-it}$$

Inserting in Equation (8.73), the equations for f and for f^* become

$$\dot{f} = -i\left\{\left(\tfrac{1}{2}\right)\mu - \left(\tfrac{1}{64}\right)\varepsilon^2 - \left(\tfrac{1}{8}\right)\mu^2\right\}f - i\left\{\tfrac{1}{4}\varepsilon - \left(\tfrac{1}{8}\right)\varepsilon\mu\right\}f^* \quad (8.74a)$$

$$\dot{f}^* = i\left\{\tfrac{1}{4}\varepsilon - \left(\tfrac{1}{8}\right)\varepsilon\mu\right\}f + i\left\{\left(\tfrac{1}{2}\right)\mu - \left(\tfrac{1}{64}\right)\varepsilon^2 - \left(\tfrac{1}{8}\right)\mu^2\right\}f^* \quad (8.74b)$$

This set of coupled linear equations is solved by exponentials of the form $\exp(\lambda t)$ with λ satisfying

$$\lambda^2 + \left\{\left(\tfrac{1}{2}\right)\mu - \left(\tfrac{1}{64}\right)\varepsilon^2 - \left(\tfrac{1}{8}\right)\mu^2\right\}^2 - \left\{\left(\tfrac{1}{4}\right)\varepsilon - \left(\tfrac{1}{8}\right)\varepsilon\mu\right\}^2 = 0 \quad (8.75)$$

The solution is unstable when $\lambda^2 > 0$. Without loss of generality, assume $\varepsilon > 0$. In first order, the zone of instability is

$$|\mu| < \tfrac{1}{2}\varepsilon \quad (8.76)$$

Computing second order terms consistently, one finds

268 NONAUTONOMOUS OSCILLATORY SYSTEMS

$$-\tfrac{1}{2}\varepsilon - \tfrac{1}{32}\varepsilon^2 < \mu < \tfrac{1}{2}\varepsilon - \tfrac{1}{32}\varepsilon^2 \qquad (8.77)$$

II. With free functions. Studying Equation (8.72a) we find that U_{20} can be made to vanish by choosing

$$F_{10} = \left(\tfrac{1}{32}\right) v^2 u^* \qquad (8.78a)$$

On the other hand, from Equation (8.72c) one finds that no free function F_{01} can make U_{02} vanish. Hence we choose

$$F_{01} = 0 \qquad (8.78b)$$

With the choice of Equation (8.78a), the expression for U_{11} is modified. The normal form becomes

$$\dot{u} = -i\left\{1 + \left(\tfrac{1}{2}\right)\mu - \left(\tfrac{1}{8}\right)\mu^2\right\} u - i\left\{\left(\tfrac{1}{4}\right)\varepsilon - \left(\tfrac{3}{32}\right)\varepsilon\mu\right\} v^2 u^* \qquad (8.79)$$

Repeating the steps that lead to Equations (8.74a) and (8.74b), the eigenvalue constraint, Equation (8.75) is now replaced by

$$\lambda^2 + \left(\left(\tfrac{1}{2}\right)\mu - \left(\tfrac{1}{8}\right)\mu^2\right)^2 - \left(\left(\tfrac{1}{4}\right)\varepsilon - \left(\tfrac{3}{32}\right)\varepsilon\mu\right)^2 = 0 \qquad (8.80)$$

Equation (8.80) is different from Equation (8.75). However, in first order, it yields the same instability domain as Equation (8.76). Moreover, solving for the instability domain ($\lambda^2 > 0$) consistently through second order shows that Equation (8.77) is obtained again. The difference between Equations (8.75) and (8.80) affects the results only in orders higher than second (for which we have not performed the computation). Thus, the physical conclusions are unaffected by different choices of the free functions in the first-order term [which correspond to different choices of the zero-order approximation, $u(t)$] in the expansion.

8.3.2 The case $n=2$

The case $n=2$ is interesting since resonant terms occur here only in second order. Consequently, the domain of instability is confined to $\mu = O(\varepsilon^2)$, rather than $\mu = O(\varepsilon)$. Equation (8.67) now becomes

$$\dot{z} = -2i\, z - i\left(\frac{\mu}{4}\right)(z + z^*) - i\left(\frac{\varepsilon}{4}\right)(z + z^*)\cos 2t \qquad (8.81)$$

MATHIEU EQUATION

It is convenient to define a new time variable, $t'=2t$, and $v=\exp(-it')$. Equation (8.81) is now converted into an autonomous system (Throughout the remainder of this section, a dot on top of a function will denote a derivative with respect to t'.)

$$\dot{z} = -i\,z - \tfrac{1}{8}i\,\mu(z+z^*) - \tfrac{1}{16}i\,\varepsilon(z+z^*)(v+v^*) \qquad (8.82)$$

The normal form expansion yields the following:

$$T_{10} = \left(\tfrac{1}{16}\right)vu - \left(\tfrac{1}{16}\right)vu^* - \left(\tfrac{1}{16}\right)v^*u - \left(\tfrac{1}{48}\right)v^*u^* + F_{10} \qquad (8.83a)$$

$$T_{01} = -\left(\tfrac{1}{16}\right)u^* + F_{01} \qquad (8.83b)$$

$$U_{10} = 0 \qquad (8.84a)$$

$$U_{01} = -\tfrac{1}{8}i\,u \qquad (8.84b)$$

$$U_{20} = \tfrac{1}{96}i\,u + \tfrac{1}{64}i\,v^2 u^* \qquad (8.84c)$$

$$U_{11} = -\tfrac{1}{8}i\,F_{10} \qquad (8.84d)$$

$$U_{02} = \tfrac{1}{128}i\,u + \tfrac{1}{8}i\left(-F_{01} + u\frac{\partial F_{01}}{\partial u} - u^*\frac{\partial F_{01}}{\partial u^*}\right) \qquad (8.84e)$$

Introducing an $F_{10}\neq 0$ is not desired; it will make U_{11} nonzero. Also, U_{02} can be eliminated only with an F_{01} that leads to secular behavior in T_{01}. This is easy to see, writing $u=\rho\cdot\exp(-i\varphi)$. Equation (8.84e) then becomes

$$U_{02} = \tfrac{1}{128}i\,u - \tfrac{1}{8}\left(i\,F_{01} + \frac{\partial F_{01}}{\partial \varphi}\right)$$

Requiring hat U_{02} vanishes yields an F_{01} that has a secular term:

$$F_{01} = \left\{C - i\left(\tfrac{1}{16}\right)\rho\varphi\right\}\exp(-i\,\varphi)$$

Thus, we choose $F_{10}=F_{01}=0$. The normal form then becomes

$$\dot{u} = -i\left\{1 + \left(\tfrac{1}{8}\right)\mu - \left(\tfrac{1}{96}\right)\varepsilon^2 - \left(\tfrac{1}{128}\right)\mu^2\right\}u + i\,\varepsilon^2\left(\tfrac{1}{64}\right)v^{*2}u \qquad (8.85)$$

The width of the instability domain in the $\varepsilon - \mu$ plane is $O(\varepsilon^2)$:

$$-\left(\tfrac{1}{24}\right)\varepsilon^2 < \mu < \left(\tfrac{5}{24}\right)\varepsilon^2 \tag{8.86}$$

This result is a consequence of the fact that, for $n=2$, the $\cos(2t)$ multiplicative factor in Equation (8.81) resonates with the natural oscillations only in second order.

8.3.3 The general case

For general n, the computation is cumbersome. Still the trend can be seen in a qualitative manner. The μ part of the perturbation in Equation (8.68) has a resonant term already in first order, which is linear in u. In a rough approximation, the time dependence of u is $\exp(-int)$ and that of v is $\exp(-it)$. This provides an indication of the type of terms required to compete with the resonant term in Equation (8.68). The simplest term that will resonate with it will be $v^{2n} \cdot u^*$. Such a term will appear for the first time in U_{n0}, the $O(\varepsilon^n)$ term in the normal form. Therefore, the competition is between the $O(\mu)$ and $O(\varepsilon^n)$ terms. The instability domain will be

$$|\mu| \leq O(\varepsilon^n)$$

8.3.4 Do we have to convert the equations into autonomous ones?

We now reanalyze the case $n=1$ (Section 8.3.1) without converting the Mathieu equation into an autonomous system of higher dimension. Namely, we do not make the substitution $v = \exp(-it)$ and, instead, introduce an explicit time dependence in the T functions. The example will be given in first order only. Extension to higher orders is self-evident. Again, we resort to Equation (4.9b):

$$U_{01} + \frac{\partial T_{01}}{\partial t} = -i\, T_{01} + i\, u \frac{\partial T_{01}}{\partial u} - i\, u^* \frac{\partial T_{01}}{\partial u^*} - i\, \frac{\varepsilon}{2}(u + u^*) \tag{8.87a}$$

$$\begin{aligned} U_{10} + \frac{\partial T_{10}}{\partial t} &= -i\, T_{10} + i\, u \frac{\partial T_{10}}{\partial u} - i\, u^* \frac{\partial T_{10}}{\partial u^*} \\ &\quad - \tfrac{1}{4} i\, \varepsilon (u + u^*)(\exp(-2\,i\,t) + \exp(-2\,i\,t)) \end{aligned} \tag{8.87b}$$

MATHIEU EQUATION

In addition, we assume for each of the T functions the expansion:

$$T_{mn} = \sum_{pq,r} \alpha_{pqr}^{(m,n)} u^p u^{*q} \exp(i\, rt) \qquad (8.88)$$

A similar expansion is adopted for U_{mn}. We now use the T_{nm} to eliminate nonresonant terms of the interaction on the r.h.s. of Equations (8.87a,b), leaving the resonant ones as contributions to U_{mn}. In Equations (8.87a,b) each coefficient will appear as

$$\alpha_{pqr}^{(m,n)} i\,(-1+p-q-r)$$

Thus, resonant terms, for which $-1+p-q-r=0$, will be attributed to the U functions. For example, in Equation (8.87a), there are four combinations of (p,q,r): $(1,0,-2)$, $(1,0,2)$, $(0,1,-2)$, $(0,1,2)$. The resulting T and U functions are

$$T_{10} = \left(\tfrac{1}{8}\right)\exp(-2i\,t)u - \left(\tfrac{1}{8}\right)\exp(+2i\,t)u - \left(\tfrac{1}{16}\right)\exp(+2i\,t)u^* \\ + F_{10} \qquad (8.89a)$$

$$T_{01} = -\left(\tfrac{1}{4}\right)u^* + F_{01} \qquad (8.89b)$$

$$U_{10} = -i\,\left(\tfrac{1}{4}\right)\exp(-2i\,t)u^* \qquad (8.89c)$$

$$U_{01} = -i\,\left(\tfrac{1}{2}\right)u \qquad (8.89d)$$

The results are identical to Equations (8.70a,b) and (8.71a,b), respectively, with v replaced by $\exp(-2it)$. The free functions are now functions of powers of u, u^* and $\exp(-2it)$ with resonant combinations of the powers p, q, r. From here on the analysis and the results are identical to those of Section 8.3.1.

8.4 FORTUITOUS "EXPLOSIVE INSTABILITIES": EXAMPLES OF THE NONLINEAR MATHIEU EQUATION

8.4.1 One-parameter problem

In some problems, the first-order approximation in a perturbation expansion for the full solution has characteristics different from those of the full solution. The first-order approximation to the normal form exhibits an *explosive instability*: – it diverges at some finite time $t=t_0$. At the same time, the *full solution is well behaved* at $t=t_0$. The question that immediately arises is whether such a situation signifies the breakdown of the perturbation expansion.

In this section we demonstrate this phenomenon in the case of a nonautonomous version of the nonlinear Mathieu equation. We show that the occurrence of a divergence in the first-order approximation need not signify the breakdown of the perturbation expansion. When the second-order term is included in the normal form, the singularity in the approximation disappears. In particular, we show how the use of the free functions in the normal form expansion may simplify in a significant manner the explanation of the phenomenon.

We start by analyzing the following simple problem:

$$\ddot{x} + x + \delta \cos 4t \, x^3 = 0 \tag{8.90}$$

For sufficiently small δ, the solution of Equation (8.90) is well behaved. However, the first-order perturbation analysis leads to a divergence of the lowest approximation to the solution at a finite time ("finite-time blowup"). To see this, we diagonalize Equation (8.90) by substituting $z=x+idx/dt$.. The equation for z is

$$\dot{z} = -i\,z - \tfrac{1}{16} i \left(e^{-4it} + e^{4it}\right)\left(z+z^*\right)^3 \tag{8.91}$$

As previously done in this chapter, we convert Equation (8.91) to a higher (four)-dimensional problem by defining $v=\exp(-i\,t)$. We only need to expand z in a near-identity transformation, as v is fully known:

$$z = u + \delta T_1\left(u,u^*,v,v^*\right) + \cdots \tag{8.92}$$

Noting that the only resonant term in the perturbation in Equation (8.91) is $v^4 z^{*3}$, the first-order approximation for the normal form is readily obtained:

$$\dot{u} = -i\,u + \delta U_1 = -i\,u - \tfrac{1}{16} i\,\delta u^{*3} v^4 \tag{8.93}$$

It is easiest to analyze this equation in polar coordinates: $u = \rho e^{-it-i\varphi}$, where, again, we have written explicitly the part of the phase that has a fast time dependence. The equations ρ for and φ are

EXPLOSIVE INSTABILITIES

$$\dot{\rho} = \tfrac{1}{16}\delta\rho^3 \sin 4\varphi \qquad (8.94a)$$

$$\dot{\varphi} = \tfrac{1}{16}\delta\rho^2 \cos 4\varphi \qquad (8.94b)$$

Analysis of Equation (8.94). From Equation (8.94b) we see that φ has a fixed point at $4\varphi=\pm\pi/2$. A linear stability analysis yields that, for $\delta>0$, $4\varphi=+\pi/2$ is stable for $t\to\infty$; and $4\varphi=-\pi/2$, for $t\to-\infty$. For $\delta<0$, the roles of the two fixed points are interchanged. Thus, without loss of generality, we assume $\delta>0$ and concentrate on the stable fixed point at $4\varphi=\pi/2$.

To find the behavior of ρ near the fixed point, we note that Equations (8.94a,b) have a first integral. Multiplying Equation (8.94a) by $4\rho^3\cos 4\varphi$ and Equation (8.94b) by $4\rho^4\sin 4\varphi$ and then subtracting the equations from one another, we obtain

$$4\rho^3 \dot{\rho}\cos 4\varphi - 4\rho^4 \sin 4\varphi\,\dot{\varphi} = \frac{d}{dt}\{\rho^4 \cos 4\varphi\} = 0 \qquad (8.95)$$

Hence, we have (ρ_0 and φ_0 are the initial conditions)

$$\rho^4 \cos 4\varphi = \rho_0^4 \cos 4\varphi_0$$

yielding

$$\rho = \rho_0 \left\{\frac{\cos 4\varphi_0}{\cos 4\varphi}\right\}^{1/4} \qquad (8.96)$$

Therefore, in the vicinity of $4\varphi=\pm\pi/2$, ρ diverges. Clearly, this signifies that we cannot trust the results of the first-order calculation long before t reaches t_0. Let us write

$$4\varphi = \pm\frac{\pi}{2} \mp \eta \quad (0<\eta\ll 1) \Rightarrow \cos 4\varphi = \sin\eta \cong \eta \qquad (8.97)$$

and insert Equations (8.96) and (8.97) in Equation (8.94b); we find that η is given approximately by

$$\eta \cong \eta_0 \left(1 \mp \left(\tfrac{1}{8}\right)\delta\rho_0^2\, t\right)^2 \qquad (8.98)$$

Thus, near $4\varphi=+\pi/2$, η vanishes as $t\to t_0=(8/(\delta\rho_0^2))$, while near

274 NONAUTONOMOUS OSCILLATORY SYSTEMS

$4\varphi=-\pi/2$, η drifts away from 0. This shows that $4\varphi=+\pi/2$ is a stable fixed point, while $4\varphi=-\pi/2$ is unstable.

Comment. Let us focus on the fixed point at $4\varphi=+\pi/2$. We are used to an asymptotic exponential rate of approach to fixed points, $\exp(\lambda t)$ with a negative constant exponent, λ. However, here, near $4\varphi=+\pi/2$, φ reaches its fixed point in a finite time. To see how this happens, insert Equations (8.97) and (8.98) in Equation (8.96) to obtain

$$\rho = \frac{\rho_0}{\sqrt{1-\left(\frac{1}{8}\right)\delta\rho_0^2 t}} \tag{8.99}$$

Thus, ρ diverges as $t \to t_0$. As a result, the coefficient multiplying $\cos 4\varphi$ in Equation (8.94b) is not constant. It grows rapidly in time, making the rate of approach of φ to its fixed point value much more rapid than an exponential one. This is why the fixed point is reached at a finite time.

To find whether the perturbation expansion breaks down completely, or the next orders rectify what the first one spoils, we analyze the normal form through second order:

$$\dot{u} = -iu + \delta U_1 + \delta^2 U_2 \tag{8.100}$$

Employing the general formalism of Chapter 4, we find

$$T_1 = \left(\tfrac{1}{96}\right)u^3 v^4 - \left(\tfrac{1}{128}\right)u^{*3}v^{*4} - \left(\tfrac{1}{32}\right)u^3 v^{*4} + \left(\tfrac{3}{64}\right)u^2 u^* v^4 \\ - \left(\tfrac{1}{32}\right)u u^{*2} v^{*4} - \left(\tfrac{3}{64}\right)u^2 u^* v^{*4} + \left(\tfrac{3}{32}\right)u u^{*2} v^4 + F_1 \tag{8.101}$$

$$U_2 = -\left(\tfrac{61}{2048}\right) i\, u^3 u^{*2} \\ -\left(\tfrac{3}{16}\right) i\, u^{*2} v^4 F_1^* + \left(\tfrac{1}{16}\right) i\left(u^{*3}v^4 \frac{\partial F_8}{\partial u} - u^3 v^{*4}\frac{\partial F_1}{\partial u^*}\right) \tag{8.102}$$

It turns out that no useful simplification is gained by introduction of the free function, hence we choose $F_1=0$. Through second order, the normal form becomes

$$\dot{u} = -i\,u - \left(\tfrac{1}{16}\right) i\, \delta u^{*3} v^4 - \left(\tfrac{61}{2048}\right) i\, \delta^2 u^3 u^{*2} \tag{8.103}$$

In polar coordinates (with the fast time-dependent part of the phase

EXPLOSIVE INSTABILITIES

written explicitly), $u = \rho e^{-it-i\varphi}$, Equation (8.103) becomes

$$\dot{\rho} = \left(\tfrac{1}{16}\right)\delta\rho^3 \sin 4\varphi \qquad \dot{\varphi} = \left(\tfrac{1}{16}\right)\delta\rho^2 \cos 4\varphi + \left(\tfrac{61}{2048}\right)\delta^2 \rho^4 \quad (8.104)$$

The fixed points (found in the first order equations) at $4\varphi = \pm\pi/2$ are removed. Near the fixed points, $d\eta/dt$ does not vanish. Repeating the analysis that led to Equation (8.95), one finds that $\rho^4\cos 4\varphi$ is not a first integral of the motion in this order. Instead, Equations (8.104) have another first integral:

$$\rho^4 \cos 4\varphi + \tfrac{61}{192}\delta\rho^6 = \text{const} = \rho_0^4 \cos 4\varphi_0 + \tfrac{61}{192}\delta\rho_0^6 \quad (8.105a)$$

As a result, ρ does not develop a singularity at a finite time, but oscillates indefinitely with a bounded amplitude.

In Figure 8.5, we show the full solution of Equation (8.90) and $\operatorname{Re} u(t)$, the zero-order term, in two approximations: the solution of Equation (8.94) and of Equation (8.104) for the normal form through first and second orders, respectively.

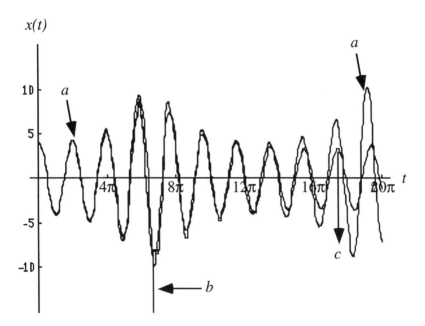

Figure 8.5 Solution and approximations for $x(t)$ of Equation (8.90): $x(0)=4$, $dx/dt(0)=0$, $\delta=0.03$. (a) Full solution; (b) zero-order approximation using first-order normal form [Equation (8.94)]; (c) zero-order pproximation using second-order normal form [Equation (8.104)].

For $t<t_0=O(1/(\delta\rho_0^2))$, both approximations agree very well with the full solution. At $t=t_0$, the first-order approximation, based on Equation (8.94), diverges. In the second-order approximation, based on Equation (8.104), this divergence disappears. The latter approximation provides a reasonable description of the full solution up to $t \approx 2t_0$, that is, far beyond the point of divergence in first order. Not surprisingly, further on, the second-order approximation breaks down. First, we have not added the T function. Second, the update of the slow phase in higher orders needs to be incorporated in the approximation, to make it a good one for longer times.

Choosing $\varphi_0=0$ as the initial condition, Equation (8.105a), implies that the amplitude at $4\varphi=\pi$ is related to the amplitude at $4\varphi=0$, by

$$\rho_\pi^4 + \rho_0^4 = \frac{\rho_\pi^6 - \rho_0^6}{\tilde{\rho}^2} \qquad \tilde{\rho}^2 \equiv \tfrac{192}{61}\delta^{-1} \qquad (8.105b)$$

From Equation (8.105b) one finds that, to a very good approximation, $\rho_\pi^2 = \tilde{\rho}^2$ for $\rho_0^2 \ll \tilde{\rho}^2$. In Figure 8.5 we use $\rho_0 \approx 4$ and $\delta = 0.03$, which yield $\rho_0^2 \approx 16 \ll \tilde{\rho}^2 = 105.5$. Therefore, we can use the approximate value $\rho_\pi^2 = \tilde{\rho}^2$, yielding $\rho_\pi \approx 10.2$. This result agrees very well with the maximal value of the amplitude, indicated by Figure 8.5.

8.4.2 Two-parameter problem

We now focus our attention on a more complicated problem:

$$\ddot{x} + x + (\gamma + \delta\cos 4t)x^3 = 0 \qquad (8.106)$$

There is a range of values of γ and δ, for which the solution of Equation (8.106) is bounded. In the same domain, the first-order approximation for $x(t)$ in a normal form analysis has an explosive instability. Again, the latter disappears in a second-order analysis. Owing to the more complex nature of the present problem, the analysis becomes simple only when specific free functions are used.

Writing $v=\exp(-it)$, the complex version of Equation (106) is

$$\dot{z} = -i\,z - \tfrac{1}{8}i\,\gamma\left(z+z^*\right)^3 - \tfrac{1}{16}i\,\delta\left(v^4 + v^{*4}\right)\left(z+z^*\right)^3 \qquad (8.107)$$

We now write the following two-parameter expansions:

$$z = u + \gamma T_{10} + \delta T_{01} + \cdots \qquad (8.108)$$

$$\dot{u} = -i\,u + \gamma U_{10} + \delta U_{01} + \gamma^2 U_{20} + \gamma\delta U_{11} + \delta^2 U_{02} + \cdots \quad (8.109)$$

Using the formalism of Chapter 4, we find that T_{01} is identical to T_1 of Equation (8.101) and U_{01}, to U_1 of Equation (8.93). In addition, we find

$$T_{10} = \left(\tfrac{1}{16}\right)u^3 - \left(\tfrac{3}{16}\right)uu^{*2} - \left(\tfrac{1}{32}\right)u^{*3} + F_{10} \quad (8.110)$$

$$U_{10} = -\left(\tfrac{3}{16}\right)i\,u^2 u^* \quad (8.111)$$

$$U_{20} = \left(\tfrac{51}{256}\right)i\,u^3 u^{*2} - \left(\tfrac{3}{4}\right)i\,uu^* F_{10} - \left(\tfrac{3}{8}\right)i\,u^2 F_{10}^*$$
$$+ \left(\tfrac{3}{8}\right)i\left(u^2 u^* \frac{\partial F_{10}}{\partial u} - uu^{*2}\frac{\partial F_{10}}{\partial u^*}\right) \quad (8.112)$$

$$U_{11} = \left(\tfrac{3}{512}\right)i\,u^5 v^{*4} + \left(\tfrac{15}{512}\right)i\,uu^{*4} v^4$$
$$- \left(\tfrac{3}{16}\right)i\,u^{*2} v^4 F_{10}^* - \left(\tfrac{3}{8}\right)i\left(2uu^* F_{01} + u^2 F_{01}^*\right)$$
$$+ \left(\tfrac{3}{8}\right)i\left(u^2 u^* \frac{\partial F_{01}}{\partial u} - uu^{*2}\frac{\partial F_{01}}{\partial u^*}\right) \quad (8.113)$$
$$+ \left(\tfrac{1}{16}\right)i\left(u^{*3} v^4 \frac{\partial F_{10}}{\partial u} - u^3 v^{*4}\frac{\partial F_{10}}{\partial u^*}\right)$$

Using, again, polar coordinates, with the fast time dependence of the phase written explicitly, $u = \rho e^{-it-i\varphi}$, we write the free functions as

$$F_\alpha = u\,G_\alpha(\rho,\varphi) \qquad (\alpha = 10 \text{ or } 01) \qquad (8.114)$$

Comment. A priori, G_α should depend on all nonresonant combinations: $u\cdot u^*$, $u\cdot v^*$, $u^*\cdot v$ and $v\cdot v^*$. However, in polar coordinates, with $v = \exp(-it)$, they depend only on ρ, and φ.

In first order, the normal form Equation (8.109) yields the following equations for ρ and φ:

$$\dot{\rho} = \left(\tfrac{1}{16}\right)\delta\rho^3 \sin 4\varphi \quad (8.115)$$

$$\dot{\varphi} = \left(\tfrac{1}{16}\right)\rho^2 (6\gamma + \delta\cos 4\varphi) \quad (8.116)$$

For

$$6\left|\frac{\gamma}{\delta}\right|<1 \qquad \delta>0 \qquad (8.117)$$

the solution of Equation (8.116) for φ has a stable fixed point [5] at

$$\varphi^* = \cos^{-1}\left(-6\frac{\gamma}{\delta}\right) \qquad 8.118)$$

leading to an asymptotically divergent solution of Equation (8.115):

$$\rho = \frac{\rho_0}{\sqrt{1-(t/t_0)}} \qquad t_0 = \frac{8}{\delta \sin 4\varphi^* \rho_0^2} \qquad (8.119)$$

The second order normal form is

$$\dot{\rho} = \left(\tfrac{1}{16}\right)\delta\rho^3 \sin 4\varphi - \gamma^2 \left(\tfrac{3}{8}\right)\rho^3 \frac{\partial \operatorname{Re} G_{10}}{\partial \varphi}$$

$$+\gamma\delta \left\{ \begin{array}{l} -\left(\tfrac{3}{128}\right)\rho^5 \sin 4\varphi \\ +\left(\tfrac{1}{16}\right)\rho^3 \left(2\operatorname{Re} G_{10} \sin 4\varphi - 4\operatorname{Im} G_{10} \cos 4\varphi\right) \\ -\left(\tfrac{1}{16}\right)\rho^4 \dfrac{\partial \operatorname{Re} G_{10}}{\partial \rho} \sin 4\varphi \\ -\left(\tfrac{1}{16}\right)\rho^3 \dfrac{\partial \operatorname{Re} G_{10}}{\partial \varphi} \cos 4\varphi - \left(\tfrac{3}{8}\right)\rho^3 \dfrac{\partial \operatorname{Re} G_{01}}{\partial \varphi} \end{array} \right\}$$

$$+\delta^2 \left\{ \begin{array}{l} \left(\tfrac{1}{16}\right)\rho^3 \left(2\operatorname{Re} G_{01} \sin 4\varphi - 4\operatorname{Im} G_{01} \cos 4\varphi\right) \\ -\left(\tfrac{1}{16}\right)\rho^4 \dfrac{\partial \operatorname{Re} G_{01}}{\partial \rho} \sin 4\varphi - \left(\tfrac{1}{16}\right)\rho^3 \dfrac{\partial \operatorname{Re} G_{01}}{\partial \varphi} \cos 4\varphi \end{array} \right\}$$

(8.120)

$\dot{\varphi} = \left(\tfrac{1}{16}\right)\rho^2(6\gamma + \delta\cos 4\varphi)$

$+ \gamma^2\left\{-\left(\tfrac{51}{256}\right)\rho^4 + \left(\tfrac{3}{8}\right)\rho^2\left(2\operatorname{Re}G_{10} + \dfrac{\partial\operatorname{Im}G_{10}}{\partial\varphi}\right)\right\}$

$+ \gamma\delta\left\{\begin{array}{l}-\left(\tfrac{9}{256}\right)\rho^4\cos 4\varphi + \left(\tfrac{3}{8}\right)\rho^2\left(2\operatorname{Re}G_{01} + \dfrac{\partial\operatorname{Im}G_{01}}{\partial\varphi}\right) \\ + \left(\tfrac{1}{16}\right)\rho^2(2\operatorname{Re}G_{10}\cos 4\varphi + 4\operatorname{Im}G_{10}\sin 4\varphi) \\ + \left(\tfrac{1}{16}\right)\rho^3\dfrac{\partial\operatorname{Im}G_{10}}{\partial\rho}\sin 4\varphi + \left(\tfrac{1}{16}\right)\rho^2\dfrac{\partial\operatorname{Im}G_{10}}{\partial\varphi}\cos 4\varphi\end{array}\right\}$

$+ \delta^2\left\{\begin{array}{l}\left(\tfrac{61}{2048}\right)\rho^4 + \left(\tfrac{1}{16}\right)\rho^2(2\operatorname{Re}G_{01}\cos 4\varphi + 4\operatorname{Im}G_{01}\sin 4\varphi) \\ + \left(\tfrac{1}{16}\right)\rho^3\dfrac{\partial\operatorname{Im}G_{01}}{\partial\rho}\sin 4\varphi + \left(\tfrac{1}{16}\right)\rho^2\dfrac{\partial\operatorname{Im}G_{01}}{\partial\varphi}\cos 4\varphi\end{array}\right\}$

(8.121)

The singular behavior disappears when the normal form is computed through second order. The numerical study of the solutions of Equations (8.120) and (8.121) shows that the second-order contributions, with or without (bounded) free functions, eliminate the finitetime blowup that occurs in first order.

Without free functions, the equations become

$$\dot{\rho} = \left(\tfrac{1}{16}\right)\delta\left\{1 - \left(\tfrac{3}{8}\right)\gamma\rho^2\right\}\rho^3\sin 4\varphi \qquad (8.122)$$

$$\dot{\varphi} = \left(\tfrac{1}{16}\right)\rho^2(6\gamma + \delta\cos 4\varphi) + \left\{\left(\tfrac{61}{2048}\right) - \left(\tfrac{51}{256}\right)\gamma^2\right\}\rho^4 \\ - \left(\tfrac{9}{256}\right)\gamma\delta\{\rho^4\cos 4\varphi\} \qquad (8.123)$$

An important point emerges here. Comparing terms in Equations (8.122) and (8.123) that compete with the lower-order ones [as in Equations (8.115) and (8.116)], we see that the relative size of the new terms is always of the form

$$(\text{Small parameter})\cdot\rho^2$$

Thus, the *effective* expansion parameters are not γ and δ, but $\gamma\rho^2$ and $\delta\rho^2$, respectively. If ρ tends to become large in the first-order calculation, these two effective expansion parameters may become of order unity (rather than remaining much smaller than unity). To see whether this can be avoided, we must repeat the calculation in the domain where ρ is *still* $O(1)$ rather than $O(1/\delta)$ (so that the effective expansion parameters are small) and include the second-order terms in the calculation. If we are lucky, the second-order terms may tame the behavior of the first order. This turns out to be the case in the present problem.

In a numerical calculation, using Equations (8.122) and (8.123) the divergence of ρ is avoided. However, it is difficult to see how this happens analytically. With an appropriate choice of the free functions, the structure of the equations through second order can be simplified into a form that makes the disappearance of the divergent behavior understandable.

Guided by intuition, we seek free functions that simplify the ρ equation so that, through second order, that equation remains the same as Equation (8.115), and that also simplify the φ equation. Inspection of Equations (8.120) and (8.121) shows that introducing a G_{10} for that purpose makes Equation (8.121) more complicated and less amenable to simple analysis. Thus, we choose $G_{10}=0$ (i.e., $F_{10}=0$). On the other hand, the choice

$$G_{01} = \left(\tfrac{1}{64}\right)e^{4i\varphi} \qquad \left(F_{01} = \left(\tfrac{1}{64}\right)u^{*3}v^{4}\right)$$

does simplify the equations. The ρ equation remains unchanged in this order:

$$\dot\rho = \left(\tfrac{1}{16}\right)\delta\rho^{3}\sin 4\varphi + O\!\left(\gamma^{3},\gamma^{2}\delta,\gamma\delta^{2},\delta^{3}\right) \qquad (8.124)$$

and the φ equation becomes

$$\dot\varphi = \left(\tfrac{1}{16}\right)\rho^{2}(6\gamma + \delta\cos 4\varphi) + \left(\left(\tfrac{73}{2048}\right)\delta^{2} - \left(\tfrac{51}{256}\right)\gamma^{2}\right)\rho^{4} \\ + O\!\left(\gamma^{3},\gamma^{2}\delta,\gamma\delta^{2},\delta^{3}\right) \qquad (8.125)$$

The second-order term in Equation (8.125) prevents the occurrence of the fixed point at $\varphi=\varphi_0$. [See Equation (8.118)]. Within the "dangerous" range of γ and δ [Equation (8.117)], the coefficient of the ρ^4 term is positive. Hence, the second-order term in Equation (8.125) is always positive, and prevents the vanishing of $d\varphi/dt$ near the prospective fixed point. As a result, φ drifts away from the fixed point by a positive contribution that grows monotonically in time. Consequently, no blowup occurs in the solution of Equation (8.124): ρ oscillates indefinitely with a finite oscillation amplitude.

This is demonstrated in Figure 8.6, where we show the full numerical solution, and the zero-order approximations to $x(t)$ [$=\mathrm{Re}\,u(t)$] for two cases: (1) $u(t)$ - computed from the first-order normal form [Equations (8.115) and (8.116)], yielding a blowup at a finite time, and (2) $u(t)$ - computed with the second-order correction included in the normal form [Equations (8.124) annd (8.125)]. It is quite remarkable that $u(t)$, when computed from the first-order normal form, diverges at a finite time, but, when computed from the second-order normal form, provides a reasonable description of the solution far beyond the prospective "blowup" time.

At longer times, even the better approximation deviates from the full solution. This is to be expected, as we have approximated $x(t)$ by the zero order term, $u(t)$, ommitting the first-order terms ($\gamma T_{10} + \delta T_{01}$).

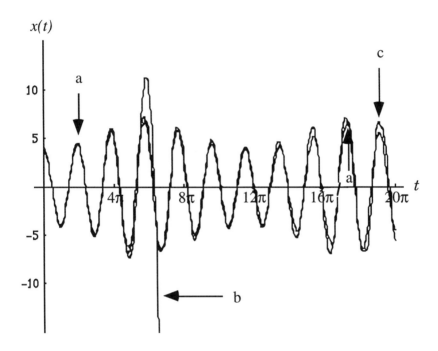

Figure 8.6 Solution and approximations for $x(t)$ of Equation (8.106): $x(0)=4$, $dx/dt\,(0)=0$, $\delta=0.03$, $\gamma=0.003$. (a) Full solution of Equation (8.106); (b) zero-order approximation via first-order normal form [Equations (8.115) and (8.116)]; (c) zero-order approximation via second-order normal form [Equations (8.124) and (8.125)].

We now return to he first order normal form equations, Equations (8.115) and (8.116). They have an integral of motion:

$$\rho^4\left(\cos 4\varphi + 6\frac{\gamma}{\delta}\right) = \text{const} = \rho_0{}^4\left(\cos 4\varphi_0 + 6\frac{\gamma}{\delta}\right) \quad (8.126a)$$

Consider what happens if Equation (8.117) is obeyed and a fixed point exists in 4φ in the first-order equations. Then, as 4φ reaches the fixed point, the term $(\cos 4\varphi + 6\gamma/\delta)$ on the l.h.s. of Equation (8.126a) vanishes, forcing ρ to diverge. If Equation (8.117) is not obeyed, this term cannot vanish. There is, then, no divergence in ρ, which oscillates between a minimal and a maximal value over a timescale close to t_0. Again, adopting $4\varphi_0=0$ as the initial point, the amplitudes at $4\varphi_0=0$ and $4\varphi_0=\pi$ are related by

$$\rho_\pi{}^4\left(6\frac{\gamma}{\delta} - 1\right) = \rho_0{}^4\left(6\frac{\gamma}{\delta} + 1\right) \quad (8.126b)$$

In that situation, as the first-order term is bounded, the second-order one constitutes a mere small correction.

8.5 HARMONIC OSCILLATORS WITH PERIODICALLY MODULATED FRICTION

Until now, all the examples we have studied in this chapter did not involve friction. In the present section we study two examples of harmonic oscillators that are subjected to friction that is modulated by a periodic coefficient. In particular, we show how the behavior of a simple nonautonomous system is affected by the initial conditions.

8.5.1 Linear friction

The example, drawn from Ref. 6, is:

$$\ddot{x} + x + \left(4\varepsilon \sin^2 t\right)\dot{x} = 0 \quad (8.127a)$$

We study two possible initial conditions:

(A) $\quad x(0) = a \quad \dfrac{dx}{dt}(0) = 0 \quad (8.127b)$

(B) $\quad x(0) = 0 \quad \dfrac{dx}{dt}(0) = a \quad (8.127c)$

OSCILLATORS WITH MODULATED FRICTION

Transforming to $z = x + i\,(dx/dt)$ and $v = \exp(-it)$, the system becomes a nonlinear one with two degrees of freedom. Equations (8.127a–c) then become

$$\dot{z} = -i\,z + \tfrac{1}{2}\varepsilon(v - v^*)^2(z - z^*) \qquad \dot{v} = -i\,v \qquad (8.128\mathrm{a})$$

with the corresponding initial conditions:

(A) $\qquad z(0) = a \qquad$ (real number) $\qquad (8.128\mathrm{b})$

(B) $\qquad z(0) = i\,a \qquad$ (imagiary number) $\qquad (8.128\mathrm{c})$

The first-order normal form is readily obtained by noting that the friction term contributes two resonant contributions: a $v^2 u^*$ term and a $(vv^*)u$ term, yielding equations for u and u^*:

$$\dot{u} = -i\,u - \tfrac{1}{2}\varepsilon v^2 u^* - \varepsilon v v^* u \qquad (8.129\mathrm{a})$$

$$\dot{u}^* = i\,u^* - \tfrac{1}{2}\varepsilon v^{*2} u - \varepsilon v v^* u^* \qquad (8.129\mathrm{b})$$

We now write

$$u = f e^{-it} \qquad (8.130)$$

Note that in the equations for u and u^* the real and imaginary components transform in a different manner. Hence, we cannot assume that $f(t)$ is real. The equations for f and f^* become

$$\dot{f} = -\varepsilon f - \tfrac{1}{2}\varepsilon f^* \qquad (8.131\mathrm{a})$$

$$\dot{f}^* = -\tfrac{1}{2}\varepsilon f - \varepsilon f^* \qquad (8.131\mathrm{b})$$

The solution of this set of equations is

$$f(t) = G \exp\!\left(-\left(\tfrac{1}{2}\right)\varepsilon\,t\right) + H \exp\!\left(-\left(\tfrac{3}{2}\right)\varepsilon\,t\right) \qquad (8.132)$$

Equations (8.131a,b) also require that

$$G = -G^* \qquad\qquad H = H^* \qquad (8.133)$$

Thus, G has to be purely imaginary ($G = i\tilde{G}$) and H, purely real. With an error $O(\varepsilon)$ for times of $O(1/\varepsilon)$, we can therefore write

$$x = \tfrac{1}{2}(u + u^*) = \tilde{G}\exp(-\tfrac{1}{2}\varepsilon t)\sin t + H\exp(-\tfrac{3}{2}\varepsilon t)\cos t \quad (8.134)$$

Now we impose the initial conditions. For the sake of convenience, we choose to chhose the free functions so that the initial conditions are fully obeyed by the zero-order approximation. This requires that the free functions have specific forms in all orders of the expansion of the near-identity transformation for $z(t)$. From Equation (8.134) we obtain

$$x(0) = H \qquad \dot{x}(0) = \tilde{G} - \left(\tfrac{3}{2}\right)\varepsilon H \quad (8.135)$$

However, we must be careful in the application of the initial condition in this way, as Equation (8.134) already constitutes an $O(\varepsilon)$ approximation for x. Hence, for consistency, we must drop the ε term in the dx/dt. Doing so, we find for the initial condition of Equation (8.127b):

$$\tilde{G} = 0 \qquad H = a$$
$$\Rightarrow x = a\exp(-\tfrac{3}{2}\varepsilon t)\cos t \quad (8.136a)$$

For the initial condition of Equation (8.127c), we find

$$\tilde{G} = a \qquad H = 0$$
$$\Rightarrow x = a\exp(-\tfrac{1}{2}\varepsilon t)\sin t \quad (8.136b)$$

Clearly, the trap we almost fell into could have been avoided had we applied the initial condition of Equations (8.128b) or (8.128c) to the complex function $u(t)$ [the zero-order approximation for $z(t)$]. The reason is that, for timescales of $O(1/\varepsilon)$, $\text{Re}\,u(t)$ and $\text{Im}\,u(t)$ are $O(\varepsilon)$ approximations for $x(t)$ and (dx/dt), respectively. Thus, the initial conditions significantly affect the temporal evolution of the system.

8.5.2 The modified Rayleigh oscillator

Consider now a modified Rayleigh oscillator (Equation (7.26):

$$\ddot{x} + x = \varepsilon\dot{x}\left(1 - \cos^2 t\,\dot{x}^2\right) \qquad \varepsilon > 0 \quad (8.137a)$$

OSCILLATORS WITH MODULATED FRICTION

Transforming to $z=x+i(dx/dt)$ and $v=\exp(-it)$, Equation (8.137a) is transformed into the following equation:

$$\dot{z} = -i\,z + \varepsilon\left[\tfrac{1}{2}(z-z^*) + \tfrac{1}{32}\left((v+v^*)^2(z-z^*)^3\right)\right] \quad (8.137b)$$

In first order, the normal form is

$$\dot{u} = -i\,u + \varepsilon\left[\tfrac{1}{2}u + \tfrac{1}{32}\left\{3uu^{*2}v^2 - 6u^2u^*vv^* + u^{*3}v^{*2}\right\}\right] \quad (8.138)$$

Using polar coordinates for u and extracting the fast time dependence explicitly

$$u = \rho\exp(-i\,t)\exp(-i\,\varphi)$$

(φ is the slowly varying phase of u), the equations for ρ and φ become

$$\dot{\rho} = \varepsilon\tfrac{1}{2}\rho\left(1 + \tfrac{1}{8}\rho^2(2\cos 2\varphi - 3)\right) \quad (8.139a)$$

$$\dot{\varphi} = -\tfrac{1}{16}\varepsilon\rho^2\sin 2\varphi \quad (8.139b)$$

The fixed points of Equations (8.139a,b) are at

1. $\quad \rho = 0$
2. $\quad \varphi = 0 \qquad \rho^2 = 8 \qquad (8.139c)$
3. $\quad \varphi = \pi/2 \qquad \rho^2 = 8/5$

Stability of the fixed points. We choose $\varepsilon>0$. (The analysis for $\varepsilon<0$ is similar, with the stability characteristics of the fixed points exchanged.) The characteristics of the three fixed points are summarized in th following:

1. Equation (8.139a) implies that $\rho=0$ is an unstable fixed point.
2. Writing $\rho^2 = 8+\xi$ and $\varphi = 0+\eta$ and retaining in Equations (8.139a,b) only terms linear in ξ and η, we find

$$\dot{\xi} \cong -\varepsilon\xi \qquad \dot{\eta} \cong -\varepsilon\sqrt{8}\,\eta$$

$\rightarrow \qquad$ The second fixed point is locally stable.

3. Writing $\rho^2 = \frac{8}{5} + \xi$ and $\varphi = (\pi/2) + \eta$, the linear approximations for Equations (8.139a,b) are

$$\dot{\xi} \cong -\varepsilon \xi \qquad \dot{\eta} \cong +\varepsilon \sqrt{\frac{8}{125}} \eta$$

→ The third fixed point is locally unstable.

As ρ and φ tend to the *stable* fixed point values (item 2, above), u approaches a circle of radius $\rho = (8)^{1/2}$ with an angular frequency that is very close to 1. Plotted against time as a third coordinate, u evolves over a cylinder of fixed radius. The higher-order terms in the expansion for z add undulations around the cylinder.

Effect of second order on fixed-point positions. Higher-order corrections cause a shift in the positions of the fixed points in u. We can see the trend of the shift by studying the $O(\varepsilon^2)$ correction to the normal form. Choosing not to include free functions in the first-order T functions, we have computed U_2, to find that Equations (8.139a,b) are modified into

$$\dot{\rho} = \varepsilon \tfrac{1}{2} \rho \{ 1 + \rho^2 (\tfrac{1}{4} \cos 2\varphi - \tfrac{3}{8}) \}$$
$$+ \varepsilon^2 \rho^3 \{ -\tfrac{1}{16} \sin 2\varphi + \rho^2 (-\tfrac{3}{512} \sin 2\varphi + \tfrac{3}{1024} \sin 4\varphi) \} \quad (8.139d)$$

$$\dot{\varphi} = -\varepsilon \tfrac{1}{16} \rho^2 \sin 2\varphi$$
$$+ \varepsilon^2 \left\{ \begin{aligned} &\tfrac{1}{8} + \rho^2 (\tfrac{3}{32} - \tfrac{5}{64} \cos 2\varphi) \\ &+ \rho^4 (\tfrac{41}{4096} - \tfrac{3}{256} \cos 2\varphi + \tfrac{3}{1024} \cos 4\varphi) \end{aligned} \right\} \quad (8.139e)$$

As a result, the correct positions of the fixed points of the normal form are shifted slightly away from the values predicted by the first-order analysis given by Equation (8.139c). Through $O(\varepsilon)$, the shift for the stable fixed point (item 2, above) is:

$$\rho^2 = 8 \Rightarrow \rho^2 = 8 + O(\varepsilon^2) \qquad \varphi = 0 \Rightarrow \varphi = 0 + \tfrac{21}{64} \varepsilon + O(\varepsilon^2)$$

and for the unstable fixed point (item 3, above):

$$\rho^2 = \tfrac{8}{5} \Rightarrow \rho^2 \tfrac{8}{5} + O(\varepsilon^2) \qquad \varphi = \frac{\pi}{2} \Rightarrow \varphi = \frac{\pi}{2} - \tfrac{741}{320} \varepsilon + O(\varepsilon^2)$$

OSCILLATORS WITH MODULATED FRICTION

Fixed point 2 is stable; i.e., it is an attractor. Suppose that at $t=0$, u is close to that fixed point. Then u is attracted toward it. Therefore, with general initial conditions $x(0)=A$, $y(0)=B$, the solution for ρ and φ will tend to the stable fixed point. Consequently, in a three-dimensional plot of x and (dx/dt) versus time, the full solution (u plus all higher-order corrections) will generate a distorted cylinder, the distortion becoming more prominent as ε is increased.

The question that arises is whether it is possible to realize the full solution that is associated with the unstable fixed point in a numerical calculation. In principle, it is possible, but in practice – very hard to achieve. We see from Equations (8.139) that if we choose the initial phase to be exactly $\pi/2$, and ρ^2 close to $\frac{8}{5}$, then φ will remain at $\pi/2$ and ρ^2 will tend to $\frac{8}{5}$. However, this statement is not obeyed by the full solution, as the fixed point is shifted away from these values. It is possible to remain on the distorted cylinder corresponding to the unstable fixed point *only if* we arrange the initial conditions so that, at $t=0$, u hits the precise, fully updated to all orders, position of that fixed point. If u is very close to the unstable fixed point, but not *on it*, u will drift away towards the stable fixed point.

The conclusions of the normal form analysis are borne out by the plots in Figures 8.7–8.9. Each figure presents (a) a plot of the time evolution of x and $y=dx/dt$ and (b) the x–y plane projection of the motion. In Figure 8.7 the initial conditions are outside the stable limit cycle; in Figure 8.8, they are close to the unstable limit cycle; and in Figure 8.9, they are close to the origin. In all cases the asymptotic motion converges onto the stable limit cycle (for which the zero-order approximation is a circle of radius $8^{1/2}$).

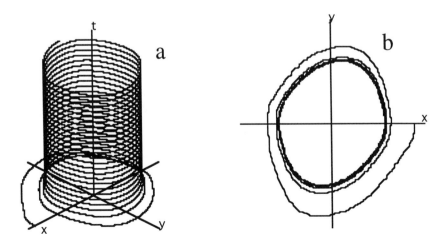

Figure 8.7 Solution of Equation (8.137a), $\varepsilon=0$; $x(0)=4.5$, $dx/dt(0)=0$.

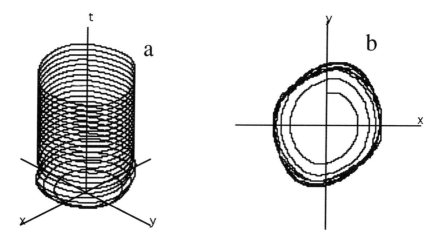

Figure 8.8 Solution of Equation (8.137a), $\varepsilon=0.2$; $x(0)=0$, $dx/dt(0)=1.5$.

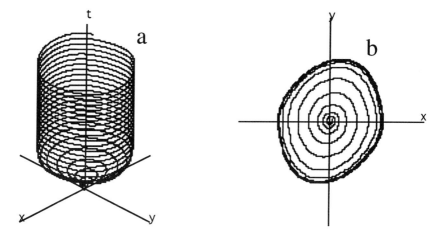

Figure 8.9 Solution of Equation (8.137a), $\varepsilon=0.2$; $x(0)=0.1$, $dx/dt(0)=0.1$.

8.6 NORMAL FORM PERTURBATION ANALYSIS OF NMR EQUATIONS

Nuclear magnetic resonance (NMR) is a very useful technique in medicine as well as a subject of intensive research. It provides an excellent example of a linear oscillatory system that is modified under the effect of an external periodic force. Resonance phenomena in which the frequency of natural oscillations and that of the external force participate are avidly sought for by scientists, as they provide

8.6.1 The model

The time dependence of the spin density matrix elements in certain models describing NMR [7,8] is governed by coupled nonautonomous linear equations:

$$\frac{d}{dt}\begin{pmatrix} x \\ y \end{pmatrix} = i \begin{pmatrix} -\delta + D & \frac{1}{2}D \\ \frac{1}{2}D & \delta + D \end{pmatrix} \begin{pmatrix} x \\ y \end{pmatrix} \qquad (8.140)$$

where x and y are the density matrix elements. In the absence of a magnetic field, δ is equal to the natural frequency, δ_0. D is the weak coupling between the nuclear spin of the sample and the magnetic field:

$$D = 2\varepsilon\omega_R(\alpha \cos\omega_R t + \beta \cos 2\omega_R t) \qquad |\varepsilon| \ll 1 \quad (8.141)$$

The experimental variables are the amplitude of the magnetic field, represented in dimensionless form by $|\varepsilon| \ll 1$, and the rotation frequency ω_R. One is particularly interested in the results near resonance: $\omega_R \approx \delta_0/N$ ($N = \frac{1}{2}, 1, \frac{3}{2}, 2, ...$). One therefore writes

$$\delta_0 = (N + \mu)\omega_R \qquad |\mu| \ll 1 \qquad (8.142)$$

In the presence of a magnetic field one has

$$\delta = \delta_0 + 2\mu\omega_R (A \cos\omega_R t + B \cos 2\omega_R t) \qquad (8.143)$$

The aim is to find the characteristics of the resulting oscillations of the nuclear spins in the magnetic field, that is, to compute the frequency (which should be slightly different from the unperturbed one) and amplitude of the resulting oscillations as well as to detect resonance effects. The usual treatment of this problem near resonance is through numerical solutions of the equations [7,8]. Here we show how resonance phenomena can be analyzed within the framework of the method of normal forms.

Experimentally, one varies ω_R. However, it is easier to analyze the problem by regarding ω_R as fixed and varying μ (which measures the deviation of δ_0 from the resonance value, $N \cdot \omega_R$) as a small parameter. In this manner, one avoids the need to expand the trigonometric functions that include ω_R in their argument (a procedure that would be far more cumbersome!).

To eliminate a trivial part in the time dependence of x and y, we write

$$\begin{pmatrix} x \\ y \end{pmatrix} = \exp\left(i \int_0^t D(t') dt' \right) \begin{pmatrix} \xi \\ \eta \end{pmatrix} \qquad (8.144)$$

yielding a 2×2 set of linear nonautonomous equations for ξ and η:

$$\frac{d}{dt} \begin{pmatrix} \xi \\ \eta \end{pmatrix} = i \begin{pmatrix} -\delta & \frac{1}{2}D \\ \frac{1}{2}D & \delta \end{pmatrix} \begin{pmatrix} \xi \\ \eta \end{pmatrix} \qquad (8.145)$$

As in previous examples, we convert Equation (8.145) into a system of autonomous nonlinear coupled equations by defining

$$z \equiv \exp(-i\,\omega_R t)$$

(At the end of this section we show that the normal form expansion can be applied to the nonautonomous problem directly, without resorting to this last substitution.) We obtain

$$\dot{\xi} = -i\,N\omega_R \xi - i\,\mu\omega_R \xi$$
$$-\tfrac{1}{2} i\, \mu\omega_R \{A(z + z^*) + B(z^2 + z^{*2})\} \xi \qquad (8.146a)$$
$$+\tfrac{1}{2} i\, \varepsilon\omega_R \{\alpha(z + z^*) + \beta(z^2 + z^{*2})\} \eta$$

$$\dot{\eta} = +i\,N\omega_R \eta + i\,\mu\omega_R \eta$$
$$+\tfrac{1}{2} i\, \mu\omega_R \{A(z + z^*) + B(z^2 + z^{*2})\} \eta \qquad (8.146b)$$
$$+\tfrac{1}{2} i\, \varepsilon\omega_R \{\alpha(z + z^*) + \beta(z^2 + z^{*2})\} \xi$$

NORMAL FORM ANALYSIS OF NMR EQUATIONS

$$\dot{z} = -i\,\omega_R z \qquad \dot{z}^* = +i\,\omega_R z^* \qquad (8.146c)$$

8.6.2 The normal form expansion

We now expand ξ and η in a near-identity transformation:

$$\xi = u + \varepsilon T_{10}^u + \mu T_{01}^u + \varepsilon^2 T_{20}^u + \varepsilon\mu T_{11}^u + \mu^2 T_{02}^u + \cdots$$

$$\eta = v + \varepsilon T_{10}^v + \mu T_{01}^v + \varepsilon^2 T_{20}^v + \varepsilon\mu T_{11}^v + \mu^2 T_{02}^v + \cdots \qquad (8.147)$$

The normal form equations for the zero-order approximations, u and v, are written as

$$\dot{u} = -i\,N\omega_R u + \varepsilon U_{10} + \mu U_{01} + \varepsilon^2 U_{20} + \varepsilon\mu U_{11} + \mu^2 U_{02} + \cdots$$

$$\dot{v} = +i\,N\omega_R v + \varepsilon V_{10} + \mu V_{01} + \varepsilon^2 V_{20} + \varepsilon\mu V_{11} + \mu^2 V_{02} + \cdots \qquad (8.148)$$

With the formalism of Chapter 4, we find for Equations (8.148)

$$U_{10} = \tfrac{1}{2} i\,\beta z^2 v \delta_{N,1} + \tfrac{1}{2} i\,\alpha z v \delta_{N,(1/2)} \qquad (8.149a)$$

$$U_{01} = -i\,u \qquad (8.149b)$$

$$U_{20} = -\left(\frac{i}{4\omega_R}\right)\left\{\alpha^2\left[\frac{(1-\delta_{N,(1/2)})}{2N-1} + \frac{1}{2N+1}\right] + \beta^2\left[\frac{(1-\delta_{N,1})}{2N-2} + \frac{1}{2N+2}\right]\right\} u \qquad (8.149c)$$

$$U_{11} = -i\left[\frac{\alpha B - 2\beta A}{4\omega_R}\right] z v \delta_{N,(1/2)} - i\left(\frac{\alpha A}{2\omega_R}\right) z^2 v \delta_{N,1}$$

$$-i\left[\frac{\alpha B + 2\beta A}{4\omega_R}\right] z^3 v \delta_{N,(3/2)} - i\left(\frac{\beta B}{4\omega_R}\right) z^4 v \delta_{N,2} \qquad (8.149d)$$

$$U_{02} = 0 \qquad (8.149e)$$

$$V_{10} = \tfrac{1}{2} i \, \alpha z^* u \delta_{N,(1/2)} + \tfrac{1}{2} i \, \beta z^{*2} u \delta_{N,1} \qquad (8.150a)$$

$$V_{01} = i \, v \qquad (8.150b)$$

$$V_{20} = \left(\frac{i}{4\omega_R}\right) \left\{ \begin{array}{l} \alpha^2 \left[\dfrac{1}{2N+1} + \dfrac{1}{2N-1}\left(1 - \delta_{N,(1/2)}\right) \right] \\ + \beta^2 \left[\dfrac{1}{2N+2} + \dfrac{1}{2N-2}\left(1 - \delta_{N,1}\right) \right] \end{array} \right\} v \qquad (8.150c)$$

$$V_{11} = -i \left[\frac{\alpha B - 2\beta A}{4\omega_R} \right] z^{*2} z u \delta_{N,(1/2)} - i \left[\frac{\alpha A}{2\omega_R} \right] z^{*2} u \delta_{N,1}$$
$$- i \left[\frac{\alpha B + 2\beta A}{4\omega_R} \right] z^{*3} u \delta_{N,(3/2)} - i \left[\frac{\beta B}{4\omega_R} \right] z^{*4} u \delta_{N,2} \qquad (8.150d)$$

$$V_{02} = 0 \qquad (8.150e)$$

$$T_{10}^{u} = \frac{\alpha}{2\omega_R(2N+1)} z^* v + \frac{\alpha}{2\omega_R(2N-1)}\left(1 - \delta_{N,(1/2)}\right) z v$$
$$+ \frac{\beta}{2\omega_R(2N+2)} z^{*2} v + \frac{\beta}{2\omega_R(2N-2)}\left(1 - \delta_{N,1}\right) z^2 v \qquad (8.151a)$$

$$T_{0,1}^{u} = -\frac{A}{2\omega_R} z^* u + \frac{A}{2\omega_R} z u - \frac{B}{4\omega_R} z^{*2} u + \frac{B}{4\omega_R} z^2 u \qquad (8.151b)$$

$$T_{10}^{v} = -\frac{\alpha}{2\omega_R(2N+1)} z u - \frac{\alpha}{2\omega_R(2N-1)}\left(1 - \delta_{N,(1/2)}\right) z^* u$$
$$- \frac{\beta}{2\omega_R(2N+2)} z^2 u - \frac{\beta}{2\omega_R(2N-2)}\left(1 - \delta_{N,1}\right) z^{*2} u \qquad (8.151c)$$

$$T_{0,1}^{v} = -\frac{A}{2\omega_R} z v + \frac{A}{2\omega_R} z^* v - \frac{B}{4\omega_R} z^2 v + \frac{B}{4\omega_R} z^{*2} v \qquad (8.151d)$$

No free functions have been introduced, as we find that, at least in first order, there is no advantage in their introduction. In Equations

NORMAL FORM ANALYSIS OF NMR EQUATIONS

(8.147) and (8.148) $\delta_{a,b}$ is the Kronecker delta:

$$\delta_{a,b} = \begin{cases} 1 & a = b \\ 0 & a \neq b \end{cases}$$

The major time dependence of $\xi(t)$ and $\eta(t)$ is governed by the rapid oscillations of $\exp(\pm iN \cdot \omega_R t)$. If these are extracted, the cumbersome expression resulting for the normal form (not given here) can be significantly simplified. We therefore write

$$\xi(t) = \exp(-i\, N\omega_R t) f(t) \qquad \eta(t) = \exp(i\, N\omega_R t) g(t) \quad (8.152)$$

Here, $f(t)$ and $g(t)$ depend only weakly on time. Using Equations (8.149) and (8.150) for U_{ij} and V_{ij}, respectively, and Equation (8.152), Equation (8.148) leads to

$$\dot{f} = -i\, Df + i\, Cg \qquad \dot{g} = i\, Cf + i\, Dg \quad (8.153)$$

In Equation (8.150) D and C are given by

$$D = \mu + \frac{\varepsilon^2}{4\omega_R}\left[\left(\frac{1}{2N-1}(1-\delta_{N,(1/2)}) + \frac{1}{2N+1}\right)\alpha^2 \right. \\ \left. + \left(\frac{1}{2N-2}(1-\delta_{N,1}) + \frac{1}{2N+2}\right)\beta^2\right] \quad (8.154)$$

$$C = \frac{\varepsilon}{2}(\alpha\delta_{N,(1/2)} + \beta\delta_{N,1})$$
$$+ \frac{\varepsilon\mu}{4\omega_R}\begin{bmatrix}(-\alpha B + 2\beta A)\delta_{N,(1/2)} - 2\alpha A\delta_{N,1} \\ -(\alpha B + 2\beta A)\delta_{N,(3/2)} - \beta B\delta_{N,2}\end{bmatrix} \quad (8.155)$$

Equation (8.153) yields a simple second order equation for f (and an identical one for g):

$$\ddot{f} + \Delta\Omega^2 f = 0 \qquad (\Delta\Omega^2 = C^2 + D^2) \quad (8.156)$$

The value of the actual oscillation frequency differs from the unperturbed one, $N \cdot \omega_R$, by $\pm\Delta\Omega$. It is important to note that the $N=\frac{1}{2}$ and $N=1$ resonances appear already in first order, while the $N=\frac{3}{2}$ and

N=2 resonances appear for the first time only in second order.

The solutions for $f(t)$ and $g(t)$ can be written as

$$f = Me^{i\Delta\Omega t} + Ne^{-i\Delta\Omega t}$$
$$g = Pe^{i\Delta\Omega t} + Qe^{-i\Delta\Omega t}$$
(8.157)

Substituting Equations (8.157) in Equations (8.153), one obtains the following equations for M, N, P, and Q:

$$(D+\Delta\Omega)M - CP = 0 \qquad (D-\Delta\Omega)P + CM = 0$$
$$(D-\Delta\Omega)N - CQ = 0 \qquad (D+\Delta\Omega)Q + CN = 0$$
(8.158)

The determinant of Equations (8.158) vanishes, so that only two of the four coefficients are independent:

$$P = \frac{D+\Delta\Omega}{C}M \qquad Q = \frac{D-\Delta\Omega}{C}N \qquad (8.159)$$

At this stage we need to impose specific initial conditions. Assume, for example, that at $t=0$ the nuclear sample is produced in a pure x spin state, e.g.

$$f(t=0) = f_0 \qquad g(t=0) = 0 \qquad (8.160)$$

The results for the coefficients are

$$M = \frac{C^2}{2\Delta\Omega(C+\Delta\Omega)}f_0 \qquad N = \frac{D+\Delta\Omega}{2\Delta\Omega}f_0 \quad (8.161a)$$

$$P = \frac{C}{2\Delta\Omega}f_0 \qquad Q = -\frac{C}{2\Delta\Omega}f_0 \quad (8.161b)$$

yielding

$$f(t) = f_0\left[\left\{\cos\Delta\Omega t - i\left(\frac{D}{\Delta\Omega}\right)\sin\Delta\Omega t\right\}\right] \qquad (8.162a)$$

$$g(t) = i\left(\frac{C}{\Delta\Omega}\right) f_0 \sin \Delta\Omega t \qquad (8.162b)$$

We now search for resonant behavior. Without loss of generality, assume $C>0$. Varying μ is equivalent to varying ω_R, the frequency of the magnetic field. $\mu<0$ corresponds to $\omega_R>\delta_0/N$ and $\mu>0$, to $\omega_R<\delta_0/N$. This affects D in $O(\mu)$ and C, in $O(\epsilon\mu)$. For $|D|\gg C$ (ω_R away from resonance), one finds

$$f(t) \cong \begin{cases} f_0 e^{+i\Delta\Omega t} & D \ll -C \\ \\ f_0 e^{-i\Delta\Omega t} & D \gg +C \end{cases}$$

$$g(t) \cong i\left(\frac{C}{D}\right) f_0 \sin \Delta\Omega t$$

(8.163)

Consequently,

$$\frac{\|g\|}{\|f\|} = O\left(\left|\frac{C}{D}\right|\right) \ll 1$$

On the other hand, for small values of D ($|D|\ll C$), one has

$$f(t) \cong f_0 \cos \Delta\Omega t \qquad g(t) \cong i\, f_0 \sin \Delta\Omega t \qquad (8.164)$$

Thus, the y spin state is now populated at the same intensity as the x spin state and the matrix elements are out of phase. $C/\Delta\Omega$, the amplitude of $g(t)$, exhibits the resonance behavior. The position of the peak of the resonance curve occurs at $D=0$. Its width, defined by the point at which the amplitude has fallen to a half of its peak value, occurs when $D=\pm\sqrt{\frac{3}{2}}\, C$. For the $N=\frac{1}{2}$ resonance, this yields a width of $\sqrt{\frac{3}{2}}\,\alpha$ for the resonance curve.

8.6.3 Results

The $N=\frac{1}{2}, 1$ resonances (resonance defined as the position of the peak in the μ-dependence of the amplitude of the y matrix element) appear already in first order, whereas the $N=\frac{3}{2}, 2$ resonances occur in second order. Ω, the frequency of oscillations of x and y, differs from the unperturbed frequency, $\delta_0 = N \cdot \omega_R$, by $\pm \Delta\Omega$. Here we provide the results for $A=B=\alpha=\beta=1$, with an initial state of $x(t=0)=1$ and $y(t=0)=0$. The μ dependence of the amplitude of y for $\varepsilon=0.02$ is shown in Figure 8.10 for $N=\frac{1}{2}, 1$.

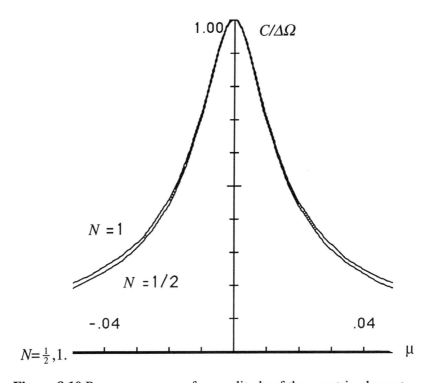

Figure 8.10 Resonance curves for amplitude of the y matrix element, $\varepsilon=0.02$.

The lowest order contributions to ω_R and $\Delta\Omega$ at resonance are given in Table 8.1.

Table 8.1 Resonance point values of ω_R and $\Delta\Omega$

N		$\tfrac{1}{2}$	1	$\tfrac{3}{2}$	2
$\dfrac{\mu}{N} \cong 1 - \dfrac{\omega_R}{(\delta_0/N)}$		$\tfrac{1}{12}\varepsilon^2$	$-\tfrac{19}{48}\varepsilon^2$	$-\tfrac{13}{60}\varepsilon^2$	$-\tfrac{3}{40}\varepsilon^2$
$\Delta\Omega$		$\tfrac{1}{2}\varepsilon$	$\tfrac{1}{2}\varepsilon$	$\tfrac{39}{160}\varepsilon^3$	$\tfrac{3}{80}\varepsilon^3$

8.6.4 Do we need to substitute $z \equiv \exp(-i\omega_R t)$?

Methodically, it is easier to analyze the present problem in the way we have done: to convert the system into an autonomous one through the substitution $z = \exp(-i\omega_R t)$. However, as the formalism of Chapter 4 indicates, one can analyze the problem as a nonautonomous one just as well. Here we show that the results are indeed the same.

Consider again Equation (8.146a), but *retain the explicit time-dependent form* of z. Substituting the expansions Equations (8.147) and (8.148) in Equation (8.146a), we embark again on the saga of obtaining the normal form. However, now we have to account for the nonautonomous nature of the equations by allowing an explicit time dependence in the T functions. Thus, for instance, the equation relating U_{10} and T_{10}^u will now be

$$U_{10} = -\frac{\partial T_{10}^u}{\partial t} + i\, N\omega_R \left\{ u \frac{\partial T_{10}^u}{\partial u} - v \frac{\partial T_{10}^u}{\partial v} - T_{10}^u \right\} \quad (8.165)$$
$$+ i\left(\alpha \cos\omega_R t + \beta\cos 2\omega_R t\right)v$$

Now write T_{10}^u as

$$T_{10}^u = \sum_{n,m\,p\,q} g_{nmpq}(t)\, u^n u^{*m} v^p v^{*q} \quad (8.166)$$

Clearly, to kill the driving term on the r.h.s. of Equation (8.165), we need in T_{10}^u a term with $n=m=q=0$ and $p=1$. The equation for g_{0010} is obtained as

$$\dot{g}_{0010} = -2i\, N\omega_R\, g_{0010} + i\left(\alpha\cos\omega_R t + \beta\cos 2\omega_R t\right) \quad (8.167)$$

Writing

$$g_{0010} = e^{-2i\, N\omega_R t}\, \tilde{g}$$

and the cosines in terms of exponentials, we obtain

$$\dot{\tilde{g}} = -i\, e^{2i\, N\omega_R t}\left(\tfrac{1}{2}\alpha\left(e^{i\,\omega_R t} + e^{-i\,\omega_R t}\right) + \tfrac{1}{2}\beta\left(e^{2i\,\omega_R t} + e^{-2i\,\omega_R t}\right)\right) \quad (8.168)$$

For $N=\tfrac{1}{2}$, one can eliminate all the terms on the r.h.s. of Equation (8.168) without introducing a secular behavior in g_{0010}, except for the $\exp(-i\omega_R t)$ term. The latter can be eliminated only with a g_{0010} that develops a secular behavior. Similarly, for $N=1$ one can eliminate, without introducing a secular behavior in g_{0010}, all the contributions on the r.h.s., except for the $\exp(-2i\omega_R t)$ term. Hence, the requirement that the near-identity transformation is composed of bounded terms only leads to the following division of the contributions to g_{0010} between the U and T functions:

$$U_{10} = \tfrac{1}{2}i\,\alpha e^{-i\,\omega_R t}\, v\delta_{N,(1/2)} + \tfrac{1}{2}i\,\beta e^{-2i\,\omega_R t}\, v\delta_{N,1} \quad (8.169)$$

$$\tilde{g} = \frac{\alpha}{2\omega_R(2N+1)} e^{i\,\omega_R t} + \frac{\alpha}{2\omega_R(2N-1)}\left(1-\delta_{N,(1/2)}\right)e^{-i\,\omega_R t}$$
$$+ \frac{\beta}{2\omega_R(2N+2)} e^{2i\,\omega_R t} + \frac{\beta}{2\omega_R(2N-2)}\left(1-\delta_{N,1}\right)e^{-2i\,\omega_R t} \quad (8.170)$$

These yield for U_{10} and T_{10}^u the same expressions as in Equations (8.149a) and (8.151a), respectively. It is not difficult to show that the same holds for all other U and T functions.

REFERENCES

1. Thompson, J. M. T., and H. B. Stewart, *Nonlinear Dynamics and Chaos*, Wiley, New York (1986).

2. Conte, M., and W. M. MacKay, *An Introduction to the Physics of Particle Accelerators*, World Scientific, Singapore (1991).
3. Nayfeh, A. H., and D. T. Mook, *Nonlinear Oscillations*, Wiley, New York (1979).
4. Arnold, V. I., "Remarks on the Perturbation Theory for Problems of Mathieu Type," *Russian Math. Surveys* **38**, 215 (1983).
5. Mond, M., G. Cederbaum, P. B. Kahn, and Y. Zarmi, "Stability Analysis of the Nonlinear Mathieu Equation," *J. Sound Vib.* **167**, 77 (1993).
6. Sanders, J. A., and F. Verhulst, *Averaging Methods in Nonlinear Dynamical Syatems*, Springer-Verlag, New York (1985).
7. Raleigh, D. P., M. H. Levitt, and R. G. Griffin, "Rotational Resoance in Solid State NMR," *Chem. Phys. Lett.* **146**, 71 (1988).
8. Levitt, M. H., D. P. Raleigh, F. Creuzet, and R. G. Griffin, "Theory and Simulation of Homonuclear Spin Pair Systems in Rotating Solids," *J. Chem. Phys.* **92**, 6347 (1990) and references cited therein.

9

PROBLEMS WITH A ZERO EIGENVALUE

Problems with one zero eigenvalue, with the remaining ones having negative real parts, present new options. In previous chapters, we studied systems for which the eigenvalues all had nonzero real parts. We introduced a near-identity transformation as a means for finding an approximation for the evolution of a dynamical system, enabling us to remain near the linear problem, which is presumably solvable. However, in that case, we depended on the justification provided by the Hartman–Grobman (HG) theorem for the use of near-identity transformations. By contrast, if the system has some eigenvalues with zero real part, then one *cannot* obtain information regarding its behavior for small amplitudes by a linear stability analysis. One *cannot* invoke the HG theorem.

In such cases, if one is close enough to the fixed point at the origin, the coordinates corresponding to the eigenvalues with negative real parts decay exponentially onto a manifold (the *center manifold*) that is constructed from the coordinate corresponding to the zero eigenvalue. The subsequent motion takes place there. One then examines the stability of the origin. If the origin is stable, then all the modes will decay. If, on the other hand, the origin is an unstable fixed point, the system will either grow until it reaches a fixed point or exhibit finite-time blowup. (See Chapter 4.) In the latter case, the perturbation expansion will not provide a valid approximation for the evolution of the system over a long time.

We will study this class of problems, using normal forms. We will obtain the time dependence of the approximation to the full solution in the form of a power series in the small parameter. Although there is no linear term to guide the computation, as far as applications are concerned, one proceeds precisely as with problems in which there is a nonzero linear term. The important difference is that in the new class of problems, the decay or growth is not of exponential character.

For problems in this class, many investigators introduce the metod of the *center manifold* as a way to reduce the dimensionality of the system and to study the resulting motion on the center manifold. The theory and applications of the method of the center manifold are thoroughly described in many texts [1–4]. We prefer the method of normal forms as it allows one to follow the transients at the cost of some complexity in the analysis. With this in mind, we restrict our exposition of the center manifold method to showing its connection with the results obtained by the method of normal forms. In particular, we show that the equations derived in a center manifold perturbation analysis are *precisely* the equations for the time dependence of the zero-order approximation in the method of normal forms. The T functions carry all the explicit exponential time dependence of the solution. Once the exponential decay has taken its effect, one is left with the equation for the zero-order term, obtained also in the asymptotic limit of the center manifold method.

We emphasize that, in this chapter, we restrict the discussion to problems in which there is at least one zero eigenvalue, at least one eigenvalue with negative real part, and *no* pure imaginary eigenvalues. The connection that we exhibit between the center manifold and normal form methods is valid only for this class of problems.

9.1 ANALYSIS USING THE NORMAL FORM EXPANSION

We proceed to develop the normal form expansion in a straightforward manner and apply it to a variety of simple problems.

Statement of the problem and discussion. Consider an m-dimensional system that has r zero eigenvalues and s eigenvalues with negative real part. The basic equations are

$$\frac{dx_i}{dt} = 0 + \varepsilon X_i(x_1,\ldots,x_r,y_1,\ldots,y_s;\varepsilon); \quad i = 1,\ldots,r$$

$$(s+r) = m \quad (9.1)$$

$$\frac{dy_j}{dt} = \lambda_j y_j + \varepsilon Y_j(x_1,\ldots,x_r,y_1,\ldots,y_s;\varepsilon); \quad j = 1,\ldots,s$$

$$\left(\operatorname{Re}\lambda_j < 0\right)$$

Here X_i and Y_j are nonlinear. We write the normal form expansion as

$$x_i = u_i + \sum_n \varepsilon^n T^u_{i,n}(u,v) \quad (9.2a)$$

NORMAL FORM EXPANSION

$$\frac{du_i}{dt} = \sum_{n \geq 1} \varepsilon^n U_{i,n} \tag{9.2b}$$

$$y_j = v_j + \sum_n \varepsilon^n T^v_{j,n}(u,v) \tag{9.2c}$$

$$\frac{dv_j}{dt} = V_{j,0} + \sum_{n \geq 1} \varepsilon^n V_{j,n} \tag{9.2d}$$

$$V_{j,0} = \lambda_j v_j \tag{9.2e}$$

To see the structure of the various terms in the expansion, it is sufficient to develop a second-order calculation for problems in which the interaction term is proportional to ε only [i.e., no additional ε dependence of X_i and Y_j in Equations (9.1)]. Higher-order computations follow the same pattern.

1. Computing $U_{i,1}$

$$U_{i,1} = -\sum_k V_{k,0} \frac{\partial T^u_{i,1}}{\partial v_k} + X_{i,1}(u,v) = -\sum_k \lambda_k v_k \frac{\partial T^u_{i,1}}{\partial v_k} + X_{i,1}(u,v) \tag{9.3}$$

The U_1 equation will be of the form

$$U_{i,1} = f_i(u) \tag{9.4}$$

because interaction terms of the form $f_i(u)$ cannot be killed by a T function, whereas any term of the form $g_i(u,v)$ can be killed.

2. Equations for the v variables. We have

$$V_{i,1} = \sum_k T^v_{k,1} \frac{\partial Y_{k,0}}{\partial v_k} - \sum_k V_{k,0} \frac{\partial T^v_{i,1}}{\partial v_k} + Y_{i,1}(u,v)$$

$$= \lambda_i T^v_{i,1} - \sum_k \lambda_k v_k \frac{\partial T^v_{i,1}}{\partial v_k} + Y_{i,1}(u,v) \tag{9.5}$$

On the basis of our studies in previous chapters, we now examine the structure of the r.h.s. of Equation (9.5) and reach the following conclusions:

a. One can eliminate all terms in Equation (9.5) that are of the form $f(u)$. One can also eliminate terms of the form $g(u,v)$ except when they resonate with the variable v_i. As these terms also decay, the equation for $V_{i,n}$ will consist entirely of exponentially decaying terms.
b. It follows then that, in the asymptotic limit, the structure of the equations for y_j will be determined entirely by those T^v functions that are of the form $f(u)$, as all others will have a decay factor.

We now compute the second-order terms. The general form of the equation for U_2 is

$$U_{i,2} = -\sum_k V_{k,0} \frac{\partial T^u_{i,2}}{\partial v_k} + T^u_{j,1} \frac{\partial X_{i,1}}{\partial u_j} + T^v_{j,1} \frac{\partial X_{i,1}}{\partial v_j}$$

$$- U_{j,1} \frac{\partial T^u_{i,1}}{\partial u_j} - V_{j,1} \frac{\partial T^u_{i,1}}{\partial v_j}$$

$$= -\sum_k \lambda_k v_k \frac{\partial T^u_{i,1}}{\partial v_k} + T^u_{j,1} \frac{\partial X_{i,1}}{\partial u_j} + T^v_{j,1} \frac{\partial X_{i,1}}{\partial v_j}$$

$$- U_{j,1} \frac{\partial T^u_{i,1}}{\partial u_j} - V_{j,1} \frac{\partial T^u_{i,1}}{\partial v_j}$$

(9.6)

We need concentrate only on those terms that originate from the v equations and have the capacity to be free from v variables. With this noted, the only term that cannot be killed in $U_{i,2}$ is of the form

$$T^v_{j,1} \frac{\partial X_{i,1}}{\partial v_j}$$

This type of term will enter the U_2 equation if there is a component in the T^v equation of the form $f(u) \cdot v$. This occurs if the interaction term has a term that is linear in the variable y.

9.1.1 Illustrative examples

Example 9.1. We return to Example 5.9, with one eigenvalue, considered in Chapter 5. We discuss this one-dimensional example extensively, because of its simplicity: (Often the long-time motion of

NORMAL FORM EXPANSION

the approximate, or "reduced" system takes place on a one-dimensional manifold.) In the analysis of Example 5.9, we required the eigenvalue to have negative real part. Here we choose $\lambda = 0$. Hence, our basic equation becomes

$$\frac{dx}{dt} = \sum_{k=1}^{4} x \{ \varepsilon^k c_k x^k \} \qquad (9.7)$$

where $\varepsilon > 0$, and the coefficients c_k are real.

The normal form expansion is straightforward. However, to analyze it, we need to have some information about the coefficients, c_i.

Case 1: All $c_i < 0$. Given any initial amplitude, we will observe decay to the origin.

Case 2: All $c_i > 0$. Given any initial amplitude, we will have finite-time blowup, the behavior of which is characterized by the leading power, i.e., x^5. The normal form expansion will lose its validity before the blowup time and hence may, at best, provide a faithful characterization of the solution only for short times.

Case 3: Some $c_i > 0$ and other $c_i < 0$. If $c_1 < 0$, then the origin is an attractor, and sufficiently small initial displacements will decay. If $c_1 > 0$, the origin is unstable. Then, the result depends on the sign of the nonzero c_i. A small initial disturbance from the origin may grow until it saturates. This happens, for example, if $c_1 > 0$ whereas c_2, c_3, $c_4 < 0$. Other relationships may lead to finite-time blowup.

Analysis. The original equation is itself a valid normal form since all its terms are resonant. However, one can simplify the normal form equation by constructing a formal expansion. (The word "formal" is used, because the resulting perturbation expansion may not properly characterize the long-time behavior of the system.) As all terms in the original equation are resonant, the T functions can contain only *free functions,* that is, they can be constructed from resonant terms only. Choosing the free functions to vanish (which makes the whole T functions vanish), the normal form is simply the original equation. On the other hand, in each order, k, of the expansion, we can include in T_k a term of the same monomial character as the resonant monomial in that order in the normal form. In the present case, this encompasses all possible monomials. Choosing each generating function as $T_k = a_k u^{k+1}$ produces the following normal form:

$$\frac{du}{dt} = 0 \cdot u + \varepsilon c_1 u^2 + \varepsilon^2 c_2 u^3 + \varepsilon^3 \left(c_3 + c_2 a_1 - c_1 a_2 + c_1 a_1^2\right) u^4$$

$$+ \varepsilon^4 \left(c_4 + 2 c_3 a_1 + c_2 a_1^2 - 2 c_1 a_3 + 4 c_1 a_1 a_2 - 2 c_1 a_1^3\right) u^4$$

$$+ O(\varepsilon^5)$$

(9.8)

From Equation (9.8) we see that the coefficients in the generating function of orders $k \leq 3$ can be used to eliminate contributions to the normal form in orders $k=3,4$. In fact, with the following choice:

$$a_1 = 0; \quad a_2 = \frac{c_3}{c_1}; \quad a_3 = \frac{c_4}{2 c_1}$$

the normal form equation reduces to a minimal normal form given by

$$\frac{du}{dt} = \varepsilon c_1 u^2 + \varepsilon^2 c_2 u^3 \tag{9.8a}$$

(There is freedom in the choice of the coefficients.) This result is true in all orders, as the pattern persists in all orders higher than second.

Discussion. The result of Equation (9.8a) may be deceptive.

In case 1, the solution of the minimal normal form decays to the origin. Once the amplitudes are sufficiently small so that the effect of the quartic and quintic terms can be neglected in Equation (9.8), Equation (9.8a) faithfully describes the full solution.

In case 2, the structure of the minimal normal form leads one to the *wrong* conclusion that the blowup is characterized by the cubic term, rather than the quintic one, which obviously dominates the behavior of the solution of the full Equation (9.8) for large amplitudes. When the amplitude has grown sufficiently, the terms in the perturbation expansion are no longer in their "proper order." For small amplitudes the quadratic term dominates. Thus, for sufficiently short times and sufficiently small amplitudes, Equation (9.8a) provides a good approximation for the full solution. However, for sufficiently large amplitudes, the quintic term dominates in Equation (9.8). Then the normal form misrepresents the original problem.

In case 3, the amplitude of the full solution reaches saturation. So does the solution of Equation (9.8a), since $c_1 > 0$ and $c_2 < 0$. However, the value of the saturation amplitude obtained through the minimal normal form does not coincide with that obtained in the full equation. Therefore, the minimal normal form provides a reasonable

NORMAL FORM EXPANSION

description only for sufficiently small initial displacement.

Example 9.2. We continue the study of Example 9.1, noting that Equation (9.8a) can be solved in closed form. For simplicity of notation, we rewrite Equation (9.8a) as

$$\frac{dx}{dt} = ax^2 + bx^3 \tag{9.8b}$$

Case 1: a<0, b<0. For any initial amplitude, the system decays until its behavior is dominated by the quadratic term. Then it decays like $1/t$. The MNF provides a faithful zero-order approximation to the full solution.

Case 2: a>0, b>0. Equation (9.8b) is solved by

$$t = \frac{1}{a}\left[\frac{1}{x_0} - \frac{1}{x}\right] - \frac{b}{a^2}\ln\left[\frac{x}{x_0}\right] + \frac{b}{a^2}\ln\left[\frac{a+bx}{a+bx_0}\right] \tag{9.8c}$$

As the coefficients are positive, blowup will occur at a finite time:

$$t \to \frac{1}{a}\frac{1}{x_0} + \frac{b}{a^2}\ln\left[\frac{bx_0}{a+bx_0}\right] \tag{9.8d}$$

Perturbation theory is valid for times sufficiently far from the blowup time.

Case 3: a>0, b<0. The origin is unstable and the system evolves until it reaches the fixed point at $x^* = -a/b > 0$. One studies the stability of this fixed point by writing $x = -a/b + z$, obtaining the equation

$$\frac{dz}{dt} = \frac{a^2}{b}z - 2az^2 + bz^3 \tag{9.8e}$$

All the terms have negative coefficients. Hence, the z motion is a monotonic decay. Again, the minimal normal form equation provides a faithful zero-order approximation to the full solution.

Case 4: a<0, b>0. The fixed point at $x^* = -a/b$ is unstable. Thus, the system will decay to the origin when the initial amplitude x_0 obeys $x_0 < x^*$ and an unbounded solution for $x_0 > x^*$. Here, the initial conditions determine the time validity of the normal form expansion.

Example 9.3. Consider the following set of coupled equations:

$$\frac{dx}{dt} = -\tfrac{1}{2}\varepsilon x y \tag{9.9a}$$

$$\frac{dy}{dt} = -9y + \tfrac{1}{2}\varepsilon x^2 \tag{9.9b}$$

We introduce the near-identity transformations

$$x \to u + \sum_n \varepsilon^n T_n(u,v) \qquad y \to v + \sum_n \varepsilon^n S_n(u,v)$$

and normal form equations

$$\dot{u} = U_0 + \sum_{n\geq 1} \varepsilon^n U_n \qquad \dot{v} = V_0 + \sum_{n\geq 1} \varepsilon^n V_n$$

and obtain, through $O(\varepsilon^2)$:

$$T_1 = \frac{uv}{18} + f_1(u); \quad S_1 = \frac{1}{18}u^2 + g_1(u)v$$

$$T_2 = \frac{1}{2\cdot 18^2}uv^2 + f_2(u); \quad S_2 = g_2(u)v$$

$$U_0 = 0; \quad V_0 = -9v \tag{9.10}$$

$$U_1 = V_1 = 0$$

$$U_2 = -\frac{1}{36}u^3; \quad V_2 = \frac{1}{18}u^2 v$$

The functions $f_n(u)$ and $g_n(u)$ are arbitrary. With a proper choice of these functions, one can have $U_n = V_n = 0$, for all $n > 2$. For instance, choosing $f_1 = g_1 = 0$, yields $U_3 = V_3 = 0$.

The equations for the zero-order approximation are then

$$\frac{du}{dt} = -\tfrac{1}{36}\varepsilon^2 u^3 \qquad \frac{dv}{dt} = -9v + \tfrac{1}{18}\varepsilon^2 u^2 v \tag{9.11}$$

NORMAL FORM EXPANSION

Equations (9.11) are solved by

$$u(t) = \frac{u(0)}{\sqrt{1 + \frac{1}{18}\varepsilon^2 u(0)^2 t}}$$

$$v(t) = v(0)\exp(-9t)\ln\{[1 + \frac{1}{18}\varepsilon^2 u(0)^2 t]\}$$

(9.12)

Observe that $u(t)$ decays as an inverse power, *not* exponentially, while $v(t)$ decays exponentially. These results follow immediately from the structure of the original equations.

We can verify that we do indeed have a valid approximate solution to our original set of equations by substituting for x and y the expressions for u, v and the T and S functions. We write

$$x(t) = u + \tfrac{1}{18}\varepsilon u v + \tfrac{1}{648}\varepsilon^2 u v^2 + O(\varepsilon^3)$$

$$y(t) = v + \tfrac{1}{18}\varepsilon^2 u^2 + O(\varepsilon^3)$$

(9.13)

The verification involves some algebra but is straightforward.

Comment. At this level of approximation, we have chosen the free functions to be equal to zero. The solutions are displayed explicitly in Equation (9.13) to enable comparison with those obtained in Example 9.4 by the method of the center manifold.

9.2 CENTER MANIFOLD METHOD: CONNECTION WITH METHOD OF NORMAL FORMS

In this section, we establish a connection between the method of normal forms and the method of the center manifold, for problems in which the unperturbed system has one eigenvalue equal to zero and the remaining one negative. We show that the zero-order equation derived for the zero-eigenvalue component within the method of normal forms is *precisely* the same as the equation obtained through the method of the center manifold. We begin with an equation of the form

$$\frac{dx}{dt} = \sum_{n=2}^{N} \varepsilon^{n-1} f_n(x, y) = \sum_{n=2}^{N} \varepsilon^{n-1} \alpha_{n;p,q} x^p y^q; \quad p + q = n \quad (9.14a)$$

$$\frac{dy}{dt} = \sigma y + \sum_{n=2}^{N} \varepsilon^{n-1} g_n(x,y) = \sum_{n=2}^{N} \varepsilon^{n-1} \beta_{n;p,q} x^p y^q; \quad p+q = n \quad (9.14b)$$

(The eigenvalue σ is negative.) Before embarking on the detailed comparison, we make the following comments.

Method of the center manifold. In the system modeled by Equations (9.14) there are two time regimes. One is associated with the decay of the coordinate with the negative eigenvalue. The other is determined by the growth or decay of the zero eigenvalue coordinate. The exponential decay of y is called a *transient*. The initial conditions are restricted to "sufficiently small" excitations. One waits a time long enough for the exponential decay of the effect of the initial conditions to have fully expressed itself so that the time regime determined by the zero-eigenvalue coordinate is reached. In that time regime, although the transient component of y has decayed, y still survives, owing to its coupling to x. This coupling, in turn, affects the stability of x on its *center manifold*. One studies this stability.

Normal form analysis. We proceed with a straightforward analysis, introducing a near-identity transformation. To connect with the method of the center manifold, we separate the exponentially decaying terms from the others. As the algebra is messy, we have limited the study to this simple two-dimensional system. Furthermore, we have only carried out the expansions through $O(\varepsilon^3)$. The extension to the general case is cumbersome but straightforward.

9.2.1 Method of normal forms

We write the near-identity transformation as

$$x = u + \sum_{n=1}^{M} \varepsilon^n T_n(u,v) \qquad y = v + \sum_{n=1}^{M} \varepsilon^n \{\hat{S}_n(u) + S_n(u,v)\} \quad (9.15)$$

[The upper limit M in Equation (9.15) just indicates that series is truncated at the order ε^M; otherwise it has no significance.]

In anticipation of the structure of the expansion, we separate the generating functions for y into two parts:

1. Terms with a pure u dependence, denoted by $\hat{S}(u)$, that serve for establishing the connection with the method of the center manifold
2. Terms that have a v dependence, denoted by $S(u,v)$, leading to exponential decay

CONNECTION WITH METHOD OF CENTER MANIFOLD

We first substitute the near-identity transformation for x into the l.h.s. of Equation (9.14) to obtain

$$\frac{dx}{dt} = \varepsilon \left[U_1 + \frac{\partial T_1}{\partial v} \sigma v \right] + \varepsilon^2 \left[U_2 + \frac{\partial T_1}{\partial u} U_1 + \frac{\partial T_1}{\partial v} V_1 + \frac{\partial T_2}{\partial v} \sigma v \right]$$

$$+ \varepsilon^3 \left[U_3 + \frac{\partial T_1}{\partial u} U_2 + \frac{\partial T_2}{\partial u} U_1 + \frac{\partial T_1}{\partial v} V_2 + \frac{\partial T_2}{\partial v} V_1 + \frac{\partial T_3}{\partial v} \sigma v \right]$$

$$+ O(\varepsilon^4)$$

(9.16a)

Now substitute the near-identity transformation, Equation (9.15), into the r.h.s. of Equation (9.14) and obtain

$$\frac{dx}{dt} = \varepsilon \left[\alpha_{20} u^2 + \alpha_{11} u v + \alpha_{02} v^2 \right]$$

$$+ \varepsilon^2 \begin{bmatrix} \alpha_{30} u^3 + \alpha_{21} u^2 v + \alpha_{12} u v^2 + \alpha_{30} v^3 + 2\alpha_{20} u T_1 \\ + \alpha_{11} \{ u(S_1 + \hat{S}_1) + v T_1 \} + 2\alpha_{20} v (S_1 + \hat{S}_1) \end{bmatrix}$$

(9.16b)

$$+ \varepsilon^3 \begin{bmatrix} \alpha_{40} u^4 + \alpha_{31} u^3 v + \alpha_{22} u^2 v^2 + \alpha_{13} u v^3 + \alpha_{04} v^4 \\ + \alpha_{20} (2 u T_2 + T_1^2) + \alpha_{11} \{ u(S_2 + \hat{S}_2) + v T_2 + 2 T_1 (S_1 + \hat{S}_1) \} \\ + \alpha_{02} \{ 2 v (S_2 + \hat{S}_2) + (S_1 + \hat{S}_1)^2 \} \\ + 3\alpha_{30} u^2 T_1 + \alpha_{21} \{ u^2 (S_1 + \hat{S}_1) + 2 u v T_1 \} \\ + \alpha_{12} \{ v^2 T_1 + 2 u v (S_1 + \hat{S}_1) \} + 3\alpha_{03} v^2 (S_1 + \hat{S}_1) \end{bmatrix}$$

Study of Equations (9.16a,b): $O(\varepsilon)$ analysis. We first concentrate on the T_1 terms keeping in mind that $U_0 = 0$. Resonant terms are "pure u" and cannot be killed by T_1. Thus, we have

$$U_1 = \alpha_{20} u^2 \tag{9.17a}$$

On the other hand, T_1 can kill the remaining terms. To this end, it must contain a v component. Solving, one finds

$$T_1 = \frac{\alpha_{11}}{\sigma}uv + \frac{\alpha_{02}}{2\sigma}v^2 \qquad (9.17b)$$

As we will be trying to identify pure-u terms, we need not display the structure of the higher-order T terms but just note that each of them contains some v dependence.

With these observations made, we display the structure of the u-dependent terms in Equations (9.16a,b). The u dependence of the y terms is expressed by the $\hat{S}$ generators. Through $O(\varepsilon^3)$, we have

$$\frac{du}{dt} = \varepsilon U_1 + \varepsilon^2 U_2 + \varepsilon^3 U_3$$

$$= \varepsilon \alpha_{20} u^2 + \varepsilon^2 \left[\alpha_{30} u^3 + \alpha_{11} u \hat{S}_1 \right]$$

$$+ \varepsilon^3 \left[\alpha_{40} u^4 + \alpha_{21} u^2 \hat{S}_1 + \alpha_{11} u \hat{S}_2 + \alpha_{02} \hat{S}_1^{\,2} \right] \qquad (9.18)$$

We now wish to obtain the $\hat{S}$ functions. Through $O(\varepsilon^3)$, the l.h.s. of the y equation is

$$\frac{dy}{dt} = \sigma v + \varepsilon \left[V_1 + \frac{\partial S_1}{\partial v} \sigma v \right]$$

$$+ \varepsilon^2 \left[V_2 + \frac{d\hat{S}_1}{du} U_1 + \frac{\partial S_1}{\partial u} U_1 + \frac{\partial S_1}{\partial v} V_1 + \frac{\partial S_2}{\partial v} \sigma v \right]$$

$$+ \varepsilon^3 \left[\begin{array}{c} V_3 + \dfrac{d\hat{S}_1}{du} U_2 + \dfrac{\partial S_1}{\partial u} U_2 + \dfrac{d\hat{S}_2}{du} U_1 + \dfrac{\partial S_2}{\partial u} U_1 \\[6pt] + \dfrac{\partial S_1}{\partial v} V_2 + \dfrac{\partial S_2}{\partial v} V_1 + \dfrac{\partial S_3}{\partial v} \sigma v \end{array} \right] \qquad (9.19a)$$

The right-hand side is constructed from terms that are identical to those on the r.h.s. of Equation (9.16b), with α_{pq} replaced everywhere by β_{pq}, with an additional contribution coming from the σ term:

CONNECTION WITH METHOD OF CENTER MANIFOLD

$$\frac{dy}{dt} = \sigma\left[v + \varepsilon\left(\hat{S}_1 + S_1\right) + \varepsilon^2\left(\hat{S}_2 + S_2\right) + \varepsilon^3\left(\hat{S}_3 + S_3\right)\right]$$
(9.19b)
+ r.h.s. of Equation (9.16b) with $\alpha_{pq} \to \beta_{pq}$

Examining the structure of Equations (9.19a,b), one sees that the $\hat{S}$ functions can be used to kill the pure u dependence entirely. Hence, all the V_j terms have v dependence. One finds

$$\hat{S}_1 = -\frac{\beta_{20}}{\sigma}u^2 \qquad S_1 = \frac{\beta_{02}}{\sigma^2}v^2$$

$$V_1 = \beta_{11}uv \qquad (9.20)$$

$$\hat{S}_2 = \left[-2\frac{\beta_{20}\alpha_{20}}{\sigma^2} + \frac{\beta_{11}\beta_{20}}{\sigma^2} - \frac{\beta_{30}}{\sigma}\right]u^3$$

These terms will be sufficient for us to establish the desired connection to the center manifold equations.

9.2.2 Method of the center manifold

To motivate the procedure we follow, we quote from page 2 of Carr [1], as it relates to Equation (9.14): "The theory that we develop tells us that this equation has an invariant manifold, $y=h(x)$, |x| small, with $h=O(x^2)$ as $x \to 0$. Furthermore, the asymptotic stability of the zero solution can be proved by studying a first order equation." By construction (or by assumption), the y dependence is, therefore, expressed completely in terms of x. With these remarks noted, the procedure is to substitute for y, everywhere in Equation (9.14), a power series in x:

$$y = h(x) = \sum_{n=1}^{M} \varepsilon^n A_n x^{n+1} \qquad (9.21a)$$

Comment. In the usual treatment, one chooses the form of $h(x)$ to mimic the interaction terms. It simply makes sense. We could have done this also in the normal form expansion. Instead, we did not assume a specific form for the S and $\hat{S}$ functions. As our goal is to establish the connection to the normal form equations, we have found it better to write Equation (9.21a) in the form

$$y = h(x) = \sum_{n=1}^{M} \varepsilon^n A_n(x) \tag{9.21b}$$

The dy/dt part of Equation (9.14) becomes

$$\frac{dy}{dt} = \frac{dh(x)}{dx}\frac{dx}{dt} = \frac{dh(x)}{dx}\varepsilon X(x,h(x);\varepsilon) \tag{9.22}$$

where $X(x,y;\varepsilon)$ represents the nonlinear r.h.s. of Equation (9.14a). We substitute Equation (9.21b) into both sides of Equation (9.22) and collect all terms through $O(\varepsilon^3)$, to obtain

$$\left[\varepsilon\frac{dA_1}{dx}+\varepsilon^2\frac{dA_2}{dx}\right]\left[\begin{array}{l}\varepsilon\alpha_{20} x^2 + \varepsilon\alpha_{11} x(\varepsilon A_1 + \varepsilon^2 A_2) \\ + \varepsilon^3 \alpha_{02} A_1^2 + \varepsilon^2 \alpha_{30} x^3\end{array}\right]$$

$$= \sigma\left[\varepsilon A_1 + \varepsilon^2 A_2 + \varepsilon^3 A_3\right] \tag{9.23}$$

$$+ \left[\begin{array}{l}\varepsilon\beta_{20} x^2 + \varepsilon\beta_{11} x(\varepsilon A_1 + \varepsilon^2 A_2) + \varepsilon^3\beta_{02} A_1^2 \\ + \varepsilon^2 \beta_{30} x^3 + \varepsilon^3 \beta_{21} x^2 A_1 + \varepsilon^3 \beta_{40} x^4\end{array}\right]$$

The coefficients α_{pq} and β_{pq} are known. Substituting Equation (9.23) into Equations (9.22) and solving for A_n, we have

$$O(\varepsilon): \left[\sigma A_1 + \beta_{20} x^2\right] = 0$$

$$O(\varepsilon^2): \left[\frac{dA_1}{dx}\alpha_{20} x^2 = \sigma A_2 + \beta_{11} x A_1 + \beta_{30} x^3\right] \tag{9.24a}$$

$$O(\varepsilon^3): \left[\begin{array}{l}\frac{dA_1}{dx}(\alpha_{11} x A_1 + \alpha_{30} x^3) + \frac{dA_2}{dx}\alpha_{20} x^2 \\ = \sigma A_3 + \beta_{11} x A_2 + \beta_{02} A_1^2 + \beta_{21} x^2 A_1 + \beta_{40} x^4\end{array}\right]$$

yielding

$$A_1 = -\frac{\beta_{20}}{\sigma}x^2 \qquad A_2 = \left[-\frac{\beta_{30}}{\sigma} - 2\frac{\alpha_{20}\beta_{20}}{\sigma^2} + \frac{\beta_{11}\beta_{20}}{\sigma^2}\right]x^3 \tag{9.24b}$$

This leads to an expression for $y = h(x)$. We are now ready to substitute for the y variable in Equation (9.14a) and obtain, through $O(\varepsilon^3)$, the desired form for dx/dt: [we have written the $O(\varepsilon^3)$ equation for y, although we do not need it to obtain dx/dt through $O(\varepsilon^3)$]:

$$\frac{dx}{dt} = \begin{bmatrix} \varepsilon \alpha_{20} x^2 + \varepsilon^2 (\alpha_{30} x^3 + \alpha_{11} x A_1) \\ + \varepsilon^3 (\alpha_{40} x^4 + \alpha_{21} x^2 A_1 + \alpha_{11} x A_2 + \alpha_{02} \varepsilon^2 A_1^2) \end{bmatrix} \quad (9.25)$$

Observations. If one equates everywhere in Equation (9.18), $\hat{S}$ with A and u with x, using Equation (9.20), the result is identical to Equation (9.25). The center manifold equations for dx/dt are the same as the zero-order normal form equations for u. However, we see explicitly how the normal form method generates the transient behavior as well as the center manifold results. For this reason we prefer it. (See Example 9.5.)

9.3 ADDITIONAL EXAMPLES

Example 9.4. We return to Example 9.3 [Equation (9.9)] and analyze it by the method of the center manifold. To obtain an $O(\varepsilon^2)$ approximation for $x(t)$, it is sufficient to restrict the analysis to the leading term in $y=h(x)$. Thus, we write $h(x)=\varepsilon \cdot x^2$. In $O(\varepsilon)$, one need only look at the dy/dt equation:

$$\frac{dy}{dt} = 2\varepsilon A x \frac{dx}{dt} = 2\varepsilon A x \{-\tfrac{1}{2} x \varepsilon A x^2\} = -9\varepsilon A x^2 + \tfrac{1}{2}\varepsilon x^2 \quad (9.26)$$

Solving Equation (9.26) in $O(\varepsilon)$ for A, we find

$$y = \tfrac{1}{18}\varepsilon x^2$$

Substitute this expression for $y(x)$ in the equation for dx/dt and neglecting terms of $O(\varepsilon^3)$ and obtain:

$$\frac{dx}{dt} = -\tfrac{1}{2}\varepsilon x y = -\tfrac{1}{36}\varepsilon^2 x^3 \quad (9.27)$$

Finally, we solve Equation (9.27) and obtain

$$x(t) = \frac{x(0)}{\sqrt{1 + \tfrac{1}{18}\varepsilon^2 x(0)^2 t}} \quad (9.28)$$

The expression for $x(t)$ in Equation (9.28) is *identical* to the one for $u(t)$ of Equation (9.12). Comparing Equations (9.28) and (9.13), we observe that the center manifold equations do not incorporate the T functions and thus lack terms that depend on v, that constitute the exponential component of the solution. Losing the transients is characteristic of the method of the center manifold.

Comment. The center manifold method provides an asymptotic solution valid only for long times, after the transients have decayed. Therefore, one cannot follow the development of the system from the initial time, $t = 0$. For example, in Equation (9.28), the constant $x(0)$ is not the initial amplitude. In the method of normal forms, the transient terms provide the connection with the initial time, so that $u(0)$ in Equation (9.12) is indeed the initial amplitude of the zero-order term.

Example 9.5. Consider the system

$$\frac{dx}{dt} = -\varepsilon x^2 + \varepsilon x y \qquad \frac{dy}{dt} = -\lambda y + \varepsilon x^2 \qquad (9.29)$$

where $\lambda > 0$. In the method of the center manifold, we substitute $y = h(x)$ and construct dy/dt. We obtain

$$\frac{dy}{dt} = \frac{dh}{dx}\frac{dx}{dt} = \frac{dh}{dx}\left[-\varepsilon x^2 + \varepsilon x h(x)\right] = -\lambda h(x) + \varepsilon x^2 \qquad (9.30)$$

Writing $h(x)$ as a power series in x

$$h(x) = a x^2 + b x^3 + c x^4 + \cdots$$

we substitute this series for $h(x)$ in Equation (9.30), equate terms order by order, and obtain the first three coefficients:

$$a = \frac{\varepsilon}{\lambda} \qquad b = 2\left[\frac{\varepsilon}{\lambda}\right]^2 \qquad c = 4\left[\frac{\varepsilon}{\lambda}\right]^3$$

(We will only use the leading one.) Return to Equation (9.29) and construct the equation for dx/dt:

$$\frac{dx}{dt} = -\varepsilon x^2 + \varepsilon^2 \frac{1}{\lambda} x^3 + O(\varepsilon^3) \qquad (9.31)$$

Now construct an approximate solution to Equation (9.29) using the method of normal forms. We obtain, through $O(\varepsilon^2)$

$$x = u - \varepsilon \frac{1}{\lambda} uv + \varepsilon^2 \left[\frac{1}{\lambda^2} u^2 v - \frac{1}{2\lambda^2} uv^2 \right] \tag{9.32a}$$

$$y = v + \varepsilon \frac{1}{\lambda} u^2$$

and to the same order of accuracy

$$\frac{du}{dt} = -\varepsilon u^2 + \varepsilon^2 \frac{1}{\lambda^3} u^3 \qquad \frac{dv}{dt} = -\lambda v + O(\varepsilon^2) \tag{9.32b}$$

Comment. Observe that the du/dt component of Equation (9.32b) is precisely the same as the x term in Equation (9.31). This follows because all u terms are resonant. The T functions, apart from any free functions, will contain the v dependence.

Example 9.6. To further amplify on the connection between the center manifold and normal form equations, we exhibit a model system on page 5 of the book by Carr [1].

$$\frac{dx}{dt} = xy + ax^3 + by^2 x$$

$$\frac{dy}{dt} = -y + cx^2 + dx^2 y \tag{9.33}$$

We consider small amplitudes and scale the variables as $x \to \varepsilon x$, $y \to \varepsilon y$. The center manifold method yields the following equation that governs the stability of the zero eigenvalue solution (following Carr, we denote the dependent variable by u):

$$\frac{du}{dt} = \varepsilon^2 (a+c) u^3 + \varepsilon^4 (cd + bc^2) u^5 + O(\varepsilon^6) \tag{9.34a}$$

This is precisely the equation for the zero-order approximation, obtained in the normal form expansion. However, in the latter, one can uncover much more information about the solution. Invoking the near-identity transformation and solving for the generating functions, one finds in addition to Equation (9.34a):

$$x = u - \varepsilon uv + \varepsilon^2 \tfrac{1}{2}[1-b]uv^2$$
$$- \varepsilon^3 \left[\tfrac{1}{6}(1-3b)uv^3 + (2a+2c-2bc-d)u^3v\right] + O(\varepsilon^4)$$

$$y = v + \varepsilon c u^2 + \varepsilon^3 \left[\begin{array}{l}(bc-2c+2d)u^2v^2 \\ +(dc-2c^2-2ac)u^4\end{array}\right] + O(\varepsilon^4) \qquad (9.34\text{b}$$

$$\frac{dv}{dt} = -v + \varepsilon^2[d-2c]u^2v$$
$$+ \varepsilon^4 4\left[ac+c^2-c-bc^2\right]u^4v + O(\varepsilon^6)$$

Clearly, it takes some algebra to arrive at the results given in Equation (9.34b). However, a simple program written in a symbolic computer language such as MATHEMATICA or MAPLE does the computation.

Example 9.7. To finish our discussion, we consider an example given by the set of equations:

$$\frac{dx}{dt} = \varepsilon xy \qquad \frac{dy}{dt} = -y(1+\varepsilon \alpha x^2) \qquad (9.35)$$

In the method of the center manifold, we start by substituting $y=h(x)$ and introduce a power-series expansion for $h(x)$:

$$h(x) = ax^2 + bx^3 + cx^4 + \cdots$$

Equations (9.35) yield $a=b=c=\cdots=0$, namely, $y=h(x)\equiv 0$. As a result, the solution for x is $x\equiv$const. The center manifold is the line

$$y \equiv 0$$

The usual recipe of the method of center manifold may lead to confusion, unless one makes sure to write $h(x)$ as a series in powers of x. In fact, assuming $y=h(x)$, Equation (9.35) can be solved for $y(x)$ by simply dividing dy/dt by dx/dt, obtaining an equation for $y(x)$:

$$\frac{dy}{dx} = -\frac{1+\varepsilon \alpha x^2}{\varepsilon x} \qquad (9.36)$$

The solution of Equation (9.36) is

ADDITIONAL EXAMPLES

$$y(x) = y_0 - \frac{1}{\varepsilon}\ln\left(\frac{x}{x_0}\right) - \frac{1}{2}\alpha\left(x^2 - x_0^2\right) \qquad (9.37)$$

Obviously, $y(x)$ of Equation (9.37) cannot be expanded in a power series in ε. Nor can it be expanded in powers of x, because Equation (9.36) is not valid at $x=0$. The solution that passes through $x=0$ is

$$x \equiv 0 \qquad y = y(0)\exp(-t)$$

Once the exponential transient has decayed, the solution degenerates to the origin.

The center manifold can be also obtained by simply inserting in Equations (9.35) $y=0$. The solution is

$$x = x_0 \qquad y \equiv 0 \qquad (9.38)$$

where x_0 can take any value. Thus, the center manifold is the line

$$y \equiv 0$$

A straightforward application of the method of normal forms yields an expansion for $x(t)$ and $y(t)$ in powers of ε that provides the transient behavior *as well* as with the center manifold. For instance, through $O(\varepsilon^2)$, one obtains

$$x = u - \varepsilon u v + \varepsilon^2 \left(\frac{1}{2}uv^2 + \alpha u^3 v\right) + O(\varepsilon^3)$$

$$y = v - \varepsilon^2 \alpha u^2 v^2 + O(\varepsilon^3)$$

$$\dot{u} = 0 + O(\varepsilon^3) \qquad \dot{v} = -v - \varepsilon \alpha u^2 v + O(\varepsilon^3)$$

Thus, through $O(\varepsilon^2)$, u is any constant; hence, v decays exponentially and the asymptotic form for (x,y) is as in Equation (9.38).

9.4 CONCLUDING REMARKS

We have shown, through a variety of examples, how the method of normal forms accommodates problems that have some zero eigenvalues and some with negative real part. The formalism is unchanged and one obtains a faithful approximation to the solution.

One of our goals was to illustrate the connection between normal forms and the method of the center manifold. The conclusion is that the equations for the zero-order approximation in the normal form expansion, if one sets all the free functions equal to zero, are *identical* to those obtained by the method of center manifold. As the approximation to the center manifold is not unique, the freedom introduced by the free functions is to be expected. The v-dependent generating functions carry the time dependence of the full solution. Thus, the approximate solution obtained by the method of normal forms contains far more information about the evolution of the solution in time from an initial point onto the center manifold.

We remind the reader, in passing, that the intimate connection we have obtained between the center manifold and normal form equations is *not valid* for systems that have any pure imaginary eigenvalues. However, a generalization of the result to such cases is possible.

A fundamental characteristic of the method of the center manifold is that it enables one to reduce the dimensionality of the system of equations by solving the equations after the transients associated with the nonzero eigenvalues have decayed. This often brings with it a simpler picture of the long-time motion of the system. Thus, for example, if one has a two-dimensional problem in x and y, where the y variable has a negative eigenvalue and the x variable has a zero eigenvalue, it is possible, on the invariant manifold, to write $y=h(x)$, thereby reducing the dimension of the vector space.

It appears that the center manifold method is useful in the study of the stability of the zero solution. However, it cannot follow the time dependence of the solution from an initial time. By contrast, the normal form expansion provides a systematic procedure that allows one to follow, through a perturbation expansion, the time development of systems that have some zero eigenvalues with the remaining ones having negative real part from $t=0$. It does not enjoy the dimensional reduction associated with the decay onto a center manifold, lacking thereby possible simplifications in the analysis. However, with the ability to use symbolic manipulation programs such as MAPLE or MATHEMATICA to do the algebra, the advantages of the method of the center manifold no longer seem important. In the end, which method one uses depends on the questions one wants answered.

REFERENCES

1. Carr, J., *Applications of Centre Manifold Theory*, Springer-Verlag, New York (1981).
2. Guckenheimer, J., and P. J. Holmes, *Nonlinear Oscillations, Dynamical Systems, and Bifurcations of Vector Fields*. Springer-Verlag, New York (1988).

3. Wiggins, S., *Introduction to Applied Nonlinear Dynamical Systems and Chaos*, Springer-Verlag, New York (1990).
4. Rand, R. H., and D. Armbruster, *Perturbation Methods, Bifurcation Theory and Computer Algebra*, Springer-Verlag, New York (1987).

10

HIGHER-DIMENSIONAL HAMILTONIAN SYSTEMS

10.1 INTRODUCTION

Higher-dimensional systems exhibit a wealth of phenomena that are discussed extensively in the literature. These include systems in which the unperturbed part of the equations is already nonlinear, resonance effects, and the various aspects of chaotic behavior. In the spirit of keeping this text as simple as possible, and limiting its scope to systems that are amenable to a perturbation analysis, we discuss in this chapter only systems of more than one harmonic oscillator that are affected by nonlinear coupling forces.

In Section 10.2 we provide a detailed analysis of a problem that has been studied extensively in the literature, the betatron problem. It is a system of two harmonic oscillators with quadratic coupling forces. We use this simple case to expose new features that characterize the analysis of higher-dimensional systems within the framework of perturbation theory.

Action–angle variables are introduced in Section 10.3. These are the standard variables used in the literature in the analysis of Hamiltonian systems. They are then employed in Section 10.4 in the normal form analysis of systems of harmonic oscillators. The result is a powerful formalism that yields, in a natural manner, the notion of slow relative phases and the idea of formal constants of the motion. Finally, a specific choice of the free functions, in the near-identity transformation, that generate an extension of the idea of *minimal normal forms* (MNF, discussed in Chapter 6 in the case of a single oscillator) to higher-dimensional systems is studied.

In Section 10.5, the formalism is applied in detail to a system of two harmonic oscillators that are coupled by cubic coupling terms. In the numerical part of the study, the perturbation approximation for the solution is calculated through first order and the normal form equations are solved through second order. The notion of formal constants of the motion is studied, and the emergence and importance of the slow relative phase is exemplified. The approximation to the full solution, generated by the MNF choice, yields a better approximation to the solution than the usual *no free functions* choice.

In Section 10.6 we discuss the question of the extension of MNF to discrete maps.

10.2 BETATRON OSCILLATIONS

Consider a system of two coupled harmonic oscillators:

$$\ddot{\xi} + \xi = -\tfrac{1}{2}\beta\eta^2 \tag{10.1a}$$

$$\ddot{\eta} + \alpha\eta = -\beta\eta\xi \qquad \alpha > 0 \tag{10.1b}$$

Equations (10.1) provide in dimensionless quantities an approximate model for the coupling of longitudinal oscillations (along the beam path), described by the ξ coordinate, to transversal ones (η) of particles in a circular accelerator. The parameters α and β characterize the accelerator. If the transversal oscillations become too violent, the particle beam hits the walls of the accelerator tube and is destroyed. The aim is to find when this instability can occur.

This problem has been discussed extensively in the literature. We note that the system is Hamiltonian and follow, in part, the analysis presented in Ref. 1. Equations (10.1a,b) are obtained from the Hamiltonian:

$$\overline{H} = \tfrac{1}{2}\dot{\xi}^2 + \tfrac{1}{2}\xi^2 + \tfrac{1}{2}\dot{\eta}^2 + \tfrac{1}{2}\alpha\eta^2 + \tfrac{1}{2}\beta\eta^2\xi \tag{10.2}$$

We wish to study the onset of oscillations, namely, situations where the overall energy is small:

$$\overline{H} = \mu^2 H \qquad\qquad \mu^2 \ll 1$$

Without loss of generality, we choose $\mu > 0$ and rescale the amplitudes:

$$\xi \to \mu\xi \qquad \eta \to \mu\eta \tag{10.3}$$

The rescaled Hamiltonian becomes

$$H = \tfrac{1}{2}\dot{\xi}^2 + \tfrac{1}{2}\xi^2 + \tfrac{1}{2}\dot{\eta}^2 + \tfrac{1}{2}\alpha\eta^2 + \tfrac{1}{2}\mu\beta\eta^2\xi \tag{10.4}$$

Equations (10.1) become

BETATRON OSCILLATIONS

$$\ddot{\xi} + \xi = -\tfrac{1}{2}\beta\mu\eta^2 \qquad (10.5a)$$

$$\ddot{\eta} + \alpha\eta = -\beta\mu\eta\xi \qquad (10.5b)$$

As a result of energy conservation, motion in the four-dimensional space spanned by $\xi,\dot{\xi},\eta,\dot{\eta}$ is confined onto a three-dimensional constant energy subspace.

The unperturbed frequencies are $\omega_1=1$ and $\omega_2=\alpha^{1/2}$. If they are not resonant, the coupling between the two modes does not enter the normal form equations, making the problem formally equivalent to that of two uncoupled harmonic oscillators. However, when the two frequencies are incommensurate, the near-identity transformation suffers from the small-denominator problem, making the perturbation expansion useless. (This is why the word "formally" was used.) Therefore, we focus on situations when ω_1 and ω_2 almost obey a resonance relation. The small deviation from resonance will be treated as an additional expansion parameter.

10.2.1 Qualitative arguments

Pure longitudinal oscillations are allowed and do not lead to the destruction of the beam. This solution is obtained from

$$\eta = 0 \Rightarrow \ddot{\xi}_0 + \xi_0 = 0$$

yielding

$$\xi_0 = \rho_0 \cos t$$

where $\rho_0 = O(1)$ is the oscillation amplitude.

Now let the solution deviate from the pure longitudinal mode by a small deviation:

$$\xi = \xi_0 + x \qquad \eta = 0 + y \qquad |x|,|y| \ll 1$$

In a crude approximation, the x mode introduces a linear, time-dependent term in the equation for y, through Equation (10.5b):

$$\ddot{y} + (\alpha + \beta\mu\rho_0 \cos t)y = 0 \qquad (10.6)$$

Equation (10.6) is the linear Mathieu equation, discussed in Section

8.3.1. There, the cosine time-dependent term was $\cos 2t$. Resonance occurred when the unperturbed frequency was close to 1,2,3,.... Since here the driving term has unit frequency, resonance between the unperturbed part of the equation and the driving term occurs when $\alpha^{1/2}$ is near $\frac{1}{2}, 1, \frac{1}{2},....$ This may cause the emergence of unstable oscillations. We focus on the case $\alpha^{1/2} \approx \frac{1}{2}$ and write

$$\alpha = \tfrac{1}{4} + \delta \qquad |\delta| \ll 1 \qquad (10.7)$$

From Equation (8.77), we find through second order that, for $(\beta \cdot \mu) > 0$, the y oscillations become unstable for

$$-\tfrac{1}{2}\beta\mu\rho_0 - \tfrac{1}{8}(\beta\mu\rho_0)^2 < \delta < \tfrac{1}{2}\beta\mu\rho_0 - \tfrac{1}{8}(\beta\mu\rho_0)^2 \qquad (10.8)$$

Thus, a combination of accelerator parameters and of the amplitude of the longitudinal mode, can lead to unstable transverse oscillations.

10.2.2 The unperturbed system

For $\mu=0$ we have

$$\ddot{\xi}_0 + \xi_0 = 0 \qquad \ddot{\eta}_0 + \tfrac{1}{4}\eta_0 = 0 \qquad (10.9)$$

with the unperturbed Hamiltonian:

$$H_0 = \tfrac{1}{2}\dot{\xi}_0^2 + \tfrac{1}{2}\xi_0^2 + \tfrac{1}{2}\dot{\eta}_0^2 + \tfrac{1}{8}\eta_0^2 \qquad (10.10)$$

The solution of Equations (10.9) is given by

$$\xi_0 = \rho_0 \cos(t + \varphi_0) \qquad \eta_0 = r_0 \cos(\tfrac{1}{2}t + \psi_0) \quad (10.11)$$

From Equations (10.10 & 11) we obtain

$$\rho_0^2 + \tfrac{1}{4}r_0^2 = H_0 = \text{const} \qquad (10.12a)$$

$$\rho_0 = \text{const} \qquad r_0 = \text{const} \qquad (10.12b)$$

Equation (10.12a) implies that the motion of the unperturbed system is confined onto a three-dimensional constant energy subspace. In that subspace, motion is further confined onto a torus, because the

BETATRON OSCILLATIONS

amplitudes ρ_0 and r_0 are constant (Equation 10.12b).

10.2.3 First-order perturbation analysis

We now turn to the analysis of Equations (10.5), diagonalizing them by changing into the complex variables:

$$z = \xi + i\,\dot\xi \qquad\qquad w = \eta + 2i\,\dot\eta$$

We employ Equation (10.7). Equations (10.5a,b) become

$$\dot z = -i\,z - \tfrac{1}{8}i\,\mu\beta(w+w^*)^2 \tag{10.13a}$$

$$\dot w = -\tfrac{1}{2}i\,w - i\,\delta(w+w^*) - i\,(\tfrac{1}{2})\beta\mu(w+w^*)(z+z^*) \tag{10.13b}$$

The near-identity transformations for z and w are two-parameter expansions:

$$z = u + \mu T_{10}^u + \delta T_{01}^u + \mu^2 T_{20}^u + \mu\delta T_{11}^u + \delta^2 T_{02}^u + \cdots \tag{10.14a}$$

$$w = v + \mu T_{10}^v + \delta T_{01}^v + \mu^2 T_{20}^v + \mu\delta T_{11}^v + \delta^2 T_{02}^v + \cdots \tag{10.14b}$$

So are the normal form equations for u and v:

$$\dot u = -i\,u + \mu U_{10} + \delta U_{01} + \mu^2 U_{20} + \mu\delta U_{11} + \delta^2 U_{02} + \cdots \tag{10.15a}$$

$$\dot v = -\tfrac{1}{2}i\,v + \mu V_{10} + \delta V_{01} + \mu^2 V_{20} + \mu\delta V_{11} + \delta^2 V_{02} + \cdots \tag{10.15b}$$

The first-order terms in Equations (10.14) and (10.15) are

$$T_{10}^u = -\tfrac{1}{4}\beta vv^* - \tfrac{1}{16}\beta v^{*2} + uF_{10}^u \qquad\qquad T_{01}^u = uF_{01}^u \tag{10.16a}$$

$$T_{10}^v = \tfrac{1}{2}\beta uv - \tfrac{1}{2}\beta u^*v - \tfrac{1}{4}\beta u^*v^* + vF_{10}^v$$
$$T_{01}^v = -v^* + vF_{01}^v \tag{10.16b}$$

$$U_{10} = -\tfrac{1}{8}i\,\beta v^2 \qquad\qquad U_{01} = 0 \tag{10.16c}$$

$$V_{10} = -\tfrac{1}{2}i\,\beta uv^* \qquad\qquad V_{01} = -i\,v \tag{10.16d}$$

The explicit u and v factors in the free functions take care of their resonant character. The first-order normal form equations are

$$\dot{u} = -i\,u - \tfrac{1}{8}i\,\mu\beta v^2 \tag{10.17a}$$

$$\dot{v} = -\tfrac{1}{2}i\,v - \tfrac{1}{2}i\,\mu\beta uv^* - i\,\delta v \tag{10.17b}$$

In polar coordinates, Equations (10.17) become

$$u = \rho\exp(-i\,\varphi) \qquad v = r\exp(-i\,\psi)$$

The normal form equations become

$$\dot{\rho} = \tfrac{1}{8}\mu\beta r^2 \sin\Omega \tag{10.18a}$$

$$\dot{r} = -\tfrac{1}{2}\mu\beta\rho r \sin\Omega \tag{10.18b}$$

$$\dot{\varphi} = 1 + \tfrac{1}{8}\mu\beta \frac{r^2}{\rho}\cos\Omega \tag{10.18c}$$

$$\dot{\psi} = \tfrac{1}{2} + \delta + \tfrac{1}{2}\mu\beta\rho\cos\Omega \tag{10.18d}$$

$$\Omega = \varphi - 2\psi \tag{10.18e}$$

The slow relative phase Ω. Ω is a slowly varying relative phase. The rapidly varying parts of the phases cancel in Ω. Also, the r.h.s. of Equations (10.18a–d) do not depend on the individual phases, φ and ψ, but only on Ω. Thus, the original four-dimensional system is reduced in the normal form equations, where only resonant terms are retained, to a three-dimensional one, in the space defined by ρ, r, and Ω. Equations (10.18c,d) yield for Ω one independent equation:

$$\dot{\Omega} = \mu\beta\rho\bigl(\tfrac{1}{8}(r/\rho)^2 - 1\bigr)\cos\Omega - 2\delta \tag{10.18f}$$

Once ρ, r, and Ω are solved for, the individual phases, φ and ψ, are obtained by direct quadrature.

This phenomenon recurs in higher orders. It is not particular to the present problem, but is typical of the structure of the normal form equations in multidimensional Hamiltonian systems.

To see how the slow phase appears, we note that the unperturbed eigenvalues in the present problem are: $\pm i$ and $\pm(i/2)$. Resonant combinations of the "old" type that we have encountered in two-dimensional systems occurred between $+i$ and $-i$, and between $+(i/2)$ and $-(i/2)$. In the two-dimensional case, the phases are then obtained in a straightforward manner. Here, there is a new type of resonance combination, connecting the two degrees of freedom. The eigenvalues $\lambda_1 = -i$ and $\lambda_2^* = +(i/2)$ are connected by resonance relations:

$$\lambda_1 = (n+1)\lambda_1 + 2n\lambda_2^* \qquad (10.19)$$

with n any nonnegative integer. This corresponds to resonant monomials of the form $u^{n+1}v^{*2n}$, where Ω appears naturally. There are additional resonance relations: generating λ_1^* from a combination of λ_1^* and λ_2, or λ_2 (λ_2^*) from combinations of λ_1 (λ_1^*) and λ_2^* (λ_2). All these relations can be derived from Equation (10.19) and are summarized by one relationship between the frequencies, $\omega_1 = 1$, corresponding to $\pm\lambda_1$ and $\omega_2 = \frac{1}{2}$, corresponding to $\pm\lambda_2$:

$$n\omega_1 - 2n\omega_2 = 0$$

The $n=1$ fundamental relation, yields the pair of integers $(1, -2)$ used in the definition of Ω. We address this point again later in this chapter.

Constants of motion. Equations (10.18a,b) have an integral:

$$K \equiv \rho^2 + \tfrac{1}{4}r^2 = \text{const} \qquad (10.20)$$

However, Equations (10.18) are *first-order* equations. Namely, terms of $O(\mu^2)$, $O(\mu\delta)$ and $O(\delta^2)$ have been neglected. Therefore, the r.h.s. of Equation (10.20) is constant only approximately, with possible errors of the type $O(\mu^2 t)$, $O(\mu\delta t)$, and $O(\delta^2 t)$. Thus, constancy may be destroyed over timescales of $O(1/\mu)$ or $O(1/\delta)$. The question that arises is whether the inclusion of higher-order terms extends the constancy of this "constant of motion" to longer timescales. As we shall see, this depends on the choice of the free functions.

In the unperturbed system, ρ and r were constant. One may wonder whether a solution with fixed values for ρ and r exists here as well. One trivial case, emerging from Equations (10.18a,b), is $r=0$ and $\rho=$const, corresponding to pure ξ oscillations. Its stability domain is the complement of the instability domain for the η oscillations, given in Equation (10.8).

The other possibility in Equations (10.18a,b) is

$$\sin\Omega = 0 \Rightarrow \Omega = 0, \pi$$

Since Ω must be constant, Equation (10.18f) implies

$$\pm \mu \beta \rho_1 \left(\frac{1}{8} \left(\frac{r_1}{\rho_1} \right)^2 - 1 \right) - 2\delta = 0 \tag{10.21}$$

where the $+$ ($-$) sign corresponds to $\Omega=0$ (π) and r_1 and ρ_1 denote the constant amplitudes. This case corresponds to a fixed point.

A detailed linear stability analysis shows that the $\Omega=0$ fixed point is neutrally stable, so that any slight deviation from the point (r_1, ρ_1) yields a possible solution. The $\Omega=\pi$ fixed point is unstable. That the fixed points are not attractive and hence do not lead to stable fixed amplitudes, is not surprising. As a result of resonance, the coupling between z and w causes efficient energy transfer between the two modes, leading to variations in their amplitudes.

10.2.4 Second-order analysis

Using the first-order T functions, the second-order contributions to the normal form are found to be

$$U_{20} = \frac{i}{16} \beta^2 u v v^* + \frac{i}{8} \beta v^2 \left(F_{10}^u - 2 F_{10}^v \right)$$

$$+ i \beta u \left(\frac{1}{8} \left(v^2 \frac{\partial F_{10}^u}{\partial u} - v^{*2} \frac{\partial F_{10}^u}{\partial u^*} \right) \\ + \frac{1}{2} \left(u v^* \frac{\partial F_{10}^u}{\partial v} - u^* v \frac{\partial F_{10}^u}{\partial v^*} \right) \right) \tag{10.21a}$$

$$U_{11} = i \beta v^2 \left(\tfrac{1}{4} - \tfrac{1}{4} F_{01}^v + \tfrac{1}{8} F_{01}^u \right) + i u \left(v \frac{\partial F_{10}^u}{\partial v} - v^* \frac{\partial F_{10}^u}{\partial v^*} \right)$$

$$+ i \beta u \left(\frac{1}{8} \left(v^2 \frac{\partial F_{01}^u}{\partial u} - v^{*2} \frac{\partial F_{01}^u}{\partial u^*} \right) \\ + \frac{1}{2} \left(u v^* \frac{\partial F_{01}^u}{\partial v} - u^* v \frac{\partial F_{01}^u}{\partial v^*} \right) \right) \tag{10.21b}$$

BETATRON OSCILLATIONS

$$U_{02} = i\, u\left(v \frac{\partial F_{01}^u}{\partial v} - v^* \frac{\partial F_{01}^u}{\partial v^*}\right) \quad (10.21c)$$

$$V_{20} = \tfrac{9}{32} i\, \beta^2 v^2 v^* + \tfrac{1}{8} i\, \beta^2 u u^* v$$
$$+ \tfrac{1}{2} i\, \beta u v^* \left(F_{10}^v - F_{10}^{v*} - F_{10}^u\right)$$
$$+ i\, \beta v \left(\begin{array}{c} \tfrac{1}{8}\left(v^2 \dfrac{\partial F_{10}^v}{\partial u} - v^{*2} \dfrac{\partial F_{10}^v}{\partial u^*}\right) \\ + \tfrac{1}{2}\left(uv^* \dfrac{\partial F_{10}^v}{\partial v} - u^* v \dfrac{\partial F_{10}^v}{\partial v^*}\right) \end{array} \right) \quad (10.21d)$$

$$V_{11} = i\, \beta u v^* \left(1 + \tfrac{1}{2}\left(F_{01}^v - F_{01}^{v*} - F_{01}^u\right)\right)$$
$$+ i\, \beta v \left(\begin{array}{c} \tfrac{1}{8}\left(v^2 \dfrac{\partial F_{01}^v}{\partial u} - v^{*2} \dfrac{\partial F_{01}^v}{\partial u^*}\right) \\ + \tfrac{1}{2}\left(uv^* \dfrac{\partial F_{01}^v}{\partial v} - u^* v \dfrac{\partial F_{01}^v}{\partial v^*}\right) \end{array} \right) \quad (10.21e)$$
$$+ i\, v \left(v \dfrac{\partial F_{01}^v}{\partial v} - v^* \dfrac{\partial F_{01}^v}{\partial v^*}\right)$$

$$V_{02} = i\, v \left(1 + \left(v \dfrac{\partial F_{01}^v}{\partial v} - v^* \dfrac{\partial F_{01}^v}{\partial v^*}\right)\right) \quad (10.21f)$$

Constant of motion. With the choice of vanishing free functions, the second-order normal form equations in polar coordinates become

$$\dot{\rho} = \tfrac{1}{8}\mu\beta r^2 \sin\Omega - \tfrac{1}{4}\mu\delta\beta r^2 \sin\Omega \quad (10.22a)$$

$$\dot{r} = -\tfrac{1}{2}\mu\beta\rho r \sin\Omega + \mu\delta\beta\rho r \sin\Omega \quad (10.22b)$$

$$\dot{\varphi} = 1 + \tfrac{1}{8}\mu\beta\frac{r^2}{\rho}\cos\Omega - \tfrac{1}{16}\mu^2\beta^2 r^2 - \tfrac{1}{4}\mu\delta\beta r^2 \sin\Omega \quad (10.22c)$$

$$\psi = \tfrac{1}{2} + \tfrac{1}{2}\mu\beta\rho\cos\Omega + \delta$$
$$- \tfrac{9}{32}\mu^2\beta^2 r^2 \sin\Omega - \tfrac{1}{8}\mu^2\beta^2\rho^2 - \mu\delta\beta\rho\cos\Omega - \delta \qquad (10.22d)$$

From Equations (10.22a,b) we see that, with this choice, Equation (10.20) is also obeyed, extending the approximate constancy of $K=\rho^2+(1/4)r^2$ to longer timescales.

Also, U_{11} and V_{11} can be eliminated by the choice

$$F_{01}{}^u = F_{01}{}^v = 2 \qquad F_{10}{}^u = F_{10}{}^v = 0$$

As a result, the equations for ρ and r become the same as in first order [Equations (10.18a,b)], except that, now, the errors are of second order. Again, K is a "constant of the motion."

We note again that, in both first and second orders, K is not strictly a constant of motion. It is approximately so, with an error that is small over some extended timescale. As one goes from first to second order, the timescale is extended by making specific choices of the free functions. In general, the "constancy" of K is spoiled for an infinity of choices of the free functions.

The same is true in higher orders. The freedom enables us to extend the "constancy" of K to longer timescales or to have progressively smaller errors, by going to higher orders of the expansion and choosing the free functions appropriately. Seemingly, this would make K a genuine constant of motion, if all orders of the expansion were included. However, for this to be true, the near-identity transformation must converge. In most cases it may be just an asymptotic series. Hence, we can only view K as a formal constant of motion: Ignoring higher-order contributions, it is constant in every finite order of the expansion.

10.3 ACTION–ANGLE VARIABLES

The description of the problem in terms of u, u^*, and v^* is cumbersome. The resonant terms in the U functions and in the free functions are combinations of uu^*, vv^*, uv^{*2} and u^*v^2. These four terms depend only on three independent variables: ρ, r, and Ω. It is, therefore, advantageous to use the latter as variables and avoid the need to perform the transformations from the original set to the new one every time one has to analyze the effect of resonant terms. In this section, we introduce the action and angle variables. They yield a set that is equivalent to the set $\{\rho,r,\Omega\}$. They are the standard variables employed in the analysis of multi-dimensional conservative systems.

Basics. Integrable Hamiltonian systems and action–angle variables are a fundamental part of classical mechanics. We do not develop the topic here and refer the reader to the excellent expositions of the subject in classic texts [2,3]. Our goal is to use the language of *action–angle variables* so as to see the normal form formalism, in the case of conservative systems, from a point of view that is different from the one we have used until now. In this new language, resonance is expressed as a relation among the (real) unpertrubed frequencies, rather than among the (imaginary) eigenvalues. Also, as will become more obvious when we present the analysis in detail, the emergence and importance of slow relative phases (as the angle Ω in the case of two coupled oscillators, discussed above) is natural.

The first concept we need to introduce is that of *integrable Hamiltonian systems*. For an integrable system with n degrees of freedom, there are n constants of the motion.

More specifically, let the system be originally described by n coordinates q_i and n canonically conjugate momenta, p_i, with a Hamiltonian $H(\mathbf{p},\mathbf{q})$, where $\mathbf{p}$ and $\mathbf{q}$ are the n-dimensional vectors $(p_1, ..., p_n)$ and $(q_1,..., q_n)$, respectively. The equations of motion, Hamilton's equations, are

$$\dot{q}_i = \frac{\partial H}{\partial p_i} \qquad \dot{p}_i = -\frac{\partial H}{\partial q_i}$$

The system is integrable if a canonical transformation exists form the variables $(\mathbf{p},\mathbf{q})$ to a new canonical set: $(\mathbf{J},\boldsymbol{\theta})$ so that, in the new variables, $H=H(\mathbf{J})$. The fact that J_i are constants of the motion is a direct consequence of Hamilton's equations for the new variables: H depends only on the J_i. Therefore

$$\dot{J}_i = -\frac{\partial H(\mathbf{J})}{\partial \theta_i} = 0 \qquad (10.23a)$$

The equations for θ_i are

$$\omega_i = \dot{\theta}_i = \frac{\partial H(\mathbf{J})}{\partial J_i} \qquad (10.23b)$$

implying that ω_i is a function of $\mathbf{J}$ and, hence, is constant for any particular solution.

In terms of the old variables, under appropriate conditions, the constants of motion J_i can be chosen as the actions, defined as

$$J_i = \frac{1}{2\pi}\oint p_i\, dq_i \tag{10.24}$$

The integral is carried along the closed orbit that the *i*th degree of freedom performs in its two-dimensional phase space. The motion is then periodic in the angles θ_i.

10.3.1 Single harmonic oscillator

We begin by presenting the action and angle variables for an unperturbed harmonic oscillator. Although this is quite trivial, it will familiarize us with quantities that will be used when perturbations are introduced.

In the usual language, for a unit mass oscillator, we have the well-known equations:

$$H = \tfrac{1}{2} p^2 + \tfrac{1}{2}\omega^2 x^2 \tag{10.25a}$$

$$\dot{x} = \frac{\partial H}{\partial p} = p \tag{10.25b}$$

$$\dot{p} = \ddot{x} = -\frac{\partial H}{\partial x} = -\omega^2 x \tag{10.25c}$$

solved by

$$x = x_0 \cos(\omega t + \theta_0) \qquad p = \dot{x} = -\omega x_0 \sin(\omega t + \theta_0) \tag{10.26}$$

Inserting Equation (10.26) in Equations (10.24) and (10.25a), we get the well-known results:

$$J = \tfrac{1}{2}\omega x_0^2 \tag{10.27a}$$

$$H = \tfrac{1}{2}\omega^2 x_0^2 = J\omega \tag{10.27b}$$

Hamilton's equations for the action-angle variables are simple:

$$\dot{J} = -\frac{\partial H}{\partial \theta} = 0 \tag{10.28a}$$

implying (trivially) that J is a constant of the motion and

ACTION–ANGLE VARIABLES

$$\dot{\theta} = \frac{\partial H}{\partial J} = \omega$$

yielding the folowing solution for θ:

$$\theta = \omega t + \theta_0 \qquad (10.28b)$$

In terms of J and θ, x and p are expressed as

$$x = \sqrt{\left(\frac{2J}{\omega}\right)} \cos\theta \qquad p = \dot{x} = -\sqrt{2J\omega}\sin\theta$$

$$z = x + \frac{i}{\omega}p = \sqrt{\left(\frac{2J}{\omega}\right)}\exp(-i\,\theta) \qquad (10.29)$$

In this trivial case, we see explicitly that the action, J, is constant and that the motion is periodic in the canonically conjugate angle.

10.3.2 Nearly integrable systems

We now apply the action–angle formalism to the coupled oscillators problem of Equations (10.5). However, it is not clear whether the perturbed system is integrable and that action–angle variables can be defined for the full problem. This, in any case, is not our goal. We consider the system of Equations (10.5) as an example of a *nearly integrable system*. Rather than plunging into the latter subject, we develop the required concepts through this example.

Borrowing from the case of the unperturbed harmonic oscillator, we apply Equation (10.29) to both ξ and η, for $\omega_1=1$ and $\omega_2=\tfrac{1}{2}$:

$$\xi = \sqrt{2J_1}\cos\theta_1 \qquad p_\xi = \dot{\xi} = -\sqrt{2J_1}\sin\theta_1 \qquad (10.30a)$$

$$\eta = \sqrt{4J_2}\cos\theta_2 \qquad p_\eta = \dot{\eta} = -\sqrt{J_2}\sin\theta_2 \qquad (10.30b)$$

Now insert Equations (10.30) in the Hamiltonian H of Equation (10.4) to obtain for $\alpha = \omega_2^2 + \delta$:

$$H = J_1 \cdot \omega_1 + J_2 \cdot \omega_2 \\ + \delta J_2 \cos^2\theta_2 + \sqrt{8}\mu\beta\sqrt{J_1}\,J_2\cos\theta_1\cos^2\theta_2 \qquad (10.31)$$

Using Equations (10.5) for ξ and η, it is straightforward to show that J_i and θ_i ($i=1,2$) obey the equations

$$\dot{J}_1 = \sqrt{8}\,\mu\beta\sqrt{J_1}\,J_2\sin\theta_1\cos^2\theta_2 = -\frac{\partial H}{\partial \theta_1} \qquad (10.32a)$$

$$\dot{J}_2 = 2\delta J_2 \cos\theta_2 \sin\theta_2$$
$$+ 4\sqrt{2}\,\mu\beta\sqrt{J_1}\,J_2 \cos\theta_1 \cos\theta_2 \sin\theta_2 = -\frac{\partial H}{\partial \theta_2} \qquad (10.32b)$$

$$\dot{\theta}_1 = \omega_1 + \sqrt{2}\,J_1^{-1/2}\,J_2\cos\theta_1\cos^2\theta_2 = \frac{\partial H}{\partial J_1} \qquad (10.32c)$$

$$\dot{\theta}_2 = \omega_2 + \delta\cos^2\theta_2 + \sqrt{8}\,\mu\beta\sqrt{J_1}\,\cos\theta_1\cos^2\theta_2 = \frac{\partial H}{\partial J_2} \qquad (10.32d)$$

From Equations (10.32), we conclude that the system described by the new variables is Hamiltonian. Thus, the transformation $(\mathbf{x},\mathbf{p})\rightarrow(\mathbf{J},\boldsymbol{\theta})$ is canonical and J_i and θ_i are canonically conjugate variables. However, *they are not action–angle variables* in the true sense. The J_i are not constant.

The system of Equations (10.32) is an example of a *nearly integrable* system. One starts with an unperturbed integrable system, for which action and angle variables *can* be defined, so that it is described by a Hamiltonian $H_0(\mathbf{J})$ and the canonically conjugate vector of angles, $\boldsymbol{\theta}$. To the unperturbed Hamiltonian one adds a small perturbation that depends on both $\mathbf{J}$ and $\boldsymbol{\theta}$ (the dependence on the angles is periodic) yielding the Hamiltonian of the perturbed system:

$$H(\mathbf{J},\boldsymbol{\theta};\varepsilon) = H_0(\mathbf{J}) + \varepsilon V(\mathbf{J},\boldsymbol{\theta};\varepsilon) \qquad |\varepsilon|\ll 1 \qquad (10.33)$$

Here ε is representative of the small parameters of the problem. The perturbation is chosen so that Hamilton's equations [Equations (10.23)] are obeyed by the coordinates of the perturbed system. Only now, since H depends also on the angles, the J_i are not constant any more. The Equations of motion are

$$\dot{J}_i = -\frac{\partial H(\mathbf{J},\boldsymbol{\theta};\varepsilon)}{\partial \theta_i} = -\varepsilon\frac{\partial V(\mathbf{J},\boldsymbol{\theta};\varepsilon)}{\partial \theta_i} \qquad (10.34a)$$

ACTION–ANGLE VARIABLES

$$\dot{\theta}_i = \frac{\partial H(\mathbf{J},\boldsymbol{\theta};\varepsilon)}{\partial J_i} = \frac{\partial H_0(\mathbf{J})}{\partial J_i} + \varepsilon \frac{\partial V(\mathbf{J},\boldsymbol{\theta};\varepsilon)}{\partial J_i} \qquad (10.34b)$$

It is worth noting that J_i may deviate from constancy only over a timescale of $O(1/\varepsilon)$. Similarly, the deviation of the angular velocities from their constant unperturbed values occurs only over long timescales, contributing to the slow part of the phases.

10.4 COUPLED HARMONIC OSCILLATORS

We now concentrate on systems of N coupled harmonic oscillators. The unperturbed Hamiltonian, H_0, is then linear in the actions

$$H_0 = \sum_{i=1}^{N} J_i \omega_i \equiv (\mathbf{J} \cdot \boldsymbol{\omega}) \qquad (10.35)$$

(For the betatron problem, $n=2$.) The unperturbed frequencies, ω_i,

$$\omega_i = \frac{\partial H_0}{\partial J_i} \qquad (10.36)$$

are constant and equal to the frequencies of the uncoupled oscillators. (Notice that, if H_0, is not linear in J_i, the unperturbed frequencies depend on $\mathbf{J}$. They are constant for a given trajectory, but may change as one move from one trajectory to another, since the J_i, vary.)

We now generate Equation (10.33) by adding a small perturbation, such that the perturbed system is canonical, leading to

$$\dot{J}_i = -\varepsilon \frac{\partial V(\mathbf{J},\boldsymbol{\theta};\varepsilon)}{\partial \theta_i} \qquad (10.37a)$$

$$\dot{\theta}_i = \omega_i + \varepsilon \frac{\partial V(\mathbf{J},\boldsymbol{\theta};\varepsilon)}{\partial J_i} \qquad (10.37b)$$

In terms of our formalism of the normal form expansion of Chapter 4, the unperturbed parts of Equations (10.37) are

$$Z_0^{I_i} = 0 \qquad\qquad Z_0^{\varphi_i} = \omega_i \qquad (10.38)$$

We assume that V has a convergent expansion in powers of ε:

$$V(\mathbf{J}, \boldsymbol{\theta}; \varepsilon) = \sum_{n \geq 1} \varepsilon^n V_n(\mathbf{J}, \boldsymbol{\theta}) \qquad (10.39)$$

The dependence of V on the angles is periodic. We also assume that it has the following convergent Fourier expansion:

$$V_n(\mathbf{J}, \boldsymbol{\theta}) = \sum_{\mathbf{m}} V_n(\mathbf{J}; \mathbf{m}) \exp(i (\mathbf{m} \cdot \boldsymbol{\theta}))$$
$$[\mathbf{m} = (m_1, m_2, \ldots, m_N)] \qquad (10.40)$$

The near-identity transformations for J_i and θ_i are written as

$$J_i = I_i + \sum_{n \geq 1} \varepsilon^n T_n^{I_i} \qquad (10.41a)$$

$$\theta_i = \varphi_i + \sum_{n \geq 1} \varepsilon^n T_n^{\varphi_i} \qquad (10.41b)$$

The T functions depend on $\mathbf{I}$ and on $\boldsymbol{\varphi}$. The normal form equations are written as

$$\dot{I}_i = \sum_{n \geq 1} \varepsilon^n U_n^{I_i} \qquad (10.42a)$$

$$\dot{\varphi}_i = \omega_i + \sum_{n \geq 1} \varepsilon^n U_n^{\varphi_i} \qquad (10.42b)$$

Equation (10.42a) implies that I_i vary slowly in time, changing appreciably only over timescales of $O(1/\varepsilon)$. Consequently, Equation (10.42b) implies

$$\varphi_i = \underbrace{\omega_i t}_{\text{fast phase}} + \underbrace{\int_0^t \sum_{n \geq 1} \varepsilon^n U_n^{\varphi_i} \, dt'}_{\text{slow phase}} + \varphi_{i,0} \qquad (10.43)$$

Using the formalism of Chapter 4 and Equation (10.39), the equations for the first-order U functions are:

COUPLED HARMONIC OSCILLATORS

$$U_1^{I_i} = -\frac{\partial V_1}{\partial \varphi_i} + [Z_0, T_1]^{I_i} = -\frac{\partial V_1}{\partial \varphi_i} - \sum_{j=1}^{N} \omega_j \frac{\partial T_1^{I_i}}{\partial \varphi_j} \quad (10.44a)$$

$$U_1^{\varphi_i} = \frac{\partial V_1}{\partial I_i} + [Z_0, T_1]^{\varphi_i} = \frac{\partial V_1}{\partial I_i} - \sum_{j=1}^{N} \omega_j \frac{\partial T_1^{\varphi_i}}{\partial \varphi_j} \quad (10.44b)$$

In Equations (10.44), the dependence of the functions on $\mathbf{I}$ and $\boldsymbol{\varphi}$ has been omitted for the sake of economy in notation.

We assume Fourier expansions for the T and U functions with a notation similar to that in Equation (10.38). Equations (10.44) imply relations among the Fourier coefficients:

$$U_1^{I_i}(\mathbf{I};\mathbf{m}) = -i\, m_i\, V_1(\mathbf{I};\mathbf{m}) - i\, (\boldsymbol{\omega}\cdot\mathbf{m}) T_1^{I_i}(\mathbf{I};\mathbf{m}) \quad (10.45a)$$

$$U_1^{\varphi_i}(\mathbf{I};\mathbf{m}) = \frac{\partial V_1(\mathbf{I};\mathbf{m})}{\partial I_i} - i\, (\boldsymbol{\omega}\cdot\mathbf{m}) T_1^{\varphi_i}(\mathbf{I};\mathbf{m}) \quad (10.45b)$$

Note that, in language of action–angle variables, resonance means

$$(\boldsymbol{\omega}\cdot\mathbf{m}) = 0 \quad (10.46)$$

Thus, unlike the case in our usual notation in previous chapters, here the components of the integer vector, $\mathbf{m}$, cannot all be positive. Some must be negative to get a zero on the r.h.s. of Equation (10.46). Technically, this is a result of the fact that we have replaced the imaginary eigenvalues $\pm i\omega_i$, among which resonance relations exit with nonnegative integers, by the frequencies that are all positive.

We also note that if all the frequencies or some subsets of them are incommensurate, then there will be an infinite number of integer vectors, $\mathbf{m}$, for which $(\boldsymbol{\omega}\cdot\mathbf{m})$ will not vanish, but will be arbitrarily small. This will lead to the small-denominator problem, which we wish to avoid. Therefore, we assume that *all the frequencies are mutually commensurate.* [In this case, it is easy to show that the unperturbed frequencies obey $(N-1)$ independent resonance relations.]

Comment. In Equations (10.45) and (10.46) $\mathbf{m}=0$ is also allowed, corresponding to the angle-independent part of the U functions. It contributes only to $U_1^{\varphi_i}$.

The normal form solutions for the Fourier coefficients of the U and T functions are

$$U_1^{I_i}(\mathbf{I};\mathbf{m}) = \begin{cases} -i\, m_i\, V_1(\mathbf{I};\mathbf{m}) & (\boldsymbol{\omega}\cdot\mathbf{m}) = 0 \\ 0 & (\boldsymbol{\omega}\cdot\mathbf{m}) \neq 0 \end{cases} \quad (10.47\text{a})$$

$$U_1^{\varphi_i}(\mathbf{I};\mathbf{m}) = \begin{cases} \dfrac{\partial V_1(\mathbf{I};\mathbf{m})}{\partial I_i} & (\boldsymbol{\omega}\cdot\mathbf{m}) = 0 \\ 0 & (\boldsymbol{\omega}\cdot\mathbf{m}) \neq 0 \end{cases} \quad (10.47\text{b})$$

$$T_1^{I_i}(\mathbf{I};\mathbf{m}) = \begin{cases} -\dfrac{m_i}{(\boldsymbol{\omega}\cdot\mathbf{m})} V_1(\mathbf{I};\mathbf{m}) & (\boldsymbol{\omega}\cdot\mathbf{m}) \neq 0 \\ F_1^{I_i}(\mathbf{I};\mathbf{m}) & (\boldsymbol{\omega}\cdot\mathbf{m}) = 0 \end{cases} \quad (10.47\text{c})$$

$$T_1^{\varphi_i}(\mathbf{I};\mathbf{m}) = \begin{cases} \dfrac{1}{i(\boldsymbol{\omega}\cdot\mathbf{m})}\dfrac{\partial V_1(\mathbf{I};\mathbf{m})}{\partial I_i} & (\boldsymbol{\omega}\cdot\mathbf{m}) \neq 0 \\ F_1^{\varphi_i}(\mathbf{I};\mathbf{m}) & (\boldsymbol{\omega}\cdot\mathbf{m}) = 0 \end{cases} \quad (10.47\text{d})$$

In Equations (10.47c,d) the Fs are free terms. Again, $\mathbf{m}=0$ is allowed. The formalism can be now carried out to any desired order. Resonance in higher orders is also defined by Equation (10.46).

10.4.1 Structure of the normal form equations

We note that the Fourier components of the U functions contain contributions only from integer vectors $\mathbf{m}$ that obey the resonance relation, Equation (10.46). For instance

$$U_1^{I_i}(\mathbf{I},\boldsymbol{\varphi}) = \sum_{\substack{\mathbf{m} \\ (\boldsymbol{\omega}\cdot\mathbf{m})=0}} -i\, m_i\, V_1(\mathbf{I};\mathbf{m}) \exp\bigl(i\,(\mathbf{m}\cdot\boldsymbol{\varphi})\bigr) \quad (10.48)$$

with a similar expression for the U_1^φ. The fast component in the phase $(\mathbf{m}\cdot\boldsymbol{\varphi})$ is eliminated because of the resonance relation. Hence, $(\mathbf{m}\cdot\boldsymbol{\varphi})$ is a slow relative phase. Moreover, as the N frequencies, ω_i obey $(N-1)$ independent resonance relations, there are also $(N-1)$ independent slow relative phases. There is some arbitrariness in their definition. However, once they are selected, any slow phase can be written as a linear combination of the independent ones with integer coefficients.

Denoting a basis set of integer vectors that defines the $(N-1)$ basic resonance relations by $\mathbf{m}_k$, the independent slow relative phases are

COUPLED HARMONIC OSCILLATORS

$$\Omega_k = (\mathbf{m}_k \cdot \boldsymbol{\varphi}) \qquad (\mathbf{m}_k \cdot \boldsymbol{\omega}) = 0$$
$$k = 1,\ldots,(N-1) \qquad (10.49)$$

A general resonant vector $\mathbf{m}$ can be written in terms of the basis of independent $(N-1)$ resonance vectors as

$$\mathbf{m} = k_1 \mathbf{m}_1 + k_2 \mathbf{m}_2 + \cdots k_{(N-1)} \mathbf{m}_{(N-1)} \qquad (10.50)$$

Example 10.1. Consider three mutually commensurate frequencies:

$$\omega_1 = 1 \quad \omega_2 = 2 \quad \omega_3 = 3$$

The two $(N-1=3-1)$ fundamental resonance relations may be chosen as

$$\mathbf{m}_1 = (2,-1,0): \quad 2\omega_1 - \omega_2 = 0$$
$$\mathbf{m}_2 = (3,0,-1): \quad 3\omega_1 - \omega_3 = 0$$

corresponding to two fundamental slow relative phases:

$$\Omega_1 = 2\varphi_1 - \varphi_2 \quad \Omega_2 = 3\varphi_1 - \varphi_3$$

Other resonance combinations and the corresponding slow relative phases are expressible in terms of the fundamental ones, e.g.,

$$\omega_1 + \omega_2 - \omega_3 = -1 \cdot (2\omega_1 - \omega_2) + 1 \cdot (3\omega_1 - \omega_3) = 0$$
$$\varphi_1 + \varphi_2 - \varphi_3 = -1 \cdot \Omega_1 + 1 \cdot \Omega_2 \quad \mathbf{k} = (-1,1)$$

$$13\omega_1 - 5\omega_2 - \omega_3 = 5 \cdot (2\omega_1 - \omega_2) + 1 \cdot (3\omega_1 - \omega_3) = 0$$
$$13\varphi_1 - 5\varphi_2 - \varphi_3 = 5 \cdot \Omega_1 + 1 \cdot \Omega_2 \quad \mathbf{k} = (5,1)$$

The U functions depend *only* on the slow relative phases:

$$U_1^{I_i}(\mathbf{I},\boldsymbol{\Omega}) = \sum_{\substack{\mathbf{k} \\ (\mathbf{m}\cdot\boldsymbol{\omega})=0 \\ \mathbf{m}\neq 0}} -i\, m_i\, V_1(\mathbf{I};\mathbf{m}) \exp(i\,(\mathbf{k}\cdot\boldsymbol{\Omega})) \qquad (10.51a)$$

$$U_1^{\varphi_i}(\mathbf{I},\boldsymbol{\Omega}) = \frac{\partial V_1(\mathbf{I};\mathbf{m}=0)}{\partial I_i} + \sum_{\substack{\mathbf{k} \\ (\mathbf{m}\cdot\boldsymbol{\omega})=0 \\ \mathbf{m}\neq 0}} \frac{\partial V_1(\mathbf{I};\mathbf{m})}{\partial I_i} \exp(i\,(\mathbf{k}\cdot\boldsymbol{\Omega})) \qquad (10.51b)$$

where the sums run over all $(N-1)$-dimensional integer vectors, $\mathbf{k}$. The vector $\mathbf{m}$ and its components are expressed in terms of Equation (10.50). $\mathbf{\Omega}$ is the vector of $(N-1)$, independent, relative slow phases that one has chosen as a basis. In Equation (10.51b) the $\mathbf{m}=0$ contribution has been written explicitly. The free functions (appearing in the T functions) are also constructed from resonant terms only and have a similar structure as displayed in Equations (10.51).

Equations (10.51) imply that the normal form does not depend explicitly on the vector of the full individual phases, $\boldsymbol{\varphi}$, but only on the $(N-1)$ basic relative slow phases, $\mathbf{\Omega}$. As a result, from Equation (10.42b) for the full phases, one can generate $(N-1)$ equations for the independent slow relative phases:

$$\dot{\Omega}_l = \sum_{n \geq 1} \varepsilon^n \left(\sum_{i=1}^N m_{l,i} U_n^{\varphi_i} \right) \equiv \sum_{n \geq 1} \varepsilon^n U_n^{\Omega_l} \qquad (10.52)$$

The $2N$ equations for $(\mathbf{I},\boldsymbol{\varphi})$ are reduced to $(2N-1)$ equations for $(\mathbf{I},\mathbf{\Omega})$. Once these are solved, the individual phases $\boldsymbol{\varphi}$ can be found by direct quadrature.

10.4.2 "Constant" of motion

In first order, the system has a "constant of the motion," the quantity

$$\hat{I} \equiv \sum_{i=1}^N \omega_i I^i \qquad (10.53)$$

From the first-order result, Equation (10.48), we find

$$\dot{\hat{I}} = \varepsilon \sum_{i=1}^N \omega_i U_n^{I_i}(\mathbf{I},\boldsymbol{\varphi})$$

$$= \varepsilon \sum_{\substack{\mathbf{m} \\ (\omega \cdot \mathbf{m})=0}} -i\,(\mathbf{m}\cdot\boldsymbol{\omega}) V_1(\mathbf{I};\mathbf{m}) \exp\bigl(i\,(\mathbf{m}\cdot\boldsymbol{\varphi})\bigr) + O(\varepsilon^2) \quad (10.54)$$

$$= \qquad\qquad 0 \qquad\qquad + O(\varepsilon^2)$$

Hence, in first order, $\hat{I}$ is a constant of the motion.

For the case of betatron oscillations, $N=2$, where $\omega_1=1$ and $\omega_2=\frac{1}{2}$, we have $(N-1=1)$ and find

COUPLED HARMONIC OSCILLATORS

$$\hat{I} = 1 \cdot I_1 + \left(\tfrac{1}{2}\right) I_2$$

Denoting the first-order approximation for the amplitudes of ξ and η in Equations (10.30a,b) by ρ and r, respectively, we find

$$\hat{I} = 1 \cdot \tfrac{1}{2}\rho^2 + \tfrac{1}{2} \cdot \tfrac{1}{4} r^2 = \tfrac{1}{2}\left(\rho^2 + \tfrac{1}{4} r^2\right)$$

Apart from the overall factor of $\tfrac{1}{2}$, this is the "constant of motion" identified in the usual analysis of the betatron problem.

Clearly, the constancy is limited by the fact that higher-order terms may modify it. Formally, the constancy can be extended to higher orders, and hence to longer timescales, by an appropriate choice of the free functions. This ceases to be a mere formal statement, but generates a genuine constant of motion, *only* if the perturbation series converges.

10.4.3 Canonical transformation choice

Judging by the betatron example, there may be many choices of the free functions that will make $\hat{I}$ a formal constant of motion. Of the variety of choices, we mention here a particular one: the choice that makes the transformation from $(\mathbf{J},\boldsymbol{\theta})$ to $(\mathbf{I},\boldsymbol{\varphi})$ canonical. We claim that then $\hat{I}$ is a formal constant of motion.

For the transformation to be canonical, after substituting Equations (10.41) in Equation (10.33) [with Equation (10.35) for H_0], we must have Hamilton's equations for $\mathbf{I}$ and $\boldsymbol{\varphi}$:

$$\dot{I}_i = -\frac{\partial H}{\partial \varphi_i} \qquad (10.55a)$$

$$\dot{\varphi}_i = \frac{\partial H}{\partial I_i} \qquad (10.55b)$$

In Equations (10.55), H is now a function of $\mathbf{I}$ and $\boldsymbol{\varphi}$, of the form

$$H = H_0(\mathbf{I}) + \sum_{n \geq 1} \varepsilon^n H_n(\mathbf{I},\boldsymbol{\varphi}) \qquad (10.56)$$

Here H_0 is the same as in Equation (10.35), computed at $\mathbf{I}$ rather than at $\mathbf{J}$.

Consider Equation (10.55a). When each H_n is written in terms of its Fourier expansion, we obtain

$$\dot{I}_i = -\sum_{n\geq 1}\varepsilon^n \sum_{\mathbf{m}} i\, m_i\, H_n(\mathbf{I};\mathbf{m})\exp\!\left(i\,(\mathbf{m}\cdot\boldsymbol{\varphi})\right) \qquad (10.57)$$

At the same time, $\mathbf{I}$ and $\boldsymbol{\varphi}$ must obey the normal form equations. The latter contain only resonant terms. Thus, H_n can include only resonant terms. This means that the integer vectors $\mathbf{m}$ over which the summation is carried out in Equation (10.57) must obey $(\mathbf{m}\cdot\boldsymbol{\omega})=0$. In particular, the $\boldsymbol{\varphi}$ dependence of H_n can include only the slow relative phases. Therefore, the definition of $\hat{I}$, Equation (10.53), guarantees that the contribution to $(d\hat{I}/dt)$ vanishes in every order of the expansion. This makes $\hat{I}$ a formal constant of motion. It is a true constant of motion *only* if the series in Equation (10.57) converges.

Mutually incommensurate frequencies: more formal constants of motion. As already noted in Section 10.4.1, when some of the frequencies are incommensurate, the small-denominator problem exists. We address this case briefly, since the issue of formal constants of the motion is of interest. In reality, they turn out to be slowly varying quantities and may be useful in obtaining approximations for the solution. In this more general case, the number of formal constants of motion is greater than one.

Assume that the N frequencies are divided into k mutually incommensurate subsets. That is, all the frequencies within each subset are mutually commensurate, hence they obey resonance relations, but frequencies in different sets are incommensurate. Denoting the size of each sunset by n_r ($r=1,2,...,k$), we have

$$\sum_{r=1}^{k} n_r = N$$

We observe that the unperturbed frequencies in each subset will obey (n_r-1) independent resonance relations, and will hence have (n_r-1) slow relative phases. Since the frequencies of different subsets do not resonate, the normal form equations for each subset will depend *only* on the slow relative phases of that subset. Repetition of the analysis for the case in which *all* eigenvalues resonate (the "number of subsets" k is 1), presented earlier, shows that each subset now has its own formal constant of motion:

$$\hat{I}_r \equiv \sum_{j_r=1}^{n_r} \omega_{j_r} I_{j_r} \qquad r = 1,\ldots,k$$

Thus, the number of formal constants of motion equals the number of subsets, k. The number of independent slow relative phases is $(2N-k)$. Again, the near-identity transformation has to be canonical.

The results presented here had been originally obtained by Gustavson [4], using Birkhoff normal forms. Gustavson showed that when the overall energy of the system is low, treating the formal constants as actual constants yields a good approximation for the solution. The simplicity of the derivation outlined here is an indication of the power of action–angle variables.

10.4.4 Constraint on free functions

For the transformation $(\mathbf{J},\boldsymbol{\theta}) \Rightarrow (\mathbf{I},\boldsymbol{\varphi})$ to be canonical, the free functions must obey constraints. We develop the constraint on the free functions in first order. Substitution of Equations (10.35) and (10.41) in Equation (10.33) yields, through $O(\varepsilon)$:

$$H = \sum_{i=1}^{N} \omega_i I_i + \varepsilon \left\{ \sum_{i=1}^{N} \omega_i T_1^{I_i}(\mathbf{I},\boldsymbol{\varphi}) + V_1(\mathbf{I},\boldsymbol{\varphi}) \right\} \qquad (10.58)$$

Requiring that the transformation is canonical, we have to implement Equations (10.55). In parallel, we have to obey the normal form Equations (10.42). Comparing the two sets of equations, we find

$$\dot{I}_i = -\varepsilon \left\{ \sum_{k=1}^{N} \omega_j \frac{\partial T_1^{I_j}(\mathbf{I},\boldsymbol{\varphi})}{\partial \varphi_i} + \frac{\partial V_1(\mathbf{I},\boldsymbol{\varphi})}{\partial \varphi_i} \right\}$$
$$= -\varepsilon \left\{ \sum_{j=1}^{N} \omega_j \frac{\partial T_1^{I_i}(\mathbf{I},\boldsymbol{\varphi})}{\partial \varphi_j} + \frac{\partial V_1(\mathbf{I},\boldsymbol{\varphi})}{\partial \varphi_i} \right\} \qquad (10.59a)$$

$$\dot{\varphi}_i = \omega_i + \varepsilon \left\{ +\sum_{k=1}^{N} \omega_j \frac{\partial T_1^{I_j}(\mathbf{I},\boldsymbol{\varphi})}{\partial I_i} + \frac{\partial V_1(\mathbf{I},\boldsymbol{\varphi})}{\partial I_i} \right\}$$
$$= \omega_i + \varepsilon \left\{ -\sum_{j=1}^{N} \omega_j \frac{\partial T_1^{\varphi_i}(\mathbf{I},\boldsymbol{\varphi})}{\partial \varphi_j} + \frac{\partial V_1(\mathbf{I},\boldsymbol{\varphi})}{\partial I_i} \right\} \qquad (10.59b)$$

Using Equations (10.47) in Equations (10.59a,b), relations are derived among the Fourier coefficients of the free functions:

$$\sum_{k=1}^{N} \omega_j F_1^{I_j}(\mathbf{I};\mathbf{m}) = 0 \quad (\boldsymbol{\omega}\cdot\mathbf{m}) = 0 \quad (10.60)$$

No constraints are obtained for $F_1^{\varphi_j}(\mathbf{I};\mathbf{m})$ in this order. Equation (10.60) is solved, for instance, by

$$F_1^{I_j}(\mathbf{I};\mathbf{m}) = m_j F_1'(\mathbf{I};\mathbf{m})$$

with $F_1'(\mathbf{I};\mathbf{m})$ arbitrary.

Case of two degrees of freedom. In that case, we have

$$\omega_1 F_1^{I_1} + \omega_2 F_1^{I_2} = 0 \quad (10.60a)$$

Thus, already in first order, there is ample freedom in the choice of the free functions that obey the condition for a canonical transformation. This freedom has been studied extensively in the literature [5–8].

10.4.5 Minimal normal forms?

In the case of a single harmonic oscillator that is subjected to a small nonlinear perturbation, we have shown in Chapters 6 and 7 that, for certain types of perturbations, one can choose the free functions so that the normal form is rigorously terminated (i.e., this is not an ad hoc truncation) at a finite order in the expansion, without introducing singularities in the T functions. We called this the *minimal normal form* (MNF) choice. In Chapter 6, for conservative systems, we showed that, with a prescribed order of precision in the approximation, MNFs generate smaller errors in the next order, so that the evolution of the secular terms is much slower that in other choices of the free functions. These qualitative statements were demonstrated numerically in the case of the Duffing oscillator, where we found that MNFs gave a better approximation for the solution over longer times than the approximations generated by other common choices of the free functions.

The first proof of the possibility to always generate a MNF by exploiting the freedom in the expansion was presented within the framework of the method of averaging [9]. There it was shown for a general, n–dimensional system. We have found in Chapters 6 and 7 that, already for $n=2$ (i.e., one degree of freedom), the required free functions may be singular and that, only for a specific class of perturbations are they simple polynomials or entire functions.

Extending the notion of "well behaved" MNF (i.e., with free functions that are devoid of singularities) to more than one degree of freedom is still an open problem. We have studied the limited case of two harmonic oscillators, with frequencies both equal to unity, that are coupled by the most general quartic perturbation potential in the Hamiltonian. In complex notation ($z = x + i\,\dot{x}$), H can be written as

$$H = \tfrac{1}{2}\left(z_1 z_1^* + z_2 z_2^*\right) + \tfrac{1}{2}\varepsilon \sum_{k+l+m+n=4} h_{klmn}\, z_1^{k} z_1^{*l} z_2^{m} z_2^{*n}$$

$$\left(h_{klmn} = \left(h_{lknm}\right)^*\right) \tag{10.61}$$

The constraint in parentheses guarantees that H is real. Using Equation (10.29) with $\omega=1$, H can be rewritten in terms of action–angle variables as

$$H = J_1 + J_2$$
$$+ \varepsilon \sum_{k+l+m+n=4} g_{klmn}\, (J_1)^{(k+l)/2} (J_2)^{(n+m)/2} \exp\!\left(i\left((l-k)\theta_1 - (n-m)\theta_2\right)\right)$$

$$\left(g_{klmn} = \left(g_{lknm}\right)^*\right) \tag{10.62}$$

In this conservative system, Hamilton's equations yield the most general cubic coupling. Equality of the frequencies means that there is resonance in the system. Replacing them by unequal but mutually rational frequencies [ω and $(r/s)\omega$, with r and s integers], leads to the same conclusions and only complicates the computation.

We have limited our study even further, by addressing only the second order in the normal form. Let us write the expansion as

$$J_p = I_{p1} + \varepsilon T_1^{I_p} + \cdots \qquad \dot{I}_p = \varepsilon U_1^{I_p} + \varepsilon^2 U_2^{I_p} + \cdots$$

$$\theta_p = \varphi_p + \varepsilon T_1^{\varphi_p} + \cdots \qquad \dot{\varphi}_p = 1 + \varepsilon U_1^{\varphi_p} + \varepsilon^2 U_2^{\varphi_p} + \cdots \tag{10.63}$$

$$p = 1, 2$$

An elaborate analysis, the details of which are omitted here, shows that, in general, one cannot eliminate all U_2 functions with simple polynomials of trigonometric functions of the angles. However, for a wide class of interactions, it is easy to achieve:

$$U_2^{I_1} + U_2^{I_2} = 0 \tag{10.64}$$

Equation (10.64) is nothing but the $O(\varepsilon^2)$ reiteration of the conclusions of Sections 10.4.2 and 10.4.3 that there are choices of free functions that make the quantity

$$\omega_1 I_1 + \omega_2 I_2 = I_1 + I_2$$

a formal constant of the motion.

More interesting is the observation that a choice of nonsingular free functions exists for which

$$U_2^{\Omega} \equiv U_2^{\varphi_1} - U_2^{\varphi_2} = 0 \tag{10.65}$$

In general, it is not possible to find nonsingular free functions that yield both Equations (10.64) and (10.65).

Equation (10.65) is new. It seems to be the only feasible generalization of the MNF idea in systems with more than one degree of freedom. It means that there is a choice of simple free functions that eliminate the $O(\varepsilon^2)$ term in the equation obeyed by the slow phase:

$$\dot{\Omega} = \varepsilon \left(U_1^{\varphi_1} - U_1^{\varphi_2} \right) + O(\varepsilon^3) \tag{10.66}$$

10.5 EXAMPLE OF TWO COUPLED OSCILLATORS

We provide the results of a detailed perturbation analysis for a system of two coupled harmonic oscillators with cubic coupling terms:

$$\dot{z}_1 = -i\, z_1 - i \left(\tfrac{1}{8}\right) \varepsilon \left(z_1 + z_1^*\right)^2 \left(z_2 + z_2^*\right) \tag{10.67a}$$

$$\dot{z}_2 = -i\, z_2 - i \left(\tfrac{1}{24}\right) \varepsilon \left(z_1 + z_1^*\right)^3 \tag{10.67b}$$

The Hamiltonian of this system is

$$H = \tfrac{1}{2} z_1 z_1^* + \tfrac{1}{2} z_2 z_2^* + \tfrac{1}{48} \varepsilon \left(z_1 + z_1^*\right)^3 \left(z_2 + z_2^*\right)$$

Employing the action–angle variables [Equation (10.29)] for both z_1 and z_2, the Hamiltonian becomes

$$H = J_1 + J_2 + \tfrac{4}{3} \varepsilon J_1^{3/2} J_2^{1/2} \cos^3 \theta_1 \cos \theta_2$$

EXAMPLE OF TWO COUPLED OSCILLATORS

Using Hamilton's equations [Equations (10.34) or (10.37)], the equations of motion are found to be

$$\dot{J}_1 = 4\varepsilon J_1^{3/2} J_2^{1/2} \cos^2\theta_1 \sin\theta_1 \cos\theta_2$$

$$\dot{J}_2 = \tfrac{4}{3}\varepsilon J_1^{3/2} J_2^{1/2} \cos^3\theta_1 \sin\theta_2$$

$$\dot{\theta}_1 = 1 + 2\varepsilon J_1^{1/2} J_2^{1/2} \cos^3\theta_1 \cos\theta_2 \qquad (10.68)$$

$$\dot{\theta}_2 = 1 + \tfrac{2}{3}\varepsilon J_1^{3/2} J_2^{-1/2} \cos^3\theta_1 \cos\theta_2$$

10.5.1 Perturbation analysis

Equations (10.63) (the near-identity transformation and the normal form) and the formalism, described in Section 10.4.1, yield

$$U_1^{I_1} = \tfrac{1}{2} I_1^{3/2} I_2^{1/2} \sin\Omega$$

$$U_1^{I_2} = -\tfrac{1}{2} I_1^{3/2} I_2^{1/2} \sin\Omega$$

$$U_1^{\varphi_1} = \tfrac{3}{4} I_1^{1/2} I_2^{1/2} \cos\Omega \qquad (10.69)$$

$$U_1^{\varphi_2} = \tfrac{1}{4} I_1^{3/2} I_2^{-1/2} \cos\Omega$$

$$T_1^{I_1} = I_1^{3/2} I_2^{1/2} \begin{Bmatrix} -\tfrac{1}{4}\cos(\varphi_1+\varphi_2) - \tfrac{1}{4}\cos(3\varphi_1-\varphi_2) \\ -\tfrac{1}{8}\cos(3\varphi_1+\varphi_2) \end{Bmatrix} + F_1^{I_1}(I_1, I_2, \Omega)$$

$$T_1^{I_2} = I_1^{3/2} I_2^{1/2} \begin{Bmatrix} -\tfrac{1}{4}\cos(\varphi_1+\varphi_2) + \tfrac{1}{12}\cos(3\varphi_1-\varphi_2) \\ -\tfrac{1}{24}\cos(3\varphi_1+\varphi_2) \end{Bmatrix} + F_1^{I_2}(I_1, I_2, \Omega)$$

$$(10.70)$$

$$T_1^{\varphi_1} = I_1^{1/2} I_2^{1/2} \begin{Bmatrix} \tfrac{3}{8}\sin(\varphi_1+\varphi_2) + \tfrac{1}{8}\sin(3\varphi_1-\varphi_2) \\ +\tfrac{1}{16}\sin(3\varphi_1+\varphi_2) \end{Bmatrix} + F_1^{\varphi_1}(I_1, I_2, \Omega)$$

$$T_1^{\varphi_2} = I_1^{3/2} I_2^{-1/2} \begin{Bmatrix} \tfrac{1}{8}\sin(\varphi_1+\varphi_2) + \tfrac{1}{24}\sin(3\varphi_1-\varphi_2) \\ +\tfrac{1}{48}\sin(3\varphi_1+\varphi_2) \end{Bmatrix} + F_1^{\varphi_2}(I_1, I_2, \Omega)$$

$$U_2^{I_1} = -\tfrac{1}{4}I_1^2 I_2 \sin 2\Omega + \text{f.f.c.}$$

$$U_2^{I_2} = +\tfrac{1}{4}I_1^2 I_2 \sin 2\Omega + \text{f.f.c.}$$

$$U_2^{\varphi_1} = -\tfrac{17}{192}I_1^2 - \tfrac{9}{32}I_1 I_2 - \tfrac{1}{4}I_1 I_2 \cos 2\Omega + \text{f.f.c.}$$

$$U_2^{\varphi_2} = -\tfrac{9}{64}I_1^2 - \tfrac{1}{8}I_1^2 \cos 2\Omega + \text{f.f.c.}$$

(10.71)

In Equations (10.70) the Fs are the free functions. In Equations (10.71), "f.f.c." stands for "free function contributions."

From Equations (10.69) we note that in first order, we have

$$U_1^{I_1} + U_1^{I_2} = 0$$

Thus, I_1+I_2 is, automatically, a formal constant of motion in $O(\varepsilon)$. We also note from Equations (10.71) that if the free functions vanish, then

$$U_2^{I_1} + U_2^{I_2} = 0$$

Thus, without free functions, I_1+I_2 is also a formal constant of motion through $O(\varepsilon^2)$. We also know that if the free functions $F_1^{I_1}$ and $F_1^{I_2}$ do not vanish, but obey the first order condition for a canonical near-identity transformation [see Equation (10.60a) in Section 10.4.4]:

$$F_1^{I_1} + F_1^{I_2} = 0 \qquad (10.72)$$

However, the example of Equations (10.67) was chosen because it is a case for which it is impossible to impose Equations (10.64) and (10.65) simultaneously and obtain nonsingular free functions. If we impose Equation (10.65) alone, and require that the free functions are nonsingular, we find the following:

$$F_1^{\varphi_2} = F_1^{\varphi_1} + \left\{ \tfrac{1}{32} I_1^{1/2} I_2^{1/2} + \tfrac{17}{288} I_1^{3/2} I_2^{-1/2} \right\} \sin\Omega \qquad (10.73)$$

with $F_1^{\varphi_1}$ arbitrary, and

$$F_1^{I_1} = \left\{ \begin{array}{l} a I_1^{5/2} I_2^{-1/2} + b I_1^{3/2} I_2^{1/2} \\ + \left(-\tfrac{41}{16} - 9a + 3b \right) I_1^{1/2} I_2^{3/2} \end{array} \right\} \cos\Omega \qquad (10.74a)$$

EXAMPLE OF TWO COUPLED OSCILLATORS

$$F_1^{I_2} = \begin{cases} \left(\tfrac{17}{144} + 3a\right) I_1^{3/2} I_2^{1/2} \\ + \left(-\tfrac{33}{16} - 12a + 3b\right) I_1^{1/2} I_2^{3/2} \\ + \left(\tfrac{41}{16} + 9a - 3b\right) I_1^{5/2} I_2^{-1/2} \end{cases} \cos\Omega \quad (10.74b)$$

(Note that the monomial $I_1^{3/2} I_2^{-1/2}$ appears naturally in $T_1^{\varphi_2}$ and is, therefore, accptable also in $F_1^{\varphi_2}$.) Equations (10.74) do not yield Equation (10.72). Thus, the first-order condition for the near-identity transformation to be canonical is not satisfied. Consequently, with this choice of the free functions, I_1+I_2 is not a constant of the motion in $O(\varepsilon^2)$.

If we insist on imposing Equation (10.72), so that Equations (10.64) and (10.65) are both obeyed, Equation (10.73) for $F_1^{\varphi_1}$ and $F_1^{\varphi_2}$ is obtained again. However, for $F_1^{I_1}$ and $F_1^{I_2}$ we find

$$F_1^{I_2} = -F_1^{I_1} = \tfrac{1}{144} I_1^{3/2} I_2^{1/2} \frac{\left(-17 I_1^2 + 246 I_1 I_2 - 585 I_2^2\right)}{\left(I_1^2 + 6 I_1 I_2 - 3 I_2^2\right)}$$

The denominator vanishes along the line

$$I_1 - \left(2\sqrt{3} - 3\right) I_2 = 0$$

Thus, the free functions have a singularity along this line.

We now return to the nonsingular solution of Equations (10.73) and (10.74), for which the near-identity transformation is *not* canonical. As the equations indicate, there is still some freedom left. To simplify the free functions, we make the following choice for $F_1^{\varphi_1}$

$$F_1^{\varphi_1} = -\tfrac{1}{32} I_1^{1/2} I_2^{1/2} \sin\Omega \quad (10.75a)$$

leading, through Equation (10.73), to

$$F_1^{\varphi_2} = \tfrac{17}{288} I_1^{3/2} I_2^{-1/2} \sin\Omega \quad (10.75b)$$

In addition, in Equation (10.74), we choose $a=0$ and $b=\tfrac{41}{48}$, yielding

$$F_1^{I_1} = \tfrac{41}{48} I_1^{3/2} I_2^{1/2} \cos\Omega \quad (10.76a)$$

$$F_1^{I_2} = \left\{\tfrac{17}{144} I_1^{3/2} I_2^{1/2} + \tfrac{1}{2} I_1^{1/2} I_2^{3/2}\right\} \cos\Omega \quad (10.76b)$$

With this choice of the free functions, the second-order U functions of Equations (10.71) become rather simple:

$$U_2^{I_1} = \tfrac{1}{8} I_1^2 I_2 \sin 2\Omega \qquad (10.77a)$$

$$U_2^{I_2} = \tfrac{1}{8} I_1 I_2^2 \sin 2\Omega \qquad (10.77b)$$

$$U_2^{\varphi_1} = U_2^{\varphi_2} = -\tfrac{5}{96} I_1^2 \qquad (10.77c)$$

Note that Equation (10.77c) automatically yields Equation (10.65).

10.5.2 Numerical results

Equations (10.67) were solved numerically for

$$\begin{array}{ll} J_1(0) = J_2(0) = 1 & \theta_1(0) = \theta_2(0) = 0 \\ \left(\Rightarrow x_1(0) = x_2(0) = \sqrt{2}\right. & \left.\dot{x}_1(0) = \dot{x}_2(0) = 0\right) \end{array} \qquad (10.78)$$

$$\varepsilon = 0.2$$

Figure 10.1 presents the time dependence of J_1 and J_2. Neither are constant. Resonance between the unperturbed frequencies generates an efficient energy transfer between the two modes, with oscillations over a timescale $T \cong 34$ (close to $2\pi/\varepsilon$).

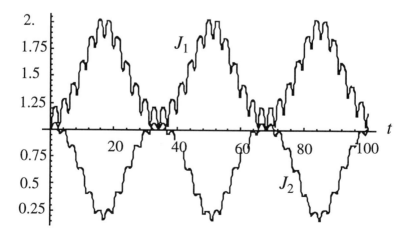

Figure 10.1 Time dependence of J_1 and J_2.

EXAMPLE OF TWO COUPLED OSCILLATORS

The fact that (I_1+I_2) can be made into a formal constant of the motion raises the question of the extent to which (J_1+J_2) deviates from constancy. This is examined in Figure 10.2. We see that (J_1+J_2) varies with time in an oscillatory manner. However, the amplitude of the oscillations is only about 5% of the average value of (J_1+J_2).

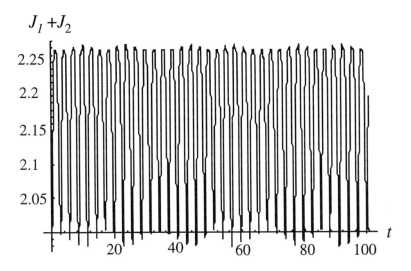

Figure 10.2 Time dependence of $(J_1 + J_2)$.

The two phases, θ_1 and θ_2, are approximately equal to t, with small $[O(\varepsilon)]$ corrections due to the interaction. In Figure 10.3 we show the slow parts, θ_1-t and θ_2-t. Over a timespan of 100, the fast part of each phase varies from 0 to 100. Hence, the slow parts constitute 10% corrections. Eeach frequency is updated roughly by a 10% correction.

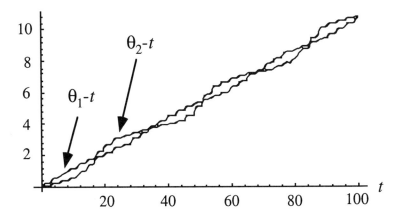

Figure 10.3 Slow parts of the phases θ_1 and θ_2.

The slow parts of the two phases oscillate around each other with a period, $T \cong 34$. This is best seen in the slow relative phase: $\theta_1 - \theta_2$, shown in Fig. 10.4. Again, $\theta_1 - \theta_2$ oscillates with a period, $T \cong 34$. (We study $\theta_1 - \theta_2$ because it is close to $\Omega = \varphi_1 - \varphi_2$, the slow relative phase that plays an important role in the perturbative analysis, with φ_i ($i=1,2$) being the zero-order approximations for θ_i, respectively.)

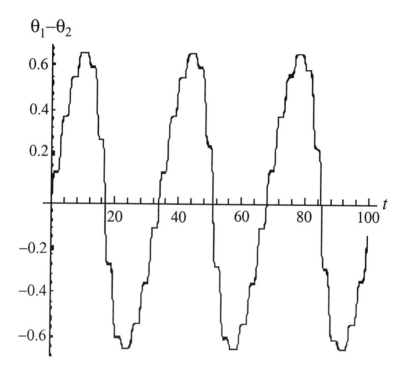

Figure 10.4 $\theta_1 - \theta_2$ versus time.

We now turn to the perturbative analysis of Equations (10.67). The expansion for J_i and θ_i was carried through first order. The normal form equations were solved through second order. The results are shown for two choices of the free functions:

1. No free functions (denoted in the figures by "*nff*"), yielding the U_2 functions of Equations (10.71) *without* the free function contributions.

2. Minimal normal form choice of Equations (10.76) (denoted by "*mnf*") yielding the U_2 functions of Equations (10.77).

A first taste of the superiority of the *mnf* choice over the *nff* one is offered in Fig. 10.5, where the full relative slow phase, $(\theta_1-\theta_2)$, is plotted together with Ω^{nff} and Ω^{mnf}. Except for oscillations caused by higher-order corrections, Ω^{mnf} reproduces $(\theta_1-\theta_2)$ rather faithfully, while Ω^{nff} gradually drifts away from $(\theta_1-\theta_2)$.

To stress this observation further, we show a succession of figures in which the evolution of secular behavior is demonstrated. Each figure depicts the difference between a quantity that is computed from the solution of the full Equations (10.67) and its approximation through first order in either the "*nff*" or the "*mnf*" choice for the free functions. The secular drift in all cases in the "*nff*" approximation is greater than in the "*mnf*" case by a factor of about 5. The absolute errors are, typically, 10–20% in the "*nff*" choice and only a few percent in the "*mnf*" approximation.

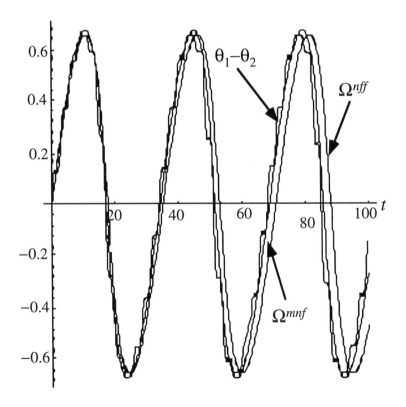

Figure 10.5 Slow relative phases.

In Figures 10.6a,b we show the errors in J_i ($i=1,2$).

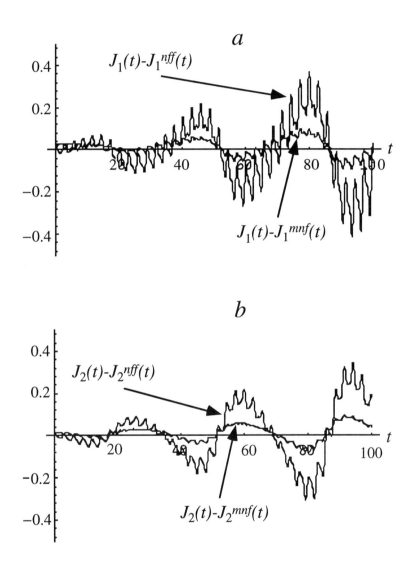

Figure 10.6 Evolution of secular terms in *nnf* and *mnf* approximations for J_1 (a) and J_2 (b).

In Figures 10.7a,b we show the errors in J_1+J_2. We note that the secular drift in J_1+J_2 is smaller than the individual drifts in J_1 (Figure 106a) and J_2 (Figure 10.6b). This is probably related to the fact that J_1+J_2 is "almost" a constant of the motion.

EXAMPLE OF TWO COUPLED OSCILLATORS

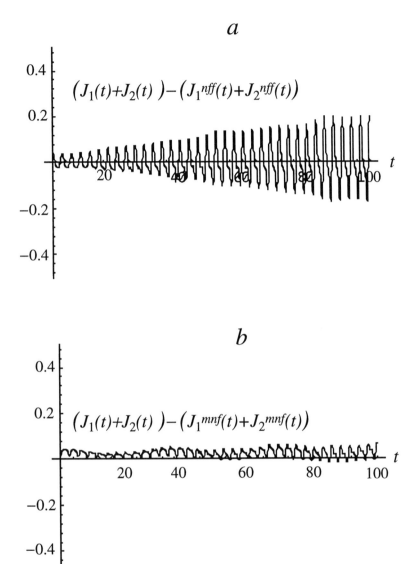

Figure 10.7 Evolution of secular term in *nff* (a) and *mnf* (b) approximations for J_1+J_2

The secular drifts in the approximations to the phases, θ_1 and θ_2, are given in Figures 10.8a,b. At long times the *"nff"* approximation yields a drift in the phases that can be of the order of $\pi/4$ to $\pi/3$. The drift in the *"mnf"* approximation reaches only about $\pi/30$ to $\pi/20$.

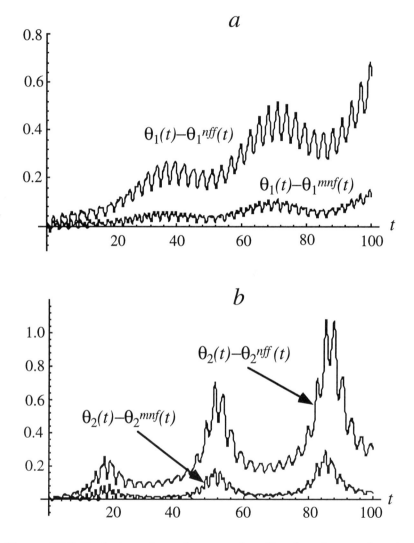

Figure 10.8 Evolution of secular terms in *nff* and *mnf* approximations for θ_1 (a) and θ_2 (b).

The tendency is similar to that found in the case of a single oscillator, analyzed in Chapter 6. MNFs yield an approximation to the full solution that is better and valid for longer times than that provided by the *no free functions* choice. In the case of a single degree of freedom we had rigorous arguments that explained the results. Here we only have the numerical results. In the case of systems with more than one oscillator, the extension of the MNF idea in general, and in particular, beyond second order, is still an open problem.

10.6 COMMENTS REGARDING SYMPLECTIC MAPS

There is a direct correspondence between normal form transformations for differential Equations (flows) and discrete maps. This is clear from the textbooks [10–14] as well as from the general literature [5,7,8,15]. We concentrate here on symplectic maps. These are the analogs of continuous Hamiltonian systems – they are energy-conserving. We study here a discrete symplectic map in which the nonlinearity is multiplied by a small parameter:

$$x_{n+1} = \cos(\mu) x_n + \sin(\mu) p_n$$
$$p_{n+1} = -\sin(\mu) x_n + \cos(\mu) p_n + \varepsilon x_{n+1}^s \quad (10.79)$$
$$|\varepsilon| \ll 1$$

where s is an integer. [Although Equation (10.79) is a two-dimensional map, the discussion is included in this chapter because many of the phenomena that characterize higher-dimensional continuous flows, e.g., bifurcations and chaotic behavior, occur in discrete maps even in one dimension.] This map is often employed in accelerator physics, as an oversimplified model for the description of harmonic oscillations of particles experiencing a magnetic octopole kick. The map is studied in Ref. 16 for the cubic case ($s=3$).

Following the formalism given in the literature [10–12], we apply the method of normal forms to Equation (10.79). The normal form is the map obeyed by the zero-order approximation to the full solution. The analysis proceeds in parallel to that developed for differential equations [17]. One begins by transforming Equation (10.79) to a diagonalized form by defining

$$z_n = x_n + i \, p_n$$

yielding

$$z_{n+1} = \lambda z_n + \tfrac{1}{8} i \, \varepsilon \left(z_{n+1} + z_{n+1}^* \right)^3$$
$$(\lambda = \exp(-i \, \mu)) \quad (10.80)$$

One now relates z_n to its zero-order approximation, u_n, by a near-identity transformation:

$$z_n = u_n + \sum_{k \geq 1} \varepsilon^k \, T_k(u_n, u_n^*) \quad (10.81)$$

The zero-order approximation, u_n, is assumed to obey a map given by

$$u_{n+1} = \lambda u_n + \sum_{k \geq 1} \varepsilon^k U_k(u_n, u_n^*) \qquad (10.82)$$

It turns out to be convenient to write

$$u_n = \rho_n \exp(-i\,\theta_n) \qquad (10.83)$$

Equation (10.82) becomes a normal form if one chooses the $T_k(u,u^*)$ so that the U_k include only resonant terms, i.e., terms of the form $[u^{k+1} \cdot u^{*k}]$. (In most of the formulas that follow, we suppress the subscript "n".)

In order to compute z_n, for any n, through $O(\varepsilon^k)$, one needs to compute $T_i(u_n, u_n^*)$ for all $i \leq k$. Furthermore, expressing z_0 by the expansion of Equation (10.81) to the same order and imposing the initial conditions determines ρ_0 to the same accuracy. Inserting Equations (10.81) and (10.82) into Equation (10.80), expanding in powers of ε, and collecting all terms in each order of ε, one finds the expressions for U_k and $T_k(u,u^*)$. This determines $T_k(u,u^*)$ up to free terms of the form $F(u_n \cdot u_n^*) \cdot u_n$. The first few terms are

$$T_1(u,u^*) = -\tfrac{1}{8} i \left(\frac{\lambda^2}{1-\lambda^2}\right) u^3 + \tfrac{3}{8} i \left(\frac{1}{1-\lambda^2}\right) u u^{*2}$$
$$+ \tfrac{1}{8} i \left(\frac{1}{1-\lambda^4}\right) u^{*3} + \alpha u^2 u^* \qquad (10.84)$$

$$U_1 = \lambda u \left(\tfrac{3}{8} i\right)(uu^*) \qquad (10.85)$$

$$U_2 = \lambda u \left\{ \tfrac{3}{8} i \left[(\alpha + \alpha^*) + i \left(\frac{9\lambda^4 + 8\lambda^2 + 9}{16(1-\lambda^4)} \right) \right] (uu^*)^2 \right\} \qquad (10.86)$$
$$+ \tfrac{3}{16} i$$

The underlined term in Equation (10.86) is real. One can, therefore, choose $\mathrm{Re}\,\alpha$ so as to eliminate this term, yielding

$$U_2 = \lambda u \tfrac{1}{2}\left(\tfrac{3}{8} i\ uu^*\right)^2 \tag{10.87}$$

The desired pattern emerges from Equations (10.85) and (10.87):

$$u_{n+1} = \lambda u_n \left\{1 + \tfrac{3}{8} i\ \varepsilon u_n u_n^* + \tfrac{1}{2}\left(\tfrac{3}{8} i\ \varepsilon u_n u_n^*\right)^2 + \cdots\right\}$$

The terms in brackets are the first terms in the expansion of

$$\exp\!\left(\tfrac{3}{8} i\ \varepsilon u_n u_n^*\right)$$

In second order, T_2 has a free term of the form $(\beta u^3 u^{*2})$. One finds that Imα (which is still free) and Reβ can be chosen so as to generate U_3 of the form

$$U_3 = \lambda \tfrac{1}{6}\left(\tfrac{3}{8} i\ \right)^3 u_n^{\ 4} u_n^{*3} \tag{10.88}$$

The extension to higher orders is straightforward, but entails a significant amount of algebra, making it a prime candidate for symbolic manipulations by computers [18,19]. Thus, with a specific choice of the free terms in $T_k(u,u^*)$, the normal form becomes

$$u_{n+1} = \exp\!\left(-i\ \mu + i\ \varepsilon \tfrac{3}{8} u_n u_n^*\right) u_n \tag{10.89}$$

Using Equation (10.83), we find

$$\rho_{n+1} = \rho_n \equiv \rho \qquad \theta_{n+1} = \theta_n + (\mu - \varepsilon \tfrac{3}{8} \rho^2) \tag{10.90}$$

The map for u_n is a rotation on a circle of radius ρ, with a constant phase shift per step.

As mentioned before, the results of this symplectic map example are analogous to those obtained for continuous single degree of freedom oscillatory systems with small nonlinearities [15]. There, the specific choice of the free functions in the expansion of the solution converted the normal forms from an infinite series in powers of the small parameter to finite sums (hence *minimal normal forms*). For an unperturbed frequency $\omega_0 = 1$ in the Duffing equation, the updated frequency is $\omega = 1 + \tfrac{3}{8}\varepsilon\rho^2$ with ρ, the radius of the zero-order term, determined by the initial conditions. Although we have not demonstrated it explicitly, it is straightforward to show that similar results are realizable for symplectic planar maps in which the unperturbed part is a simple rotation and the nonlinearity is either a polynomial or a convergent Taylor series in the unknown functions.

This elegant result suffers from one major drawback – that of small denominators. It is common to perturbative analysis of maps of this class. It originates from the fact that all terms in T_m have denominators of the form $(1-\lambda^k)$, $k \leq (2m+2)$. (See the example of T_1.)

Clearly, when λ is a root of unity, some denominator will vanish identically, making the perturbation expansion useless. Even when λ is not a root of unity, the expansion of Equation (10.81) breaks down when the worst denominator, $(1-\lambda^{2m+2})$, becomes sufficiently close to zero. At first occurrence of a small denominator, one has to truncate the expansion in the previous order. Namely, one can only write

$$z_n = u_n + \sum_{k=1}^{m-1} \varepsilon^k T_k(u_n, u_n^*) + O(\varepsilon^m) \tag{10.91}$$

As a result, the phase rotation of Equation (10.89) is valid only within the same error estimate:

$$u_{n+1} = \lambda \exp\left(\varepsilon \tfrac{3}{8} i\, u_n u_n^* + O(\varepsilon^m)\right) u_n \tag{10.92}$$

If one now uses Equation (10.92) for integrating over a large number of iterations, starting from the initial condition, one finds

$$u_n = \lambda \exp\left[\varepsilon \tfrac{3}{8} n\, i\, \rho^2 + n O(\varepsilon^m)\right] u_0 \tag{10.93}$$

Depending on how large n is, this will reduce the accuracy of the computation of u_n and, hence, of z_n. For instance, if $n = O(1/\varepsilon)$, then the error incurred is $O(\varepsilon^{m-1})$ and in Equation (10.91) one additional order, $k=m-1$, should be discarded.

There has been much progress in the area of accelerator design during the past 15 years. Lie transform techniques, used by Dragt et al. [20, 21], are rather general and are readily implemented using computer algebra techniques [19]. More important, they are not restricted to planar maps. With this in mind, it is natural to ask to what extent the techniques associated with the use of the nonuniqueness of the normal form expansion that led to MNF could be taken over into the discussion of symplectic maps. The results given in this section are the answer for the case of two-dimensional maps. The question of the usefulness of this approach as applied to simple accelerator models has been discussed in Ref. 22. A natural, correct, implementable, and numerically effective extension of minimal normal forms to symplectic maps with many degrees of freedom is generally not possible [23]. This conclusion is in the line of the statements made earlier in this chapter, that the extension, in a straightforward manner, of the results of minimal normal forms from one-degree-of-freedom Hamiltonian continuous systems, to Hamiltonian systems with more degrees of freedom is still an open problem.

REFERENCES

1. Starzhinski, V. M., *Applied Methods in the Theory of Nonlinear Oscillations*, Mir Publishers, Moscow (1980).
2. Goldstein, H., *Classical Mechanics,* 2nd Ed., Addison-Wesley, Reading, MA (1980).
3. Arnold, V. I., *Mathematical Methods of Classical Mechanics,* 2nd Ed., Springer-Verlag, New York (1989).
4. Gustavson, F., " On Constructing Formal Integrals of a Hamiltonian System Near an Equilibrium Point," *Astron. J.* **71**, 670 (1966)
5. Kummer, M., "How to Avoid Secular Terms in Classical and Quantum Mechanics," *Nuov. Cim.* **1B**, 123 (1971).
6. Van der Meer, J. C., *The Hamiltonian Hopf Bifurcation*, Springer-Verlag, New York (1985).
7. Baider, A., and R. C. Churchill, "Uniqueness and Non-Uniqueness of Normal Forms for Vector Fields," *Proc. Roy. Soc.* (Edinburgh), **A108**, 27 (1988).
8. Baider, A., "Unique Normal Forms for Vector Fields and Hamiltonians," *J. Diff. Eqns.* **78**, 33 (1989).
9. Perko, L. M., "Higher Order Averaging and Related Methods for Perturbed Periodic and Quasi-Periodic Systems," *SIAM J. Appl. Math.* **17**, 698 (1968).
10. Arnold, V. I., *Geometrical Methods in the Theory of Ordinary Differential Equations,* 2nd Ed., Springer-Verlag, New York (1988).
11. Guckenheimer, J., and P. Holmes, *Nonlinear Oscillations, Dynamical Systems and Bifurcations of Vector Fields,* Springer-Verlag, New York (1983).
12. Wiggins, S., *Introduction to Applied Nonlinear Dynamical Systems and Chaos,* Springer-Verlag, New York (1990).
13. Bruno, A. D., *Local Methods in Nonlinear Differential Equations,* Springer-Verlag, New York (1989).
14. Meyer, K. R., and G. R. Hall, *Introduction to Hamiltonian Dynamical Systems and the N-Body Problem*, Springer-Verlag, New York (1992).
15. Kahn, P. B., and Y. Zarmi, "Minimal Normal Forms in Harmonic Oscillations with Small Nonlinear Perturbations," *Physica* **D54**, 65–74, (1991).

16. La Mon, K. J., "Lie Series in an Extended Region of Phase Space," *Math. Gen.* **A23**, 3875 (1990).
17. Kahn, P. B., D. Murray, and Y. Zarmi, "Computational Aspects of Normal Form Expansions," *Proc. AIP Conf. No. 326* pp. 633-661, Y. T. Yan, J. P. Naples, and M. Syphers, eds., AIP Press, Woodbury, NY (1995).
18. Rand, R. H., and D. Armbruster, *Perturbation Methods, Bifurcation Theory and Computer Algebra,* Springer-Verlag, New York (1987).
19. Berz, M., "Differential Algebraic Description of Beam Dynamics to Very High Orders," *Particle Accelerators* **24**, 109 (1989).
20. Dragt, A. J. , and J. M. Finn, "Lie Series and Invariant Functions for Analytic Symplectic Maps", *J. Math. Phys.* **17**, 225 (1976).
21. Dragt, A. J., and E. Forest, "Computation of Nonlinear Behavior of Hamiltonian Systems Using Lie Algebraic Methods," *J. Math. Phys.* **24**, 2734 (1983).
22. Mane, S. R., and W. T. Weng, "Minimal Normal-Form Method for Discrete Maps," *Phys. Rev.* **E48**, 532 (1993).
23. Forest, E., and D. Murray, "Freedom in Minimal Normal Forms," *Physica* **D74**, 181 (1994).

11

HIGHER-DIMENSIONAL DISSIPATIVE SYSTEMS

In the realm of higher-dimensional dissipative systems, one encounters a wealth of phenomena, the climax of which is chaotic behavior and strange attractors. However, the latter are outside the scope of this book. We concentrate on a sample of phenomena that are amenable to a perturbation analysis. These include the case of a limit cycle as an asymptotic solution of a leading oscillatory coordinate, in particular, Hopf bifurcations, and the phenomenon of phase locking in a system of two coupled oscillators.

11.1 LIMIT CYCLES IN MULTIDIMENSIONAL DISSIPATIVE SYSTEMS

We have encountered limit cycles in two-dimensional dissipative systems in Chapter 7. However, in many situations, limit cycles also occur in higher-dimensional dissipative systems. In Sections 11.1.1–11.1.3 we analyze examples of three-dimensional dissipative systems for which stable limit cycle solutions exist. In Section 11.2 we present a generalization of the examples of Sections 11.1.1 and 11.1.2. This generalization is the basis of the analysis of many physical systems.

11.1.1 Center manifold with a limit cycle for leading coordinates

We consider a system that has two leading coordinates with pure imaginary eigenvalues, $\pm i$, and one nonleading coordinate:

$$\dot{x} = y \qquad \dot{y} = -x - \varepsilon y^3 z + \varepsilon^2 y^3 \qquad (11.1a)$$

$$\dot{z} = -\lambda z + \varepsilon \alpha x^2, \qquad \lambda > 0 \qquad (11.1b)$$

Without the nonlinear coupling in the equation for the nonleading coordinate, z, the latter decays to zero exponentially in time. In the full Equation (11.1b), z is sustained with a nonvanishing amplitude due to its coupling to the leading coordinates. To obtain a rough idea about the asymptotic behavior of z, we note that, in the absence of the coupling, both z and (dz/dt) vanish asymptotically in time. Insofar as the problem can be analyzed by a perturbation method, we expect that the resulting nonvanishing z and (dz/dt) to be small. Thus, in the lowest approximation, we expect (dz/dt) to satisfy

$$\frac{dz}{dt} = 0 + O(\varepsilon) \tag{11.2}$$

Since in Equation (11.1b), $\lambda = O(1)$, this requirement implies that

$$z \approx \varepsilon \left(\frac{\alpha}{\lambda}\right) x^2 \tag{11.3}$$

Equation (11.3) implies the vanishing of (dz/dt) *to leading order in* ε, namely, that the leading exponential decay of z becomes negligible compared to the time dependence induced in z through its coupling to x. Notice that this rough idea yields, for z, *values that are* $O(\varepsilon)$ compared to x. Asymptotically, the equation for x is, therefore, expected to have roughly the following form:

$$\ddot{x} + x \cong \varepsilon^2 \dot{x}^3 \left(1 - \left(\frac{\alpha}{\lambda}\right) x^2\right) \tag{11.4}$$

Equation (11.4) resembles the Van der Pol equation (with the linear friction factor replaced by a cubic one). It will have a stable limit cycle for $(\alpha/\lambda) > 0$. Hence, we expect the original three-dimensional system to decay onto a surface, the projection of which in the $x-y$ plane is a limit cycle, but where z is a function of x and y. The limit cycle is expected to be a second-order effect.

In Figure 11.1 we present the numerical solution of Equation (11.1) for $\varepsilon = 0.3$, $\lambda = 1$, and $\alpha = \frac{10}{3}$. The three-dimensional plot is shown in Figure 11.1a. Figures 11.1b–d present the two-dimensional projections; Figures 11.1e–g provide the time dependence of x, y, and z. From the figures we observe the following:

1. Figure 11.1b shows that the $x-y$ plane projection of the orbit seems to converge onto a limit cycle.

2. This observation is also demonstrated by Figures 11.1e,f, where we see that x and y converge slowly (compared to the exponential drop in z) to stable oscillatory patterns.
3. In Figure 11.1g we see that z seems to decay exponentially from its initial value to a stable oscillatory pattern.
4. From Figures 11.1c,d we see that z becomes a function of x and y.

Normal form analysis of Equation (11.1). We first diagonalize the equations for x and y by transforming to $w=x+iy$, obtaining

$$\dot{w} = -i\,w + \tfrac{1}{8}\varepsilon(w-w^*)^3 z - \tfrac{1}{8}\varepsilon^2(w-w^*)^3 \qquad (11.5\text{a})$$

$$\dot{z} = -\lambda z + \tfrac{1}{3}\varepsilon\alpha(w+w^*)^2 \qquad (11.5\text{b})$$

Now, expand w and z in near-identity transformations:

$$w = u + \varepsilon T_1 + \varepsilon^2 T_2 + \cdots \qquad z = v + \varepsilon S_1 + \varepsilon^2 S_2 + \cdots \qquad (11.6)$$

and write the normal form equations as

$$\dot{u} = -i\,u + \varepsilon U_1 + \varepsilon^2 U_2 + \cdots \qquad \dot{v} = -\lambda v + \varepsilon V_1 + \varepsilon^2 V_2 + \cdots \qquad (11.7)$$

The detailed analysis shows that $U_1=V_1=0$ and that inclusion of free functions in T_1 and S_1 does not help simplify U_2 and V_2. We therefore make the choice of no free functions. We then obtain

$$T_1 = -\tfrac{1}{8}\frac{u^3 v}{2i+\lambda} + \tfrac{3}{8}\frac{u^2 u^* v}{\lambda} - \tfrac{3}{8}\frac{u u^{*2} v}{-2i+\lambda} + \tfrac{1}{8}\frac{u^{*3} v}{-4i+\lambda} \qquad (11.8)$$

$$S_1 = \alpha\left\{-\frac{u^2}{4(2i-\lambda)} + \frac{u u^*}{2\lambda} + \frac{u^{*2}}{4(2i+\lambda)}\right\} \qquad (11.9)$$

The normal form equations become

$$\dot{u} = -i\,u$$
$$+ \tfrac{3}{8}\varepsilon^2 u^2 u^*\left\{1 - \alpha\frac{12+5\lambda^2}{6\lambda(4+\lambda^2)}u u^* - i\frac{\alpha}{3(4+\lambda^2)}u u^*\right\} \qquad (11.10\text{a})$$

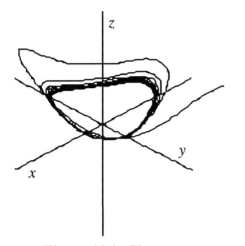

Figure 11.1a Three-dimensional plot.

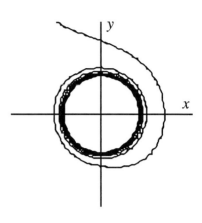

Figure 11.1b x–y plane projection; trajectory converges to a limit cycle.

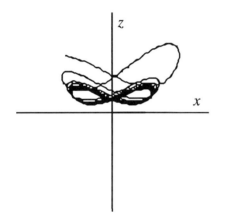

Figure 11.1c x–z plane projection.

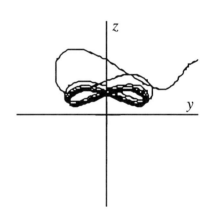

Figure 11.1d y–z plane projection.

$$\dot{v} = -\lambda v + \varepsilon^2 \tfrac{3}{2} \frac{\alpha}{\lambda(4+\lambda^2)} (uu^*)^2 v \qquad (11.10b)$$

From Equation (11.10b), we see that if the initial amplitude of u satisfies

LIMIT CYCLES

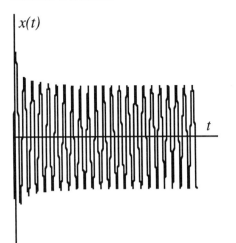

Figure 11.1e $x(t)$ versus time.

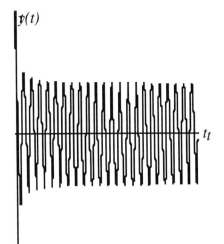

Figure 11.1f $y(t)$ versus time.

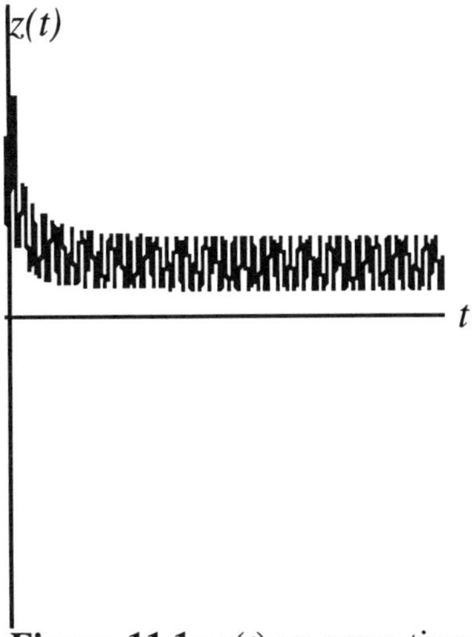

Figure 11.1g $z(t)$ vs.ersus time.

Figure 11.1 Solution of Equation (11.1), $\varepsilon=0.3$, $\lambda=1$, $\alpha=\frac{10}{3}$.

$$\varepsilon^2 \left(\tfrac{3}{2}\right) \frac{\alpha}{\lambda(4+\lambda^2)} (uu^*)^2 < \lambda \qquad (11.11)$$

then v will decay in an essentially exponential manner. Thus, in the near-identity transformation for z (see Equation (11.7)), only S_1 survives asymptotically in time: z, the nonleading coordinate, becomes a function of the leading ones only.

In addition, owing to the structure of the perturbation, as T_1 is proportional to v, it decays exponentially as well. Hence, through $O(\varepsilon)$, z remains equal to u asymptotically in time. This is *not* true in higher orders; already in $O(\varepsilon^2)$, T_2 contains nonresonant monomials in u and u^* that are not proportional to v. Therefore, the asymptotic form of z is affected by higher-order corrections.

Limiting our normal form analysis $O(\varepsilon^2)$, we write

$$u = \rho \exp(-i\varphi)$$

and obtain equations for ρ and φ

$$\dot{\rho} = \tfrac{3}{8}\varepsilon^2 \rho^3 \left(1 - \alpha \frac{12 + 5\lambda^2}{6\lambda(4 + \lambda^2)} \rho^2 \right) \qquad (11.12a)$$

$$\dot{\varphi} = 1 + \tfrac{1}{8}\varepsilon^2 \frac{\alpha}{4 + \lambda^2} \rho^4 \qquad (11.12b)$$

Thus, for $(\alpha/\lambda) > 0$, u has a limit cycle solution. The radius of the limit circle approximation is:

$$\rho^2 = \frac{6\lambda(4 + \lambda^2)}{\alpha(12 + 5\lambda^2)} \qquad (11.13)$$

With the values of the parameters used in Figure (11.1), we have $\rho = \sqrt{\tfrac{9}{17}} \cong 0.7276$, approximately where the actual limit cycle lies.

We can now obtain a rough estimate of the range of parameters for which the solution on the center manifold is stable in the vicinity of the limit cycle that we have discovered. We require that v, the zero-order approximation for z, decay in an essentially exponential manner. Equations (11.11) and (11.13) require that

$$54\varepsilon^2 \frac{\lambda(4 + \lambda^2)}{\alpha(12 + 5\lambda^2)^2} < \lambda \qquad (11.14)$$

For the parameters used in Figure 11.1, the l.h.s. of Equation (11.14) is equal to 0.0252, so that the inequality is certainly satisfied.

In Chapter 9, we analyzed systems that were also characterized by a center manifold. There, however, the eigenvalues of the linear parts of the equations for the leading coordinates vanished identically, leading

LIMIT CYCLES 371

to a much simpler structure. In particular, the asymptotic motion on the center manifold was fully described by the normal form equation, as all terms of order $n \geq 1$ in the near-identity transformation for the leading coordinates decayed exponentially in time.

This is not the case when the eigenvalues of the leading coordinates are purely imaginary, as in Equation (11.5a), so that only their real part vanishes. As mentioned earlier in this section, the higher-order corrections to $w=x+iy$ contain nonresonant terms that involve powers of u and u^* (which do not appear in the problems analyzed in Chapter 9) *in addition to* monomials that involve powers of u, u^*, and v. Terms of the second type decay asymptotically in an essentially exponential manner. However, terms of the first kind do not decay in time and modify the motion on the center manifold beyond the approximation provided by the normal form.

In summary, we have analyzed a three-dimensional system that has two leading coordinates and a single nonleading one. The eigenvalues of the linear terms in the dynamical equations of the leading coordinates have vanishing real part ($\pm i$). The linear part of the dynamical equation for the nonleading coordinate has a negative real part ($-\lambda$). The nonleading coordinate is, therefore, sustained with a nonvanishing amplitude only through its coupling to the leading ones. The solution decays onto a manifold (a two-dimensional surface in the present case) on which z is a function of the leading coordinates.

The system described by Equations (11.1) belongs to a large class of dynamical systems that decay onto a *center manifold*. In such systems, the eigenvalues of the linear part of the dynamical equations of the leading coordinates have vanishing real parts. The eigenvalues of the linear parts of the equations of the nonleading coordinates have negative real parts. As a result, the nonleading coordinates are sustained only through their coupling to the leading ones. The system decays asymptotically onto a manifold on which the nonleading coordinates are functions of the leading ones. The theory of the center manifold was studied in detail by Carr [1].

Comment. The solution of the present example converges in the three-dimensional phase space onto a line. The projection of this line on the x–y plane is a limit cycle. This feature is not a general aspect of systems that decay onto a center manifold

11.1.2 Bifurcation in a dissipative system–a scaling example

We borrow an example from the text by Rand and Armbruster [2], pages 28, 37, 39: the analysis of a three-dimensional problem with one real negative eigenvalue and a pair of complex conjugate eigenvalues.

Unlike the problem of Section 11.1.1, here there are two small parameters. One is built into the equations, and the other is discovered by the analysis. It is possible, by an appropriate scaling of the variables, to obtain a limit cycle in the "distorted" x–y plane.

The model equations

$$\frac{dx}{dt} = y + \mu x$$

$$\frac{dy}{dt} = -x + \mu y + \alpha x w \qquad (11.15)$$

$$\frac{dw}{dt} = -w + \alpha x^2$$

with $|\mu| \ll 1$.

We diagonalize the linear unperturbed matrix in Equations (11.15) by introducing $z = x + iy$ to obtain

$$\frac{dz}{dt} = -i z + \mu z + \tfrac{1}{2} i\, \alpha(z + z^*) w \qquad (11.16a)$$

$$\frac{dz^*}{dt} = +i z^* + \mu z^* - \tfrac{1}{2} i\, \alpha(z + z^*) w \qquad (11.16b)$$

$$\frac{dw}{dt} = -w + \tfrac{1}{4} \alpha(z + z^*)^2 \qquad (11.16c)$$

The origin $z=z^*=w=0$ is a fixed point. We are interested in analyzing the characteristics of the solution, in particular the stability of the origin as a fixed point, as μ is varied from negative to positive values. We begin with a qualitative analysis. Therefore, we only discuss the characteristics of the solution for sufficiently small amplitudes.

$\mu<0$. Terms proportional to i in the equation for z (and z^*) only affect the phase of z (z^*). The remaining terms (μz and μz^*, respectively), cause an essentially exponential decay of the amplitudes of z and z^* over a timescale of $O(1/|\mu|)$. As the linear term in the w equation is negative, if the amplitude of z is sufficiently small, the linear term will dominate, seemingly causing the decay of w to zero in an essentially exponential manner on a timescale of $O(1)$. However, this decay need

LIMIT CYCLES

not cause the asymptotic vanishing of w. The nonlinear coupling of w to $z+z^*$ in Equation (11.16c) implies that w can be sustained at an amplitude of the order of $|z|^2$. Thus, the actual decay of w will be slower and follow the decay of z. Still, we expect that for $\mu<0$, there will be a range of small amplitudes for z and w within which the system will always decay onto the origin. The latter will, therefore, be a stable fixed point.

$\mu>0$. The fixed point at the origin becomes unstable. Now the amplitude of z grows in an essentially exponential manner like $\exp(\mu t)$. This growth may or may not be balanced by the nonlinear coupling term in Equations (11.16a,b). To see what happens in detail, we have to delve into a more precise analysis.

Step 1: The scaling. We are interested in what happens to the system when it is close to the origin. To perform a perturbation expansion, we want to explicitly indicate that z and w are small and of the same order of magnitude. We do this by scaling them according to

$$z \to \varepsilon \cdot z \quad z^* \to \varepsilon \cdot z^* \quad w \to \varepsilon \cdot w \qquad |\varepsilon| \ll 1 \quad (11.17)$$

In our analysis we emphasize two aspects:

1. In methods usually found in the literature, one argues that the small parameters ε and μ are related from the start. Instead, we begin by treating the two as independent in a two-parameter expansion. The relation between ε and μ, which generates the desired behavior of the solution, emerges as a natural consequence of the approach.

2. In the center manifold analysis, one passes to the long-time limit early in the analysis. One follows the (asymptotic) motion after all the transients have decayed. In this approach one may erroneously conclude that the final motion takes place in the x–y plane, rather than on a surface in the three-dimensional space. Instead, we follow the transient behavior that leads to the asymptotic motion.

With the scaled variables of Equation (11.17), Equations (11.16) become

$$\frac{dz}{dt} = -i\,z + \mu z + \tfrac{1}{2} i\,\varepsilon\alpha(z+z^*)w \qquad (11.18a)$$

$$\frac{dz^*}{dt} = +i\,z^* + \mu z^* - \tfrac{1}{2}i\,\varepsilon\alpha(z+z^*)w \qquad (11.18\text{b})$$

$$\frac{dw}{dt} = -w + \tfrac{1}{4}\varepsilon\alpha(z+z^*)^2 \qquad (11.18\text{c})$$

Comment. For negative μ and sufficiently small ε, both z and w decay (essentially) exponentially. This follows directly from the Poincaré–Lyapunov theorem. We are interested in exploring limit cycle behavior, and therefore need μ > 0. With this in mind, we will watch what happens as μ crosses zero as it varies from negative to positive values. We will observe a bifurcation, or a change in stability of the origin leading to an unstable focus and then to a limit cycle.

We can capture the limit cycle only if we balance the unstable growth associated with the (μ) term in *second order*. To see this, we perform a double expansion in the two small parameters, ε and μ, and point out how the relative scales of the small parameters affect the physical picture. None of the first order εα terms in Equations (11.18) resonate with the unperturbed equations. Consequently, they do not contribute to the normal form and will only enter in $O(\varepsilon^2)$. In the scaling terminology, we would conclude that, to capture the limit cycle, we would have to require that μ scales like ε^2.

Step 2: The algebra. Introduce the two-parameter near-identity transformation with the generators having a double index. The indices i and j in T_{ij} indicate that the term is multiplied by μ^i and ε^j.

$$z = u + \mu T_{10}^u(u,u^*,v) + \varepsilon T_{01}^u(u,u^*,v) + $$
$$\mu^2 T_{20}^u(u,u^*,v) + \mu\varepsilon T_{11}^u(u,u^*,v) + \varepsilon^2 T_{02}^u(u,u^*,v) + \cdots \qquad (11.19\text{a})$$

$$z^* = u^* + \mu T_{10}^{u*}(u,u^*,v) + \varepsilon T_{01}^{u*}(u,u^*,v) + $$
$$\mu^2 T_{20}^{u*}(u,u^*,v) + \mu\varepsilon T_{11}^{u*}(u,u^*,v) + \varepsilon^2 T_{02}^{u*}(u,u^*,v) + \cdots \qquad (11.19\text{b})$$

$$w = v + \mu T_{10}^v(u,u^*,v) + \varepsilon T_{01}^v(u,u^*,v) + $$
$$\mu^2 T_{20}^v(u,u^*,v) + \mu\varepsilon T_{11}^v(u,u^*,v) + \varepsilon^2 T_{02}^v(u,u^*,v) + \cdots \qquad (11.19\text{c})$$

The superscripts, u, u^*, and v refer to the components of the vector. We have

LIMIT CYCLES

$$U_{00}^u = -i\,u \qquad U_{00}^{u*} = +i\,u \qquad U_{00}^v = -v$$

$$Z_{10}^u = u \qquad Z_{10}^{u*} = u^* \qquad Z_{10}^v = 0$$

$$Z_{01}^u = +\tfrac{1}{2}i\,\alpha(u+u^*)v \qquad Z_{01}^{u*} = -\tfrac{1}{2}i\,\alpha(u+u^*)v$$

$$Z_{01}^v = +\tfrac{1}{4}\alpha(u+u^*)^2$$

Those Z_{0j}^k that are quadratic are nonresonant, and T_{0j}^k exist that kill the corresponding U_{0j}^k ($k = u, u^*, v$). The only nonzero terms in the first-order normal form are, therefore, $U_{10}^u = u$ and $U_{10}^{u*} = u^*$. The calculations are straightforward, so we give only a few illustrative ones. For U_{01}^u, we write

$$U_{01}^u = [Z_{00}, T_{01}]^u + Z_{01}^u$$

$$= T_{01}^u \frac{\partial Z_{00}^u}{\partial u} + T_{01}^{u*} \frac{\partial Z_{00}^u}{\partial u^*} + T_{01}^v \frac{\partial Z_{00}^u}{\partial v}$$

$$- Z_{00}^u \frac{\partial T_{01}^u}{\partial u} - Z_{00}^{u*} \frac{\partial T_{01}^u}{\partial u^*} - Z_{00}^v \frac{\partial T_{01}^u}{\partial v} + \tfrac{1}{2}i\,\alpha(u+u^*)v$$

Comment. As we are only computing the lowest order, there is no need to introduce free terms in the T functions. We, therefore, write

$$T_{01}^u = \sum_{m+n+p=2} a_{mnp}\, u^m u^{*n} v^p \qquad Z_{01}^u = \sum_{m+n+p=2} b_{mnp}\, u^m u^{*n} v^p$$

with b_{mnp} known and all equal to $i(\alpha/2)$. The contribution of the Lie bracket to a given monomial is

$$[i(-1+m-n)+p]a_{mnp}$$

yielding

$$a_{mnp} = \frac{-i(\alpha/2)}{[i(-1+m-n)+p]}$$

With these results, we obtain the following:

$$T^u_{01} = -\tfrac{1}{2}i\,\alpha uv + \tfrac{1}{2}\alpha\left(\frac{2-i}{5}\right)u^*v \qquad (11.20\text{a})$$

$$T^{u*}_{01} = +\tfrac{1}{2}i\,\alpha u^*v + \tfrac{1}{2}\alpha\left(\frac{2+i}{5}\right)uv \qquad (11.20\text{b})$$

$$U^u_{01} = 0 \qquad (11.20\text{c})$$

In a similar manner, we find

$$T^v_{01} = \tfrac{1}{4}\alpha\left(\frac{1+2i}{5}\right)u^2 + \tfrac{1}{2}\alpha uu^* + \tfrac{1}{4}\alpha\left(\frac{1-2i}{5}\right)u^{*2} \quad (11.20\text{d})$$

$$U^v_{01} = 0 \qquad (11.20\text{e})$$

We now compute U^u_{02}

$$U^u_{02} = [Z_{00}, T_{02}]^u + T^u_{01}\frac{\partial Z^u_{01}}{\partial u} + T^{u*}_{01}\frac{\partial Z^u_{01}}{\partial u^*} + T^v_{01}\frac{\partial Z^u_{01}}{\partial v}$$

$$- U^u_{01}\frac{\partial T^u_{01}}{\partial u} - U^{u*}_{01}\frac{\partial T^u_{01}}{\partial u^*} - U^v_{01}\frac{\partial T^u_{01}}{\partial v}$$

Any term that is proportional to v is damped exponentially in time owing to $U^v_{00} = -v$. Thus, Z^u_{01} and T^u_{01} will decay to zero at long times.

We choose T^u_{02} to kill all non-resonant terms in U^u_{02}. We need to identify the resonant terms. Since we have

$$U^u_{00} = -i\,u$$

the resonant terms are of the form $u^m \cdot u^{*n} \cdot v^p$; with $p = 0$, $m = n+1$. These terms have zero Lie bracket since then $[i(-1+m-n)+p]=0$. Now, since Z^u_{01} and Z^{u*}_{01} both contain v as a multiplicative factor, they can only lead to the occurrence of nonresonant terms, which can be killed by a proper choice of T^u_{02}. We also have [see (11.20c&e)]

$$U^u_{01} = U^v_{01} = 0$$

LIMIT CYCLES

Thus, in the computation of U_{02}^u, we need only compute the underlined term associated with T_{01}^v to find the resonant part of the interaction:

$$T_{01}^v \frac{\partial Z_{01}^u}{\partial v} =$$

$$\left[\underline{\tfrac{1}{4}\alpha\left(\frac{1+2i}{5}\right)u^2 + \tfrac{1}{2}\alpha u u^*} + \tfrac{1}{4}\alpha\left(\frac{1-2i}{5}\right)u^{*2} \right] \left[\tfrac{1}{2} i\, \alpha(u+u^*) \right] \quad (11.21)$$

The resonant terms are those that only contain $u^2 u^*$. They are the only ones that cannot be killed by T_{02}^u as the Lie bracket vanishes when $n=m+1; p=0$. We find

$$U_{02}^u = \alpha^2 \left(\frac{11i - 2}{40} \right) u^2 u^*$$

Therefore, through second order, the normal form is

$$\frac{du}{dt} = U_{00}^u + \mu U_{10}^u + \varepsilon^2 U_{02}^u + \cdots = \\ -i\,u + \mu u + (\varepsilon\alpha)^2 u^2 u^* \left(\frac{11i-2}{40} \right) + \cdots \quad (11.22)$$

We write $u = \rho \cdot \exp(-i\theta)$. The phase correction turns out to be of no particular interest, and the equation for ρ through second order is:

$$\frac{d\rho}{dt} = \mu\rho - \tfrac{1}{20}\varepsilon^2 \alpha^2 \rho^3 \quad (11.23)$$

The fixed points (where $d\rho/dt=0$) are

$$\rho = 0 \qquad \rho = \rho^* = \left[\frac{20\mu}{(\varepsilon\alpha)^2} \right]^{1/2}$$

The origin is stable for $\mu < 0$, while $\rho = \rho^*$ is unstable. When μ becomes positive, the roles are interchanged: $\rho=0$ becomes unstable and a stable limit cycle is obtained at $\rho=\rho^*$.

Keep in mind that ε was introduced in Equation (11.18) just as a scaling parameter to help organize the calculation. Thus, $\rho^*_{true}=\varepsilon\rho^*= [20\mu/\alpha^2]^{1/2}$. Furthermore, we must require that ρ^* be $O(1)$ so as to retain the consistency of the expansion. Hence we must have $\mu=O(\varepsilon^2)$.

In the absence of the nonlinear coupling, w would have decayed exponentially on a timescale of $O(1)$. Let us see the timescale over which ρ goes to the limit cycle radius, $\rho_{lc} \equiv \rho^*$. For this we rewrite Equation (11.23) in the form

$$\frac{d\rho}{\rho(\rho_{lc}^2 - \rho^2)} = \frac{\mu\, dt}{\rho_{lc}^2}$$

which is solved by

$$\ln\frac{\rho}{\rho_i} - \tfrac{1}{2}\ln\left[\frac{\rho_{lc}^2 - \rho^2}{\rho_{lc}^2 - \rho_i^2}\right] = \mu t \qquad (11.24)$$

Thus, for instance, if the initial position is within the limit cycle, i.e., $\rho_i < \rho_{lc}$, then as $t \to \infty$, $\rho^2 \to \rho_{lc}^2$ like $\exp(-2\mu t)$.

In summary, we see that there are a few scales:

1. To get a limit cycle of radius $O(1)$ in the scaled variables, μ must scale like ε^2. This means that the contributions of U_{02} and U_{10} must be commensurate and enter in the same order.
2. If we have a limit cycle, then $\rho \to \rho_{lc}$ like $\exp(-\mu t)$, so that $1/\mu$ is the timescale of the approach.
3. The decay of the w variable is generally of a timescale of $O(1)$. Since this is fast, we reach onto the "distorted plane," where the final motion takes place very rapidly.

Discussion. For $\mu>0$, without the nonlinear coupling, Equations (11.8) lead to an exponentially outgrowing spiral and an exponentially decaying w. The coupling builds the w variable to $O(z^2)$ [$=O(x^2+y^2)$], which, in turn, produces a cubic interaction term in the z equation. As a result, a limit cycle emerges on a surface which is *not* the x–y plane. This surface is approached essentially at the rate of $\exp(-\mu t)$. The equation of the surface is

$$w = \alpha^2\left[(x+y)^2 + 2xy\right]$$

LIMIT CYCLES

Comment. Problems of this type are often analyzed by the *method of the center manifold*. We choose not to perform such an analysis. Some aspects of the method were discussed in Chapter 9.

11.1.3 Limit cycle in a three-dimensional system– no center manifold structure

In Sections 11.1.1 and 11.1.2 we analyzed systems that decayed onto a center manifold, where they evolved into a limit cycle. However, limit cycles can also occur in multidimensional systems that do not consist of leading and nonleading coordinates, where the motion relaxes asymptotically onto a center manifold. Consider, for example, the following system:

$$\ddot{x} + x = \varepsilon \dot{x}\left(1 - x^2 w\right) \quad (11.25a)$$

$$\dot{w} = -\varepsilon\left(w^2 - \alpha x^4\right) \quad (11.25b)$$

A numerical solution of Equations (11.25a,b) is given in Figures 11.2 for $\varepsilon=0.2$ and $\alpha=3$. From the figures we see that the system has a limit cycle solution. (The initial conditions were chosen so that the system is already very close to the limit cycle.) We will show that the main features of this limit cycle can be explained through a first-order normal form analysis.

We first diagonalize Equation (11.25a):

$$\dot{z} = -i\,z + \varepsilon\left(\frac{z - z^*}{2}\right)\left(1 - \frac{(z + z^*)^2}{4}w\right) \quad (11.26)$$

We denote the zero-order approximations for z and w by u and v, respectively, and note the following:

1. The w equation has no linear term. This is equivalent to saying that the eigenvalue of its linear part vanishes. In a normal form expansion, this means that any term v^n resonates with v.
2. As the unperturbed eigenvalues of the z equation are $\pm i$, neutral terms of the form $(u \cdot u^*)^n$ resonate with v for any integer n.
3. Consequently, the normal form equation for v will be composed of terms of the form $(u \cdot u^*)^n v^m$.
4. From points 1–3, we also deduce that resonant terms in the normal form equation for u will be of the general form $u^{n+1} u^{*n} \cdot v^m$.

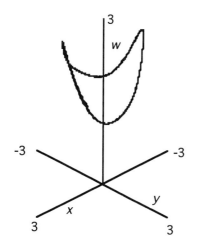

Figure 11.2a Three-dimensional plot of limit cycle solution.

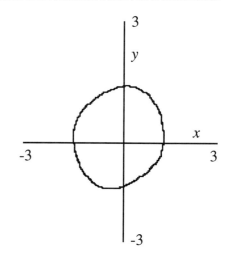

Figure 11.2b x–y plane projection.

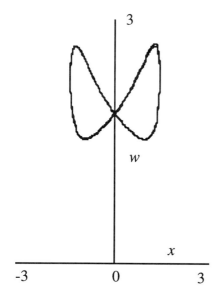

Figure 11.2c x–z plane projection.

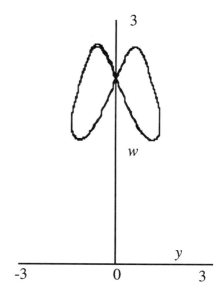

Figure 11.2d y–z plane projection.

LIMIT CYCLES

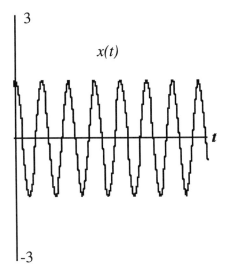

Figure 11.2e Time dependence of x.

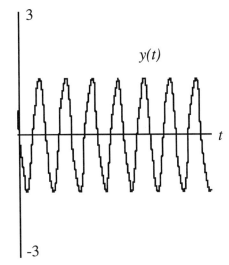

Figure 11.2f Time dependence of y.

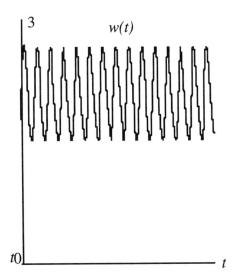

Figure 11.2g Time dependence of w.

Figure 11.2 Limit cycle solution of Equations (11.25) ($\varepsilon=0.2$, $\alpha=3$).

With points 1–4 in mind, the first-order normal form equations are readily obtained by inspecting the original equations and identifying the resonant terms:

$$\dot{u} = -i\,u + \tfrac{1}{2}\varepsilon u\left(1 - \tfrac{1}{4}(uu^*)\right)v \qquad (11.27\text{a})$$

$$\dot{v} = -\varepsilon\left(v^2 - \tfrac{3}{8}\alpha(uu^*)^2\right) \qquad (11.27\text{b})$$

Now write $u = \rho \cdot \exp(-i\varphi)$; Equations (11.27a,b) yield

$$\dot{\varphi} = 1 \qquad (11.28\text{a})$$

$$\frac{d}{dt}\rho^2 = \varepsilon\rho^2\left(1 - \tfrac{1}{4}\rho^2 v\right) \qquad (11.28\text{b})$$

$$\dot{v} = -\varepsilon\left(v^2 - \tfrac{3}{8}\alpha\rho^4\right) \qquad (11.28\text{c})$$

Equation (11.28a) indicates that in this order there is no frequency update.

Equations (11.28b,c) have a fixed point at

$$\rho^2 = \rho_0{}^2 = 2\left(\frac{3\alpha}{8}\right)^{-1/4} \qquad v = v_0 = 2\left(\frac{3\alpha}{8}\right)^{1/4} \qquad (11.29)$$

The linear stability analysis of the system of Equations (11.28b,c) is straightforward. Writing

$$\rho^2 = \rho_0{}^2 + \xi \qquad v = v_0 + \eta \qquad (|\xi|,|\eta| \ll 1)$$

we retain only the terms that are linear in ξ and η. The resulting linearized equations are

$$\frac{d}{dt}\begin{pmatrix}\xi \\ \eta\end{pmatrix} = \varepsilon \begin{pmatrix} -1 & -\left(\dfrac{3\alpha}{8}\right)^{-1/2} \\ 4\left(\dfrac{3\alpha}{8}\right)^{3/4} & -4\left(\dfrac{3\alpha}{8}\right)^{1/4} \end{pmatrix}\begin{pmatrix}\xi \\ \eta\end{pmatrix} \qquad (11.30)$$

For $\varepsilon > 0$, the eigenvalues of the matrix in Equation (11.30) both have negative real parts. Therefore, the fixed point is stable.

With $\alpha=3$, $\varepsilon=0.2$, the position of the fixed point is: $\rho_0 \cong 1.39354$, $v_0 \cong 2.05977$. Higher-order terms slightly distort the circle in the x–y plane projection, and cause oscillations in w around v_0 in the range $1.54 - 2.68$. Clearly, for smaller values of ε, the distortion of the x–y plane projection will be less pronounced; the projection will be closer to a circle of radius ρ_0. The amplitude of oscillations of w around v_0 will also be smaller then. We note that, in Figures 11.2, the oscillation frequency of w is very close to twice that of x and y.

All these characteristics can be understood already in the first-order analysis. We write the near-identity transformations for z and w as

$$z = u + \varepsilon T_1 + \cdots \qquad w = v + \varepsilon S_1 + \cdots$$

From Equation (11.25b), we find that S_1 is given by (no free functions are incorporated)

$$S_1 = \tfrac{1}{4}\alpha\rho^4\left(\sin 2\varphi + \tfrac{1}{8}\sin 4\varphi\right) \qquad (11.31)$$

Hence, on the limit cycle we have

$$W = V + \varepsilon S_1 + O(\varepsilon^2) \to v_0 + \varepsilon \tfrac{1}{4}\alpha\rho_0^4\left(\sin 2\varphi + \tfrac{1}{8}\sin 4\varphi\right) + O(\varepsilon^2)$$

In the $O(\varepsilon)$ term, the $\sin 2\varphi$ contribution dominates. Hence, to a good approximation, the oscillations of w have twice the frequency of x and y. Also, the amplitude of oscillations around v_0 is $\tfrac{1}{4}\varepsilon\alpha\rho_0^4$. For the parameter values used in Figures 11.2, this amplitude has the value 0.57, explaining the range of the oscillations in w around v_0. Finally, we observe that, relative to the zero-order term, v_0, the actual expansion parameter is not $\varepsilon=0.2$. Rather, it is $\varepsilon\alpha\rho_0^4 \approx 0.54$.

11.2 ONSET OF OSCILLATIONS IN HOPF BIFURCATIONS

The observations discussed in Sections 11.1.1-11.1.2 recur in the analysis of the onset of oscillations generated by Hopf bifurcations. These lead to "amplitude equations" (see, e.g., Kuramoto [3]).
In systems that undergo Hopf bifurcations, a change in a control parameter causes a change in the stability characteristics of a fixed point (which may be set at the origin). Below some critical value, the fixed point is a stable attractor. When the parameter crosses the critical value, the fixed point becomes unstable and the system develops an

oscillatory behavior in the form of a limit cycle. In many physical systems that involve partial-differential equations, the bifurcation signifies the onset of interesting physical phenomena. For example, in a liquid that is heated from below, the control parameter is the deviation of the Rayleigh number from a critical value. The bifurcation point, $\mu=0$, signifies the transition from pure heat conduction to convection in the form of convective rolls.

For the sake of simplicity, we consider the case of ODEs, which are amenable to a perturbative analysis through the method of normal forms. In general, one is studying an n-dimensional problem, where the various degrees of freedom are coupled:

$$\dot{\mathbf{x}} = \mathbf{A}(\mu)\mathbf{x} + \mathbf{G}(\mathbf{x};\mu) \qquad \mathbf{x} \in R^n \qquad (11.32)$$

To simplify the analysis, we assume that the $n \times n$ matrix $\mathbf{A}$ is in diagonal form. $\mathbf{G}$ is a nonlinear vector function assumed to have a Taylor expansion in $\mathbf{x}$ around the origin, the lowest power being quadratic. With these assumptions, the origin is a fixed point of the equation, since at $\mathbf{x}=0$ one has

$$\dot{\mathbf{x}} \equiv 0$$

For $\mu<0$, the we assume that the eigenvalues of $\mathbf{A}$ all have negative real parts, with a leading complex conjugate pair. By the Poincaré–Lyapunov theorem (see Chapter 2), the origin is a stable fixed point. For $\mu=0$ the leading pair of complex conjugate eigenvalues crosses the imaginary axis, and for $\mu>0$ that pair develops a positive real part. An example of this situation is shown in Figure 4.3c.

As the leading pair of eigenvalues develops a small positive real part, the origin becomes unstable. One wishes to find the behavior of the system near the origin, in particular, whether it converges onto a limit cycle or diverges in time.

We further assume that the nonleading eigenvalues all have negative real parts and are separated from the leading pair by a distance of $O(\mu^0)$ along the real axis. By the center manifold theorem (see, e.g., Carr [1]) and motivated by the analysis in Chapter 9 and in Sections 11.1.1-11.1.2, we assume that, for sufficiently long times, the dynamics of the system reduce to a two-dimensional manifold, where all $n-2$ coordinates pertaining to the nonleading eigenvalues are expressed in terms of the leading pair.

Without loss of generality, we assume

$$\lambda_\mp = \mp i + \mu(\alpha \pm i\,\beta) + O(\mu^2) \qquad (11.33)$$

ONSET OF OSCILLATIONS IN HOPF BIFURCATIONS

so that the unperturbed frequency equals unity. We then use complex notation for the leading variables, x_1 and x_2:

$$z = x_1 + i\, x_2$$

It is important that the real part of the $O(\mu)$ correction to the leading eigenvalues is strictly positive:

$$\alpha = \left.\frac{d\,\mathrm{Re}\,\lambda_\pm}{d\mu}\right|_{\mu=0} > 0 \qquad (11.34)$$

This requirement means that for

$$\mu < 0 \qquad \mathrm{Re}\,\lambda_\pm < 0 \qquad \mu > 0 \qquad \mathrm{Re}\,\lambda_\pm > 0$$

More importantly, it means that the bifurcation already occurs in $O(\mu)$. If $\alpha=0$, then the bifurcation occurs only in second order.

The generic form of the equation for the motion of the leading degree of freedom on the center manifold and near the origin is

$$\dot{z} = -i\,z + (\alpha + i\,\beta)\mu z + (\gamma + i\,\delta)\mu^2 z + \cdots$$

$$+ az^2 + bzz^* + cz^{*2} \qquad (11.35)$$

$$+ dz^3 + ez^2 z^* + f z z^{*2} + g z^{*3} + \cdots$$

In Equation (11.35), the first line presents the expansion of the eigenvalue of the linear part in powers of μ. The remainder of the equation is a Taylor expansion of the nonlinear part. Each of the coefficients a,b,c,d,e,f,g is a function of μ. For example,

$$a = a_0 + a_1 \mu + a_2 \mu^2 + \cdots \qquad (11.36)$$

11.2.1 Derivation of the normal form

Actually, the expansion is in two small parameters: μ, the bifurcation parameter and the size of the amplitude. For the sake of bookkeeping in the expansion, it is convenient to expose the latter explicitly, writing

$$z = \nu w \qquad |\nu| \ll 1 \qquad (11.37)$$

Using Equation (11.35), the equation for w becomes

$$\dot{w} = -i\,w + (\alpha + i\,\beta)\mu w + (\gamma + i\,\delta)\mu^2 w + \cdots$$

$$+ \nu(aw^2 + bww^* + cw^{*2}) \qquad (11.38)$$

$$+ \nu^2(dw^3 + ew^2w^* + fww^{*2} + gw^{*3}) + \cdots$$

Now perform a double expansion

$$w = u + \nu T_{10} + \mu T_{01} + \nu^2 T_{20} + \nu\mu T_{11} + \mu^2 T_{02} + \cdots \qquad (11.39)$$

To find for the T functions:

$$T_{10} = i\,a_0 u^2 - i\,b_0 u u^* - \tfrac{1}{3} i\,c_0 u^{*2} + F_{10}\,u \qquad T_{01} = F_{01}\,u \qquad (11.40)$$

and for the U functions:

$$U_{10} = 0 \qquad U_{01} = (\alpha + i\,\beta)u$$

$$U_{20} = \left(e_0 + i\left(b_0 b_0^* + \tfrac{2}{3} c_0 c_0^* - a_0 b_0\right)\right) u^2 u^* \qquad (11.41)$$

$$U_{11} = -2\alpha F_{10}'\,u^2 u^* \qquad U_{02} = (\gamma + i\,\delta)u - 2\alpha F_{01}'\,u^2 u^*$$

Even-powered terms are nonresonant. Hence they do not appear in the normal form in any order. Only odd ones appear. For example, T_{10} eliminates all the quadratic terms in the equation in first order.

11.2.2 Usual choice (no free functions)

With the conventional choice of $F_{10}=F_{01}=0$, the normal form becomes

$$\dot{u} = -i\,u + \left((\alpha + i\,\beta)\mu + (\gamma + i\,\delta)\mu^2\right)u + \cdots$$
$$+ \nu^2\left(e_0 + i\left(b_0 b_0^* + \tfrac{2}{3} c_0 c_0^* - a_0 b_0\right)\right)u^2 u^* + \cdots \qquad (11.42)$$

In polar coordinates, $u = \rho \cdot \exp(-i\varphi)$, we find

ONSET OF OSCILLATIONS IN HOPF BIFURCATIONS

$$\dot{\rho} = (\alpha\mu + \gamma\mu^2)\rho + (\mathrm{Re}\,e_0 + \mathrm{Im}(a_0 b_0))v^2\rho^3 + \cdots \quad (11.43a)$$

$$\omega = \dot{\varphi} = 1 - \beta\mu - \delta\mu^2 \\ - (\mathrm{Im}\,e_0 + b_0 b_0^* + \tfrac{2}{3}c_0 c_0^* - \mathrm{Re}(a_0 b_0))v^2\rho^2 + \cdots \quad (11.43b)$$

If the coefficients of ρ and ρ^3 on the r..h.s. of Equation (11.43a) have a negative relative sign, then a limit cycle exists. The radius of the limit circle is given by

$$|v|\rho_{l.c.} = \sqrt{-\frac{\alpha + \gamma\mu}{\mathrm{Re}\,e_0 + \mathrm{Im}(a_0 b_0)}}\,\mu^{1/2} \quad (11.44)$$

Thus, the competition between the linear and the cubic terms in the normal form determines the size of the amplitude of oscillations. It is $O(\mu^{1/2})$: $|v|=\mu^{1/2}$. In addition, the rescaled amplitude of oscillations at the onset of the instability ($\mu=0$) is

$$\rho_0 = \sqrt{-\frac{\alpha}{\mathrm{Re}\,e_0 + \mathrm{Im}(a_0 b_0)}}$$

The conventional choice leads here to the same results concerning limit cycles, as discussed Chapter 7. That is, besides the required rescaling, the radius of the limit circle depends on the expansion parameter μ in all orders. Hence, as in Chapter 7, updating of the asymptotic frequency is again rather cumbersome.

Amplitude equation. Equation (11.42) can be readily translated into what is called the amplitude equation, by setting $v=\mu^{1/2}$ and writing

$$u = e^{-it}\mu^{1/2} A$$

Note that the fast time dependence has been written out explicitly, so that the amplitude A is expected to have a slow time dependence. With this substitution, Equation (11.42) becomes

$$\dot{A} = ((\alpha + i\beta)\mu + (\gamma + i\delta)\mu^2)A \\ + \mu(e_0 + i(b_0 b_0^* + \tfrac{2}{3}c_0 c_0^* - a_0 b_0))|A|^2 A \quad (11.45)$$

We note that the r.h.s. of Equation (11.45) is proportional to μ, so that A has, indeed, a slow time dependence. This is made explicit by defining a slow time variable, $\tau \equiv \mu t$, leading to the well-known Stuart–Landau equation (see, e.g., Ref. 4, where the equation is derived by the method of multiple timescales):

$$\frac{dA}{d\tau} = \left((\alpha + i\,\beta) + (\gamma + i\,\delta)\mu\right)A$$
$$+ \left(e_0 + i\left(b_0 b_0{}^* + \tfrac{2}{3} c_0 c_0{}^* - a_0 b_0\right)\right)|A|^2 A$$

$$= \frac{1}{\rho_0{}^2}\left(\rho_0{}^2\left(1 + \frac{\gamma}{\alpha}\mu\right) - |A|^2\right)\alpha A \qquad (11.46)$$
$$+ i\begin{bmatrix}\beta + \mu\delta \\ + \left(\operatorname{Im} e_0 + b_0 b_0{}^* + \tfrac{2}{3} c_0 c_0{}^* - \operatorname{Re}(a_0 b_0)\right)|A|^2\end{bmatrix} A$$

In Equation (11.46), we have separated the real and imaginary parts so as to show explicitly how the radius of the limit circle is modified by the higher-order correction.

11.2.3 Special choice of free functions–limit circle radius "renormalization"

As in other limit cycle problems, here, as well, free functions can be introduced that simplify the normal form. Studying Equation (11.41), we find that a nonzero F_{10} complicates matters, so the (conventional) choice $F_{10}=0$ is made (hence $U_{11}=0$). On the other hand, if we choose

$$F_{01} = \frac{\gamma}{2\alpha\rho_0{}^2} u u^* \qquad (11.47)$$

then $\operatorname{Re} U_{02}$ becomes

$$\operatorname{Re} U_{02} = \gamma u\left(1 - \left(\frac{\rho}{\rho_0}\right)^2\right) \qquad (11.48)$$

With this choice, Equation (11.43a) is modified into

ONSET OF OSCILLATIONS IN HOPF BIFURCATIONS

$$\dot{\rho} = (\alpha\mu + \gamma\mu^2)\left(1 - \left(\frac{\rho}{\rho_0}\right)^2\right)\rho \qquad (11.49)$$

As in the analysis of limit cycles in Chapter 7, in higher orders, a choice of the free functions exists that guarantees that the term

$$\left(1 - \left(\frac{\rho}{\rho_0}\right)^2\right)$$

appears as an overall multiplicative factor in the contribution of each order to the ρ equation. Therefore, Equation (11.49) can be rewritten as

$$\dot{\rho} = (\alpha\mu + \gamma\mu^2 + O(\mu^3))\left(1 - \left(\frac{\rho}{\rho_0}\right)^2\right)\rho$$

with the $O(\mu^3)$ term representing the whole effect of higher orders in the equation.

Clearly, with this choice, unlike the case of Equation (11.46), the radius of the limit circle is not affected by higher-order corrections. Thus, except for the fact that it is proportional to $\mu^{1/2}$, *it can be calculated by the values of the dynamical parameters of the system at the bifurcation point* ($\mu=0$). Moreover, with this choice, the updating of the asymptotic frequency is again simpler; in Equation (11.43b) for the frequency, one now inserts ρ_0 for ρ, so that the computation of ω through a given order does not generate higher-order corrections in ω.

With our choice of the free functions, the amplitude equation becomes

$$\frac{dA}{d\tau} = \frac{1}{\rho_0^2}(\rho_0^2 - |A|^2)(\alpha + \gamma\mu)A$$

$$+ i\begin{bmatrix} \beta + \mu\delta \\ + (\mathrm{Im}\, e_0 + b_0 b_0^* + \tfrac{2}{3} c_0 c_0^* - \mathrm{Re}(a_0 b_0))|A|^2 \end{bmatrix} A \qquad (11.50)$$

Thus, with this choice of the near-identity transformation, the radius of the limit circle is not affected by the higher-order corrections.

HIGHER-DIMENSIONAL DISSIPATIVE SYSTEMS

In summary, the method of normal forms leads naturally to

1. Rescaling of the amplitude
2. Identification of the correct timescale for the slow variation of the amplitude
3. Choice of the near-identity transformation that fixes the radius of the limit circle at its lowest order value and simplifies the frequency updating computation

Comment. In this section we have implicitly assumed that the number of eigenvalues is finite. Often, ODEs such as Equation (11.32) are obtained from partial-differential equations. In such equations the time dependence, as well as the spatial dependence, of physical quantities (e.g., the temperature profile in a heat-transfer problem) are studied. **x** is then the vector of the amplitudes of modes obtained in a Fourier expansion of the spatial dependence of the unknown quantities. It is then an infinite-dimensional vector. Truncation to a finite number of modes is often invoked. While this is sometimes an allowed procedure, one should keep in mind the dangers that may be encountered in the truncation of nonleading coordinates. (See discussion in Chapter 5.)

11.3 PHASE LOCKING IN COUPLED OSCILLATORS

Consider the system of two coupled oscillators studied in Ref. 4:

$$\ddot{x}_1 + x_1 = \varepsilon(1 - x_1 x_2)\dot{x}_1 + \varepsilon \zeta_1 x_1 \qquad (11.51\text{a})$$

$$\ddot{x}_2 + x_2 = \varepsilon(1 - x_1^2)\dot{x}_2 + \varepsilon \zeta_2 x_2 \qquad (11.51\text{b})$$

with $\varepsilon > 0$. We think of this system as two interacting "Van der Pol-like" oscillators, with frequency mismatch through the linear ζ terms. The solution of the equations exhibits the phenomenon of "phase locking," encountered, for example, in laser physics. We study the system through first order. First, introduce the complex variables

$$z_i = x_i + i\,\dot{x}_i \qquad i = 1, 2$$

Our equations then become

PHASE LOCKING IN COUPLED OSCILLATORS

$$\dot{z}_1 = -i\, z_1 + \tfrac{1}{2}\varepsilon(z_1 - z_1^*) + \tfrac{1}{2} i\, \varepsilon\, \zeta_1(z_1 + z_1^*)$$
$$-\tfrac{1}{8}\varepsilon(z_1^2 - z_1^{*2})(z_2 + z_2^*) \qquad (11.52a)$$

$$\dot{z}_2 = -i\, z_2 + \tfrac{1}{2}\varepsilon(z_2 - z_2^*) + \tfrac{1}{2} i\, \varepsilon \zeta_2(z_2 + z_2^*)$$
$$-\tfrac{1}{8}\varepsilon(z_1^2 + 2 z_1 z_1^* + z_1^{*2})(z_2 - z_2^*) \qquad (11.52b)$$

We now introduce the near-identity transformation and develop the normal form equations for the zero order approximations, u_i, $i=1,2$. The unperturbed frequencies of both oscillators are equal: $\omega_1 = \omega_2 = 1$. Hence, resonance can occur through terms such as $u_1^{n+1} u_1^{*n}$ or $u_2^{m+1} u_2^{*m}$ as well as through mixed monomials of the form $u_1^p u_1^{*q} u_2^r u_2^{*s}$ with $(p-q+r-s)=1$.

In a first-order calculation for the normal form, it is not necessary to find the T functions. The first order equations are

$$\dot{u}_1 = -i\, u_1(1 - \tfrac{1}{2}\varepsilon\zeta_1) + \tfrac{1}{2}\varepsilon\, u_1 - \tfrac{1}{8}\varepsilon\, u_1^2 u_2^* \qquad (11.53a)$$

$$\dot{u}_2 = -i\, u_2(1 - \tfrac{1}{2}\varepsilon\zeta_2) + \tfrac{1}{2}\varepsilon u_2 + \tfrac{1}{8}\varepsilon u_1^2 u_2^* - \tfrac{1}{4}\varepsilon u_1 u_1^* u_2 \qquad (11.53b)$$

We introduce the polar representation:

$$u_1 = \rho \exp(-i\,\varphi) \qquad u_2 = r \exp(-i\,\theta) \qquad \Omega = \varphi - \theta$$

Our equations become

$$\dot{\rho} = \tfrac{1}{2}\varepsilon\rho - \tfrac{1}{8}\varepsilon\rho^2 r \cos\Omega \qquad (11.54a)$$

$$\dot{\varphi} = 1 - \tfrac{1}{2}\varepsilon\zeta_1 + \tfrac{1}{8}\varepsilon\rho r \sin\Omega \qquad (11.54b)$$

$$\dot{r} = \tfrac{1}{2}\varepsilon r - \tfrac{1}{8}\varepsilon\rho^2 r(1 + 2\sin^2\Omega) \qquad (11.54c)$$

$$\dot{\theta} = 1 - \tfrac{1}{2}\varepsilon\zeta_2 - \tfrac{1}{8}\varepsilon\rho^2 \sin 2\Omega \qquad (11.54d)$$

Comment. The equality of the unperturbed frequencies, ω_1 and ω_2, means that, for any integer n, we have $n \cdot \omega_1 - n \cdot \omega_2 = 0$. Consequently, as for Hamiltonian systems, we expect that normal form equations to involve three independent variables: ρ, r, and the relative phase Ω. In fact, from Equations (11.54c,d), we obtain a single independent equation for Ω:

$$\dot{\Omega} = -\tfrac{1}{2}\varepsilon(\zeta_1 - \zeta_2) + \tfrac{1}{8}\varepsilon\rho(r\sin\Omega + \rho\sin 2\Omega) \qquad (11.54e)$$

We are ready to find the fixed points of the system of Equations (11.54a,c,e). One fixed point is at the origin: $\rho = r = 0$. However, for $\varepsilon > 0$ (which is the interesting case), this fixed point is unstable. We, therefore, focus our attention on the second fixed point of the Equations for which one requires

$$\dot{\rho} = 0 \Rightarrow \cos\Omega = \frac{4}{\rho r} \qquad (11.55a)$$

$$\dot{r} = 0 \Rightarrow 1 + 2\sin^2\Omega = \frac{4}{\rho^2} \qquad (11.55b)$$

$$\dot{\Omega} = 0 \Rightarrow (\zeta_1 - \zeta_2) = \tfrac{1}{4}(\rho^2 \sin 2\Omega + \rho r \sin\Omega) \qquad (11.55c)$$

Equations (11.55a–c) determine specific values of ρ, r, and Ω. Hence, asymptotically the solution is characterized by two limit cycles of fixed radii; the system evolves on a torus, with energy transfer from one mode to the other. It is also characterized by a *fixed relative phase*. That is, the phases of the two oscillators vary in an identical manner, except for a constant phase difference between them. This is the phenomenon of *phase locking*.

The values of ρ, r, and Ω change as the difference in the frequency shifts, $\Delta \equiv \zeta_1 - \zeta_2$, is varied. We now deduce bounds on the allowed ranges of variation for the three quantities.

From Equation (11.55a) we easily deduce

$$0 \leq \cos\Omega \leq 1 \Rightarrow \rho r \geq 4 \qquad -\frac{\pi}{2} \leq \Omega \leq +\frac{\pi}{2} \qquad (11.56)$$

Equation (11.55b) yields a bound on ρ:

$$0 \leq \sin^2\Omega \leq 1 \Rightarrow \frac{2}{\sqrt{3}} \leq \rho \leq 2 \qquad (11.57)$$

From Equations (11.55a,b) we obtain a relation between ρ and r:

PHASE LOCKING IN COUPLED OSCILLATORS

$$3\rho^2 - \frac{32}{r^2} = 4 \tag{11.58}$$

leading to

$$r \geq 2 \tag{11.59}$$

One can also find the form of the dependence of ρ, r, and Ω on Δ for the extreme cases of $|\Delta| \ll 1$ and of very large $|\Delta|$.

$|\Delta| \ll 1$. As both ρ and r vary in ranges that are $O(1)$ relative to Δ, the smallness of the latter in Equation (11.55c) can manifest itself only if Ω is small. To lowest order in Δ, one finds

$$\Omega = \tfrac{1}{3}\Delta \qquad \rho = 2\left(1 - \tfrac{1}{9}\Delta^2\right) \qquad r = 2\left(1 + \tfrac{1}{6}\Delta^2\right) \tag{11.60}$$

Thus, as Δ moves away from 0, ρ decreases below 2 and r increases above 2.

Large Δ. As ρ is constrained to a finite range and the trigonometric functions are bounded by unity, the only way that making Δ large can avoid an inconsistency in Equation (11.55c) is by making r of $O(\Delta)$. From Equation (11.55a) we find that $\cos\Omega$ then becomes vanishingly small, hence, Ω tends to $\pm(\pi/2)$. The detailed analysis yields

$$\rho \to \frac{2}{\sqrt{3}} \qquad r \to \frac{|\Delta|}{\rho} \to \frac{\sqrt{3}}{2}|\Delta|$$

$$\Omega \to \frac{\Delta}{|\Delta|}\left(\frac{\pi}{2} - \frac{4}{|\Delta|}\right) \tag{11.61}$$

However, as r becomes large, the perturbation expansion that we have employed looses its validity.

The first-order generating functions are

$$T_1^u = \tfrac{1}{4}(\zeta_1 + i)u^* - \tfrac{1}{16}i\,u^2 v - \tfrac{1}{16}i\,u^{*2}v - \tfrac{1}{32}i\,u^{*2}v^* \tag{11.62a}$$

$$\begin{aligned}T_2^u = &\tfrac{1}{4}(\zeta_2 + i)v^* - \tfrac{1}{16}i\,u^2 v - \tfrac{1}{8}i\,uu^* v^* \\ &+ \tfrac{1}{16}i\,u^{*2}v - \tfrac{1}{32}i\,u^{2*}v^*\end{aligned} \tag{11.62b}$$

Continuing, we compute the U_2 and V_2 functions:

$$U_2 = \tfrac{1}{128} i \, \rho\left[16 + 3r^2\rho^2 + 16\zeta_1^2\right]$$
$$+ \tfrac{1}{256} i \, r\rho^2\left[-48 + \rho^2 + 8i\,(\zeta_1 + \zeta_2)\right]\cos\Omega$$
$$+ \tfrac{1}{64} i \, r^2\rho^3 \cos 2\Omega$$
$$+ \tfrac{1}{256} r\rho^2\left[16 + 7\rho^2 + 8i\,(\zeta_1 + \zeta_2)\right]\sin\Omega - \tfrac{1}{64} r^2\rho^3 \sin 2\Omega$$

$$V_2 = \tfrac{1}{256} i \, r\left[32 - 32\rho^2 + 9\rho^4 + 32\rho^2(i\,\zeta_1 + \zeta_2)\right] - \tfrac{1}{128} i \, r^2\rho^3 \cos\Omega$$
$$- \tfrac{5}{128} r^2\rho^3 \sin\Omega - \tfrac{1}{16} r\rho^2\left[1 + i\,\zeta_1\right](\sin 2\Omega + i \cos 3\Omega)$$
$$+ \tfrac{1}{64} r^2\rho^3 (\sin 3\Omega + i \cos 3\Omega)$$

REFERENCES

1. Carr, J., *Applications of Centre Manifold Theory*, Springer-Verlag New York (1981).
2. Rand, R. H., and D. Armbruster, *Perturbation Methods, Bifurcation Theory and Computer Algebra*, Springer-Verlag, New York (1987).
3. Kuramoto, Y., *Chemical Oscillations, Waves and Turbulence*, Springer-Verlag, Berlin (1984).
4. Mitropolskii, Yu. A., and A. M. Samoilenko, "Quasi-Periodic Oscillations In Linear Systems," *Ukrainian Math. J.* **24** (2), 179–193 (1972).

APPENDIX

CONSERVATIVE SYSTEM WITH ONE DEGREE OF FREEDOM: CALCULATION OF THE PERIOD

We have dwelled extensively on the analysis of conservative systems with one degree of freedom. Such systems are integrable. The motion takes place in a two-dimensional phase plane, and the frequency, ω, or the period, T, can be computed to any desired accuracy. In our normal form analysis, we often use the computed frequency as a basic element in the perturbative calculation of the time-dependent displacement. The extent to which the period depends on the amplitude is one measure of the strength of the perturbation.

In this Appendix, we describe a method for the numerical evaluation of the period, truncating the expansion in the small parameter, and for the assessment of the error incurred. We focus on a simple harmonic oscillator with symmetric perturbing potentials, as typified by the Duffing oscillator.

Denoting the total energy by E, and the potential by $V(x)$, the period is given by the integral

$$T = \oint \frac{dx}{\sqrt{2(E - V(x))}} \tag{A.1}$$

Duffing oscillator. The potential is

$$V(x) = \tfrac{1}{2}x^2 + \tfrac{1}{4}\varepsilon x^4$$

Equation (A.1) for the period becomes

$$T = 4 \int_0^A \frac{dx}{\sqrt{(A^2 - x^2) + \tfrac{1}{2}\varepsilon(A^4 - x^4)}} \tag{A.2}$$

Writing $x = A\sin\theta$, Equation (A.2) becomes

$$T = 4\int_0^{\pi/2} \frac{d\theta}{\sqrt{1 + \tfrac{1}{2}\varepsilon A^2\left[1 + \sin^2\theta\right]}} \qquad (A.3)$$

One can simply evaluate the integral in Equation (A.3) numerically. However, in that type of procedure, it is difficult to obtain good error bounds. To derive the latter, we expand the denominator and integrate, obtaining for the period the expression

$$T = 4\sum_{n=0}^{\infty} \frac{\Gamma(n+\tfrac{1}{2})}{n!\sqrt{\pi}}\left[-\tfrac{1}{2}\varepsilon A^2\right]^n \cdot \sum_{k=0}^{n} \frac{n!}{k!(n-k)!} \frac{\sqrt{\pi}\,\Gamma(k+\tfrac{1}{2})}{2k!} \qquad (A.4)$$

We define

$$F(n) \equiv \int_0^{\pi/2}\left[1+\sin^2\theta\right]^n d\theta = \sum_{k=0}^{n} \frac{n!}{k!(n-k)!}\frac{1}{2}\frac{\Gamma(\tfrac{1}{2})\Gamma(k+\tfrac{1}{2})}{k!} \qquad (A.5a)$$

Observe that, since $1 \leq [1+\sin^2\theta] \leq 2$, $F(n)$ are bounded by

$$\frac{\pi}{2} \leq F(n) \leq \frac{\pi}{2}\cdot 2^n \qquad (A.5b)$$

We now face two situations.

$\boldsymbol{\varepsilon > 0}$. In this case, the series is convergent and of alternating signs, so that the error incurred in truncation is less than the first term neglected. Expand and obtain the first few terms of the period

$$T = 2\pi\left[1 - \tfrac{3}{8}\varepsilon A^2 + \tfrac{57}{256}\varepsilon^2 A^4 - \tfrac{315}{2048}\varepsilon^3 A^6 + \cdots\right] \qquad (A.6a)$$

and the angular frequency is given as

$$\omega = \left[1 + \tfrac{3}{8}\varepsilon A^2 - \tfrac{21}{256}\varepsilon^2 A^4 + \tfrac{81}{2048}\varepsilon^3 A^6 + \cdots\right] \qquad (A.6b)$$

$\boldsymbol{\varepsilon < 0}$. The series consists entirely of positive terms. Evaluating a specified number of terms, N, in the exact expression, provides no idea about the precision of the result. To this end, we derive bounds on the remainder of the sum, using Equation (A.5b) and writing

CALCULATION OF THE PERIOD

$$T \le T_{\text{upper}} = 4\sum_{n=0}^{N}(\text{exact expression})$$
$$+ 4\sum_{n=N+1}^{\infty}(\text{upper bound}) \tag{A.7}$$

Note that when we replace $F(n)$ by their upper bounds, we obtain

$$\sum_{n=0}^{\infty}(\text{upper bound}) = \frac{2\pi}{\sqrt{1+x}} \tag{A.8}$$

where $x = \varepsilon A^2$. From Equation (A.8) we obtain

$$4\sum_{n=N+1}^{\infty}(\text{upper bound}) =$$
$$4\left\{\sum_{n=0}^{\infty}\text{upper bound} - \sum_{n=0}^{N}\text{upper bound}\right\} = \tag{A.9}$$
$$\frac{2\pi}{\sqrt{1+x}} - 2\pi\sum_{n=0}^{N}\frac{\Gamma(n+\tfrac{1}{2})}{n!\sqrt{\pi}}(-x)^n$$

Inserting Equation (A.9) in Equation (A.7), we can now compute an upper bound for the period. The lower bound, T_{lower}, is obtained by a parallel procedure, using the lower bound, $\pi/2$, for $F(n)$.

Now one chooses a suitable value for N. For example, with $N=7$, we obtain the angular frequency, ω, correct to 6 figures, with $\varepsilon = -\tfrac{1}{6}$ and $A = \pi/3$. (These values were used in Chapter 6, where we constructed Table 6.1 for the numerical analysis of the Duffing equation.) The program for the period is simple to implement, and if we choose $N = 25$, we find

$\omega_{\text{approx}} = \omega(\text{from exact expression for first 25 terms})$
$\qquad = 0.928447189058660293893\mathbf{85420064}$

$\omega(\text{From } T_{\text{upper}}) = 0.928447189058660293886\mathbf{39322946}$

$\omega(\text{From } T_{\text{lower}}) = 0.928447189058660293893\mathbf{85420054}$

We note that the upper and lower bounds differ by 7 units in the 21st digit. Also, ω_{approx} is *outside* the range of the bounds. This is not a mistake. By construction, the bounds confine the *correct* (with infinite precision) frequency on both sides. The approximate value is not forced to lie between the two bounds. Thus, the bounds provide a very good idea about the precision of the computation of the frequency, while the truncated sum may not do so.

INDEX

Action–Angle variables **332**
Amplitude equation **387**
Aperiodic 10
Autonomous system 4

Behavior near saddle 109
Betatron oscillations **324**
Bifurcation 198, 200, 371, 383
Boundary-value problems **196**

Calculation of the period 164, **395**
Center manifold **301**, 313
  limit cycle 365
  normal form
    expansion 302, 309
Choice of free functions 166
  Canonical transformations 166
  equivalence of choices 171
  killing the fundamental 170
  minimal normal forms 164, 168
  mismatch test 176
  numerical comparisons 171
  usual choice 166, 179
  "zero–zero" choice 168
Conservative planar systems **159**

Dissipative planar systems **209**
  limit cycle 209, 221
  with a fixed point 209, 211
Duffing oscillator 17, 67, **160**, 395

altered 187
minimal normal form 164, 173, 176, 179, 182
naive perturbation theory 79
normal form expansion 161
with cubic damping 218

Eigenvalues
  negative real part 59, **129**
  pure imaginary 62
  update 59, 69, 108
Energy conserving oscillations:
  behavior near a center 108
Error bounds **111**
  extension to longer times 125
  higher orders 123
  minimal normal forms 123
  naive perturbation theory 112
Explosive instabilities **271**

Finite time blowup **143**
  leading singularity analysis 148
  truncation of series 154
  two-dimensional example 148
Freedom in perturbation
  expansion 4, **73**
Frequency multiplication 11

Generators of transformation 7

Higher-dimensional dissipative systems **365**
  bifurcation 371
  limit cycles 365, 370, 379
Higher-dimensional Hamiltonian systems **323**
  canonical transformation choice 343
  constants of motion 329
    formal 332, 342, 344
  coupled harmonic oscillators 337
  integrable 333
  minimal normal forms 346
  nearly integrable 335
Higher harmonics 11
Homological equation 77
Hopf bifurcation 211, **383**
Hyperbolic fixed point 15, **24**

Lie bracket 77
  intuitive meaning 99
Limit circle 19, 211, 227, 232
  radius renormalization 228, 233, 388
Limit cycle 19
  definition 221
  first-order normal form 223
  in higher-dimensional systems **365**, 371, 379, 402
Linear oscillator with cubic damping 211
Logistic equation 145

Minimal normal form 168, 173, 179, 190, 213, 217, 219, 228
  example when not useful 193
  mixing of scales 192
Multiple limit cycles **234**

Naive perturbation theory **55**
Near-identity transformation 6
  convergence 105, 106
  divergence 107
Nonautonomous systems 4, 71, 83, **247**, 270
  accelerator design **249**
  Mathieu equation **265**
  periodically modulated friction **282**
Nonlinear Volterra equation 58
Normal form expansion 7, 14, 79, **92**
  autonomous systems 93
  convergence 102
    Poincaré domain 102
    two-dimensional case 104
  Duffing oscillator 95
  free resonant terms in $\mathbf{x}_n$ 98
  resonance condition 84
  resonant terms 82, 96
  traditional 82

Oscillator with quadratic perturbation **182**
Oscillators with cubic damping **211**

Pendulum 26, **185**
  with cubic damping 215
Perturbed linear oscillator 62
Perturbed simple harmonic oscillator 159
Phase locking **390**
Poincaré domain 14, **85**, 104
Principle of superposition 12

Rayleigh oscillator 224, **233**, 235
Resonant combinations of eigenvalues **84**

Secular terms 10, 55, **111**
Siegel domain 14, **87**, 105
Slow relative phase **328**, 340, 348, 355
Small-denominator problem **91**
Symplectic maps **359**

INDEX

Timescales 8, 57
Time validity of the truncated
 expansion **111**

Validity of expansion 178
 MNF 181
Van der Pol oscillator 19, 34,
 224, **230**
 free functions 231
 "renormalization" choice
 231

Zero-order term 4, 74, 76, 92
Zero-order updating ansatz 82,
 90, 98